FUNDAMENTALS OF LIMNOLOGY

Fundamentals of
LIMNOLOGY

By

FRANZ RUTTNER

Director, Lunz Biological Station,
Austrian Academy of Science

THIRD EDITION

Translated by

D. G. FREY and F. E. J. FRY

Indiana University *University of Toronto*

o o o

UNIVERSITY OF TORONTO PRESS

Second German edition published by
Walter de Gruyter & Co., Berlin, 1952

Reprinted, 1954, 1957, 1959

Reprinted, 1964, 1965, 1966, 1968, 1969, 1970, 1971, 1972

Printed in Canada by
University of Toronto Press,
Toronto and Buffalo

ISBN 0-8020-2028-3

LC 64-919

PREFACE

THE "FUNDAMENTALS" laid down here will in no way replace the introductions to limnology (Thienemann, Brehm, Lenz, and Welch) but complement them in certain respects. The great progress which has been made in recent years in the field of the chemical and physical properties of water and the dependence of the life processes on these makes it appear desirable to take water as an environment as the central theme, and this subject occupies half the text. This method of giving an introduction to limnology seems appropriate to me, because a complete understanding of the biological phenomena in a body of water cannot be attained without a comprehensive knowledge of the environment.

The section of this book on biotic communities will be useful merely as a review and as an illustration of causal relationships. It has been possible to make this section short because the works mentioned above contain much information on this subject.

This book has grown out of the course in Hydrobiology which has been given at the Biological Station at Lunz for some decades. This circumstance will make it clear why the text may seem to be overburdened with examples drawn from conditions in the lakes of Lunz and from work carried on at the Lunz Station.

To my old friend, my collaborator for many years at the Lunz Station, Professor V. Brehm, I owe my heartfelt thanks for many valuable suggestions. Further thanks are due Dr. F. Berger for the careful preparation of the figures, and not the least to my dear wife for the wearisome task of proof-reading.

F. R.

Biological Station, Lunz
February 1940

PREFACE TO THE SECOND EDITION

THE TREATMENT of the subject and arrangement of the text have not been essentially changed from the earlier edition. However, certain sections have been rewritten because of the progress made in the field in the last decade. Through these changes the book has been enlarged from 167 to 232 pages and the number of illustrations increased from 39 to 51. The author is again grateful to those mentioned in the preface to the first edition for further generous help. In addition he wishes to thank also Frau Dr. T. Pleskot (Vienna) for valuable references, Fräulein M. Wimmer (Vienna) for the execution of numerous illustrations and Dr. David G. Frey (Bloomington) for his contribution of American literature.

F. R.

Biological Station, Lunz
Spring 1952

PREFACE TO THE THIRD EDITION

TWO MAJOR WORKS on limnology have appeared since the publication of the second edition, in addition to records of the progress of the science in individual papers in the scientific literature. These two books are Hutchinson's *Treatise on Limnology* and Gessner's *Hydrobotanik*. Both set forth the present state of the science of limnology in a remarkably thorough fashion and have filled the long-felt need for an exhaustive treatment of the ever increasing mass of material. Hutchinson's treatise, which is a comprehensive first volume (a second volume is to follow), is limited to inland waters and offers a survey of existing knowledge of rare completeness. It is of inestimable value to every investigator in this field. Those who have a preference for the mathematical treatment of limnological problems will find this work to be entirely up to their expectations. The three-volume *Hydrobotanik* of the plant physiologist Gessner differs from Hutchinson's handbook both by having its biological chapters limited to the vegetable kingdom and by treating also the sea, thus avoiding the danger that the two cognate areas of science, limnology and oceanography, will lose contact.

When the author began to work anew on the *Fundamentals of Limnology* after the appearance of these two works, which had excited his admiration, it was particularly because he believed he could endow the book still more with the character of an introduction, referring those readers who wished to know more about a particular problem to the two handbooks.

The aim was to cover the more recent discoveries in a manner sufficiently clear for an elementary treatment while at the same time recognizing their importance as far as possible. As in the former editions, applied limnology, in particular as related to fisheries and sanitary engineering, has not been considered. The author is of the opinion that these areas, with their own special problems, should have their own limnologies.

With the rapid rise of limnology there has been an extraordinary growth not only in the number of papers but also in the number of investigators in the field. Accordingly, it appears impossible to include in the compass of an introductory treatment all investigators who are worthy of recognition, and the author must limit himself in most cases to the citation of those works that either presented the first solution to a problem or later gave it an exhaustive and comprehensive treatment. It will be understood, therefore, that the omission of so many worthy names from the book is due to its restricted scope and purpose.

In this edition as in the earlier ones I have had the benefit of valuable assistance from friendly helpers and, in particular, from my daughter-in-law Dr. Agnes Ruttner-Kolisko and my friend Dr. W. Einsele, as well as, of course, from those mentioned in the prefaces to the previous editions. I am very grateful for the assistance Professors D. G. Frey and F. E. J. Fry gave in the production of the English edition in the U.S.A. and Canada. They not only undertook the labour of preparing the English translation but also provided many valuable references. I am also grateful to Professor W. T. Edmondson for the friendly way in which he undertook to look over the revisions for the new edition. I also received valuable references from Frau Dr. Inge Dirmhirm and Dr. W. Schmitz. And finally, may I express my most sincere thanks to my publisher, Walter de Gruyter, who has given sympathetic consideration to all my wishes.

F. RUTTNER

Biological Station, Lunz
November 1960

Death softly and suddenly called away the author of this book in his eightieth year while he was in the midst of correcting the proof of the third edition.

The last months of his life were dedicated to this new edition, which required an extensive search of the literature because of the growth of limnology in the last ten years. His unwearied spirit fought a daily battle with his weakened body to keep it hard at work until his last hours.

As throughout the course which Franz Ruttner gave for almost fifty years while Director of the Biological Station at Lunz, which was the basis for *Fundamentals of Limnology*, his aim in this last revision remained the same: to draw from the vast assemblage of facts what is necessary to make the principles clear and to illustrate the causal relationships, and to present this material in a precise and clear form so that even the beginner can gain a comprehensive picture of the whole science of limnology.

May this last edition prepared by Franz Ruttner receive the same friendly reception from his colleagues all over the world and from the newcomers to limnology that its predecessors have enjoyed.

A. Ruttner-Kolisko

Biological Station, Lunz
June 1961

TRANSLATORS' PREFACE

IN SPITE OF the large number of colleges and universities in the English-speaking world which offer courses in limnology or hydrobiology, there have been woefully few books available in English that could serve as suitable texts in such courses. Furthermore, especially in recent years, progress in limnology has been so rapid along certain lines that none of the texts could claim to be up to date.

Shortly after the war the present translators became aware of the 1940 edition of Dr. Ruttner's book and independently began translating it for use in their respective classes in limnology. Soon realizing that this was perhaps the best book available in any language on the principles of limnology, they combined efforts so that the book might become generally available to beginning students as well as to those previously trained who had not kept abreast of new developments in the field. F. E. J. Fry is responsible for the physical and chemical half of the book, and D. G. Frey for the biological portion.

The translators wish to express their gratitude to Dr. Ruttner for co-operating with them in many ways, and particularly for making the manuscript of the revised edition available for translation long before it was even set in type.

This book is not just another book in limnology. It is a mature and balanced treatment of the principles of limnology, written by one of the foremost limnologists of the world, and leavened by his many years of field and laboratory experience. In fact the book is Ruttner. The translators hope that not a few readers will share their general enthusiasm for the book, and will find in it many stimulating suggestions as to the directions in which limnology will develop further.

D. G. FREY
F. E. J. FRY

Autumn 1952

TRANSLATORS' PREFACE TO THE THIRD EDITION

THE TRANSLATORS again wish to express their gratitude for the very generous co-operation they received from the late Dr. Franz Ruttner and from Dr. Agnes Ruttner-Kolisko. Again Dr. Ruttner made his revisions available in the manuscript.

They also wish to express their admiration of Dr. Ruttner as a scientist and a man, and to echo Dr. Ruttner-Kolisko's belief that limnologists will find this, his last revision, as useful and stimulating as they did his earlier ones.

D. G. FREY
F. E. J. FRY

Spring 1963

CONTENTS

BRIEF TABLE OF GERMAN-ENGLISH EQUIVALENTS

MOST OF the German terms used by Dr. Ruttner already have accepted English equivalents. A few terms were encountered, however, expressing ideas for which there are no counterpart single words in the English language. Furthermore, the German terms used in the description of the various features of bogs and also in the description of the profile of a lake bottom do not have widely used English equivalents. It was necessary, therefore, to more or less adopt words with approximately equivalent meanings, and these are listed below for the benefit of anyone wishing to refer back to the German edition of the book. The translators would be interested in learning of any English words already used for these various German terms which they have overlooked, or of any English words that would possibly be more suitable than the ones used in this translation. Words defined in the glossary are not included in this list.

Aufwuchs—*Aufwuchs* (the closest English equivalent is periphyton; see footnote, p. 183)
Austausch—eddy diffusion
Blänke—bog pool
Bülte—bog hillock
Durchflutung—inflow-outflow; flowthrough
Flachmoor—flat bog
Halde—slope (of a typical lake bottom profile)
Hochmoor—raised bog
Kampfzone—ecotone
Lagg—bog moat; marginal fossa or ditch
Lebensgemeinschaft—community
Moor—bog, moor
Pelagial—pelagial (as noun) or pelagial zone
Profundal—profundal (as noun) or profundal zone
Schlamm—ooze
Schlenke—bog puddle
Schweb—central plain (of a typical lake bottom profile)

Schwingrase—quaking bog
Sprungschicht—metalimnion, except where the temperature curve is specifically implied, then thermocline
Uferbank—shore terrace; littoral bench or platform

FUNDAMENTALS OF LIMNOLOGY

INTRODUCTION

WATER is the basis of life. Only in resting stages, such as in seeds, spores, and the like, does its proportion in the structure of plants and animals fall below 50 per cent and it normally makes up 60 to 90 per cent and even more of the total weight. The living substance of the cell, the protoplasm, is a highly complicated colloid system of which the dispersion medium is water. The complete absence of water means death.

The first life doubtless arose in water or at least in "dampness," and the first organisms were aquatic; the land was first populated after further differentiation. The organisms that embarked on this course of development were only able to do so by taking with them, as it were, their original environment in their body fluids, in the blood and cell sap. The ability to retain an indispensable amount of water, or to replace repeatedly that which is lost, spells the difference between the presence and absence of life under the various climatic conditions of the wide land spaces of the earth.

As can be imagined, since water is an essential element for the life of terrestrial as well as aquatic organisms, so that there is scarcely a single organic function in which it does not play some part or other, it would not be incorrect to consider the whole science of life as Hydrobiology. However, in the system of the sciences this subject is rather narrowly restricted. Hydrobiology is limited to the investigation of the plant and animal associations (biocoenoses) that dwell in aquatic biotopes.

The composition of plant and animal stocks of these biocoenoses is in no way an accident. They are primarily determined by geography and history. These are questions of historical plant and animal geography (Chorology). But within these limits the selection of species is primarily through the biotope, the sum of the environmental conditions impinging upon them. Of all species that reach a given place, for example a newly created body of water, only those that find their optimum near the prevailing conditions can succeed in the competitive warfare. Thus a "biocoenotic equilibrium" (Thienemann, 1918) is causally related to the conditions in the environment so that a species living in a given biotope remains there only so long as no substantial change takes place in the environment. If the environment changes, so also necessarily is there a displacement in the composition of the biocoenose. On the other hand, we find biotopes widely separated geographically which are equal in their ability to support very similar life associations (for example, the springs and brooks of the temperate latitudes, the thermal springs and the upland moors of all zones—cf. the valuable review of Macan [1961] with its extensive references).

The investigation of the causal relationship of biotic communities to their environments (when environment is taken not only to mean the physical and chemical conditions but also to include that dependence which is given by the inter-association of organisms) is recognized as the sphere of interest of Ecology, and ecological considerations thus form a fundamental of every hydrobiological investigation.

Hydrobiology is concerned with fields of science other than biology—with physics, chemistry, geology, and geography—for a comprehensive knowledge of the environment is essential to a full consideration of the subject. Thus, in our case, we must discuss the physical and chemical conditions in water and in waters, in so far as these are important to life, as well as the biological phenomena.

The part of the earth's surface which is covered with water is subdivided into two very unequal zones, the oceans and the inland waters, each of which is a special field of endeavour, Oceanography serving the one and Limnology the other. Both sciences follow somewhat parallel courses, but their subjects are in many ways so different that they must differ to a certain extent in both treatment and methods. A few of the most essential differences between the two environments can be dealt with briefly, although the fact known by all, that the seas contain salt water and the inland waters consist generally (but not always) of fresh water, will not be in the foreground.

The oceans, which cover seven-tenths of the earth's surface, are a continuum in both space and time. They have always existed since water was possible on the cooling globe of the earth and have always formed a spatially continuous unit, in spite of the great changes they have undergone in the course of the earth's history since its origin. As a consequence, the development of life in the sea has proceeded from its first beginning without ever being entirely disturbed by any catastrophe. The inland waters, on the other hand, which make up scarcely one-fiftieth of the earth's surface, are ephemeral bodies measured by the standards of geological time. Only a few large lakes (for example, Lakes Baikal and Tanganyika) extend back beyond Quaternary times into the Tertiary, and most lakes originated in the Pleistocene. Through the processes of filling and sedimentation, and through tectonic changes, the inland water surfaces disappear in an appreciably short time, and with them goes the community that populated them. Newly formed bodies of water are seeded and repeat the fate of their predecessors. Because the continual change occasioned by the formation and disappearance of biotopes can be followed only by especially adaptable organisms, the inland waters have a very restricted fauna in comparison with the sea.

Moreover, there are biotopes in the inland waters for which no parallels, or very few, can be found in the sea. Thus a parallel for the common freshwater biotope of running water and its biocoenoses with their characteristic adaptations is to be found only in the surf zone of the sea or in regions of strong currents. Biotopes comparable with bog waters scarcely occur at all in marine conditions.

Further, because of the world-wide expanse of the sea, its waters are in permanent and active interchange over all zones of the earth. The sea is very little influenced by the land masses, but often, on the other hand, it determines the climate of these. In contrast, the inland waters are relatively limited, enclosed bodies of water (see Figure 52), strongly influenced by the local climates of the land masses that surround them. Because of their small extent and depth, the regular change of the physical and chemical properties and the distribution of organisms dependent on this change are compressed into a much narrower space than in the sea and are in much less measure disturbed by currents. For the investigation of the relations between living conditions and biological phenomena the inland waters are thus more easily reviewed, and they are, in many respects, more suitable objects of study for the investigation of causal relationships than are the oceans in spite of the smaller variety of life in them.

PART A

Water as an Environment

IF AN ATTEMPT is to be made here to sketch limnological investigation in its fundamentals, it follows, as was pointed out in the Introduction, that the discussion of living processes in inland waters must be preceded by a consideration of the environment. If this is to be discussed under the title "Water as an Environment," then it must be realized that this refers not to water *per se*, but to waters in their manifold forms which produce the biotopes in question—biotopes in which the conditions for life are fixed, not by the water content alone, but by the suitability of the bottom, and the form and location of the basin, or (in running waters) of the channel. However, these factors are in general of less importance than the "hydric limitation" (Hentschel, 1923) of life in water, and we therefore first consider the peculiarities of water as an environment in contrast to the conditions under which the land (or better, aerial) organisms exist. Therefore, we shall first consider that biotope in which the "hydric limitation" finds its purest expression, *the open water of the large lakes.*

Water has a two-fold effect on the life within it: (1) through its physical properties, as a medium in which plants and animals extend their organs and move or swim; (2) through its chemical properties, as a bearer of the nutrients which produce the organic from the inorganic through the primary production of the plant kingdom.

I. PHYSICAL PROPERTIES OF THE ENVIRONMENT

It is these which above all most strikingly separate the environment of water from that of air. The vast differences in specific gravity, mobility, specific heat, and humidity—factors which have the greatest of influence on life—have many divergent effects on the plants and animals in the two environments.

1. *Density, Viscosity, Surface Tension*

The density of distilled water is 775 times greater than that of air (at 0° C., 760 mm. Hg) and, correspondingly, its buoyant effect on a body within it is also greater by the same ratio. This means a considerable saving to the organism in the energy required to support its own weight and makes possible the reduction of supporting tissue. A *Potamogeton* or *Myriophyllum*, which raises its stems and spreads its leaves in water, collapses when taken out; freshwater polyps or jellyfish become formless and motionless masses in air.

The density of water in our lakes, brooks, and rivers is not quite the same in different places and at different times. Although the differences that occur are generally small in themselves, they are nevertheless of great importance to the events in the waters under discussion. The differences in density are mostly brought about through variations in *temperature* and *salt content.*

The increase in density with increase in the content of dissolved substances is shown in the following table, which gives the relation in dilutions of sea water.

Salt content ‰ (g. per litre)	*Density* (at 4°C.)
0	1.00000
1	1.00085
2	1.00169
3	1.00251
10	1.00818
35 (Mean for sea water)	1.02822

It follows from these data that the density increases nearly linearly with increasing salt content. The figures given above are not strictly applicable to inland waters since these not only generally contain far less salt but also contain the salts in different proportions. However, we can estimate the change in density due to this factor with greater exactitude. The content of dissolved substances in normal inland waters (when we omit the saline ones) generally lies between 0.01 and 1.0 g./l., values of the order of 0.1 to 0.5 g./l. being most common. In a single lake, spatial and temporal differences in salt content are seldom greater than 0.1 g./l. Correspondingly, variations in density arising from this factor are very small (about 0.00008, i.e. 0.08 g./l.), but these cannot be wholly disregarded, as will be shown later.

The changes in density that take place through changes in temperature are of much greater importance. It is well known that water occupies a special position in this respect. Its density does not increase continuously with decreasing temperature, as is the case with all other substances, but reaches its maximum at 4° C. (more precisely 3.94°),[1] after which it decreases, at first gradually, and subsequently, on freezing, suddenly. Ice is about one-twelfth lighter than water at 0° C.

Water has this special position among the fluids because its molecules tend to form swarms or aggregations as a result of their electrical properties in a manner which is dependent on temperature. While the molecules of all other liquids arrange themselves as spheres packed compactly together somewhat as peas packed in a container, water molecules take up a tetrahedral arrangement at lower temperatures (one molecule in the centre and four in the corners of a tetrahedron). This "tridymite" structure is the sole arrangement in ice. In water in the liquid state the tetrahedra are broken down into other forms of aggregation, which with increasing temperature gradually pass through intermediate stages to that of spheres in the most compact arrangement, as is found in other liquids. The tetrahedral arrangement takes up the greatest volume of any of these states and therefore has the lowest density, while the densely packed spheres provide the greatest density. If these processes operated *alone*, the volume would thus *decrease* and the density *increase* on heating. However, as in any liquid, ordinary thermal expansion takes place at the same time. The resultant of these two opposing forces is the anomalous temperature-density curve given in Figure 1. (See Kalle [1943] and the very thorough treatment in Hutchinson, pp. 195–202.)

This anomalous behaviour of water is the cause of some very striking and, for life, important natural phenomena. Of these, the facts that our limnetic waters can only freeze on the surface (since water at 0° C. is less dense than water at 4° C.) and that the temperature in the deeper parts of lakes is generally only a little under 4° C. in winter are of prime importance. For these reasons, animals and plants in water under ice are exposed to far smaller temperature fluctuations than are land forms, and are not exposed at all to destroying frosts, whose occurrence presents an impassable barrier to the geographic distribution of many species.

[1]This is true at normal pressure. At *high pressures* the temperature of maximum density is lowered. An increase in pressure of 10 atmospheres (hence about 100 m. below the surface) decreases the temperature of maximum density by about 0.1°C. (Strøm, 1945). For this reason, in very deep lakes, as Münster-Strøm showed in Norway in 1932, temperatures below 4°C. are frequently found without the stable stratification being upset as a result.

The salt content also lowers the temperature of maximum density, there being a decrease of about 0.2°C. for each increase of 1‰. Thus, in sea water (35‰) the temperature of maximum density lies at –3.52°C. and accordingly is not reached in the liquid phase at normal pressures (the freezing point of sea water is –1.91°C.). If a volume of water is brought up from a greater depth and higher pressure to a lesser depth and lower pressure, the release from pressure results in adiabatic cooling. Thus temperatures measured at the surface in water samples drawn from the depths are somewhat lower than those actually prevailing at the points from which the samples were drawn. However, in the majority of inland waters the difference is barely measurable.

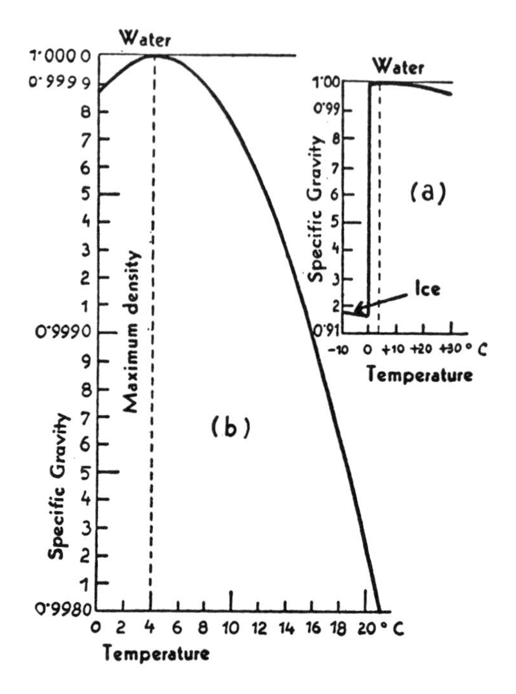

FIGURE 1. (*a*) The dependence of the specific gravity of water and ice on temperature. (*b*) Section of the curve between 0° and 20°C. with the specific gravity scale magnified 20 times and the temperature scale 5 times.

Of course, there are many small, shallow bodies of water which freeze to the bottom. The populating of such a biotope can only be by species that can protect themselves from the frost by the formation of resistant resting stages. In the case of some species, for example many phyllopods, winter eggs are produced which require freezing before they will develop further; this is also the case for seeds of certain plants.

However, even apart from this anomaly, the small differences in density with changes in temperature are of very great, and indeed of overwhelming, importance to the course of events taking place in waters. It can be said without exaggeration that the great processes that regulate the water and chemical economies of lakes are primarily a function of the differences in density. Because of the great importance of this relationship, variations in density between 0° and 20° C. are shown in Figure 1 (*b*) (which has the vertical scale of 1 (*a*) enlarged twenty times and the horizontal one, five times). The density corresponding to any temperature can be easily ascertained from this diagram. *The fact that the density changes more rapidly at higher temperatures than it does at lower ones is especially important.* Thus, the change in density between 24° and 25° C. is thirty times greater than between 4° and 5° C.

The compound H_2O normally consists of the atoms H^1 and O^{16} but H^2 (deuterium) occurs in small amounts in natural waters as does also, but still more sparsely, the radioactive H^3 (tritium). Oxygen also occurs as both O^{18} and O^{16}. Our knowledge of the amounts in which these isotopes occur is still in a state of flux.[2]

The *viscosity* of water is a physical property that is not to be underrated. This is the cause of the frictional resistance that a fluid offers to a moving body. The magnitude of this function is proportional to (1) the extent of the surface in contact with the water, (2) the speed, and (3) a constant depending on the temperature and the nature of the fluid. Since the influence of the salt content (thus the nature of the fluid) within the limits occurring in fresh water is only slight, we are chiefly interested in the effect of temperature. As the temperature rises the viscosity falls. It is twice as great at 0° C. as at 25° C. Hence at 25° C. a plankton alga under conditions which are otherwise equal will sink twice as fast (see p. 109). Since the viscosity of water is about one hundred times greater than that of air, aquatic animals must overcome much greater resistance than aerial animals are required to, and the movement of a *Cyclops*, a mayfly larva, or the lightning dart of a trout requires powerful muscular force.

The *surface tension* of water towards gaseous and solid bodies is also an important biological factor under certain conditions. It acts at the air-water interface and forms a special biotope to be mentioned later (p. 125). It affects the organs of plants and animals in a different way according to whether they are wettable or not. Young leaves of *Potamogeton* and the shells of the plankton Cladocera are water repellent. Contact of the latter with the water surface is often destructive, since the water draws away from the repellent shells and the tension of the upper surface prevents the submersion of the animal. The ecological significance of the surface tension is also to be seen in the circumstance that organs with a water-repellent surface are far less frequently the substrate for foreign growths, which limit movement and metabolism.

Under the heading of physical properties of water is a group of phenomena causally connected with each other, whose importance to life in waters is undisputed.

[2]We can now get an estimate of temperatures that prevailed in former epochs from $O^{16}:O^{18}$ ratios, in the shells of fossil molluscs for example.

2. *The Fate of Incident Solar Radiation*[3]

Solar radiation not only determines the *intensity* and *quality* of the light, the source of all life, which is available to the organisms at a given depth, but it also affects the *temperature* of waters, with its diel and annual variation through the interplay of radiation, evaporation, and conduction. Radiation phenomena, together with those phenomena that result from them, indirectly, affect almost all phases of organic and inorganic events, and it must be our task to concern ourselves with a thorough consideration of them.

(*a*) *The distribution of radiation*

The point of departure for the consideration of relations in nature is made here from the knowledge gained by physicists in the laboratory. They have discovered that a beam of light falling on a water surface at a certain angle is in part reflected and in part penetrates the water, simultaneously becoming more vertical.

The *proportion reflected* depends on the angle of incidence (calculated from the perpendicular) and is considerable when the ray is very oblique. With an angle of incidence of 60° it is only 6 per cent; at 70°, 13.4 per cent; at 80°, 34.8 per cent. In our latitudes about 2.5 per cent of the noon sunshine in summer and 14 per cent in winter is reflected. It is evident that during the change in the altitude of the sun in the course of the day there are major alterations in the proportion of light reflected. A result of this process is, for example, that—of course only when the sun is shining—the intensity of illumination in the evening decreases more rapidly under the surface of the water than above it. In addition to the greater reflection, the increasing path travelled by the light from the descending sun with its more rapid extinction (see below) plays an appreciable part in this more rapid darkening. The *diffuse light* from the sky, which according to the degree of cloudiness and the height of the sun amounts to from 8 to 100 per cent of the total radiation, strikes the water surface from all angles, and on the average 6 per cent of this is reflected. Naturally, this holds true only when the horizon is completely free. When the horizon is obscured by hills, trees, and so forth, the reflected part of the diffused light is appreciably smaller. At all but very low elevations of the sun, the spectral composition of the reflected light

[3]The diffuse radiation from the sky and clouds is considered under this heading as well as direct sunlight. This section is concerned with the total incident radiation, which is affected by latitude, altitude, season, time of day, and prevailing weather.

is substantially the same as that of the incident sunlight. This, of course, has often been noted, since images mirrored in a water surface are colour-true.

The light representing the fraction of the rays that *penetrate* the water does not pass through it unaltered; some of it is *dispersed* and some is *absorbed* and transformed into another form of energy, *heat*. The percentage held back in one metre we term the *percentile absorption*[4]; that transmitted, the *percentile transmission.*[5]

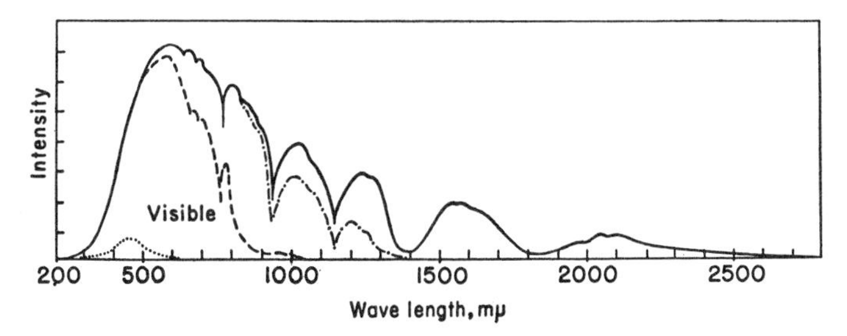

FIGURE 2. Spectral composition of solar energy: at the surface of the earth ———, after penetration of 1 cm. distilled water – · – · – ·, after penetration of 50 cm. distilled water – – –, after penetration of 100 m. distilled water · · · · ·. From Schmidt (1908).

These coefficients are by no means the same for all wave-lengths. Line *DW* in Figure 3 shows the percentages of the various wave-lengths in the region of the visible spectrum that are transmitted through one metre of distilled water when the light falls normal to the surface. It will be seen that for short wave-lengths the transparency is high and rather constant. From about 550 mμ and up, that is in the yellow, orange, and red, it decreases rapidly. Even before the *infra-red* is reached, which is beyond the visible region at wave-lengths of 900 mμ and above (heat rays), the radiations do not penetrate to any extent through a metre thickness. On the whole about 53 per cent of the total solar radiation is absorbed and turned into heat in one metre. Figure 2 also shows these relations.

[4]That is, the part lost through absorption and scattering.

[5]Expressed mathematically, the intensity of light I at a given depth may be derived from the following formula:

$$I = I_0 \cdot e^{-\epsilon h},$$

where I_0 = the intensity at the surface;
e = the base of natural logarithms;
ϵ = the extinction coefficient;
h = the length of the light path in the water column.

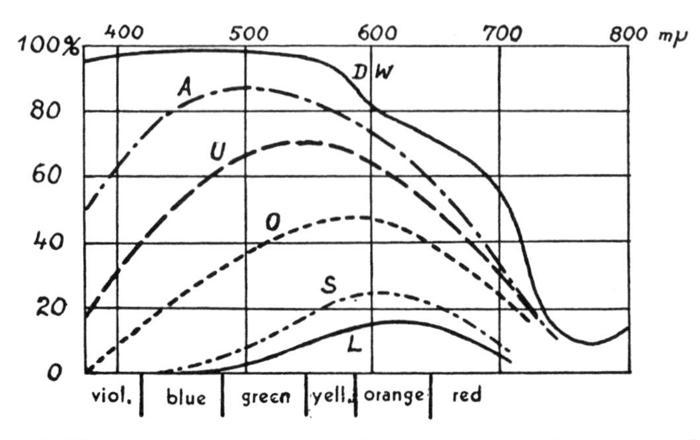

FIGURE 3. The transparency, as percentile transmission, of a stratum of water 1 m. thick with respect to different regions of the spectrum. *DW*, distilled water, from James and Birge (1938); *A*, Achensee (Tyrol); *U*, Lunzer Untersee; *O*, Lunzer Obersee (Lower Austria), from Sauberer (1939); *S*, Skärshultsjön; *L*, Lammen (South Sweden), from Åberg and Rodhe (1942).

Results of laboratory experiments cannot be applied directly to conditions in natural waters, as Aufsess (1903) showed long ago. Lakes, brooks, and springs do not contain chemically pure water but a dilute solution of inorganic and organic substances, together with suspended material of various sorts consisting of both plant and animal organisms and mineral and organic particles. All these circumstances affect the transparency of the water.

The technical difficulties that attend the exact measurement of light under water kept us from more than a meagre knowledge of the radiation climate of our waters until within the last four decades or so. Even now our knowledge is in a state of flux and is far from complete. The older investigators had to use simple methods of measurement and these are often still in use today.

One such simple method of investigation that is often employed is the determination of *the depth of visibility* by means of a *Secchi disc.* A white plate of 20 to 25 cm. diameter is lowered on a calibrated line until it just disappears. The depth thus found is the depth of visibility and gives a measure of the transparency of the water. Consideration will show that the changes the light undergoes on its way from the surface to the disc and then back to the eye are of two sorts, absorption due to the water itself or substances dissolved in it (colour) and scattering due to turbidity. Each condition can affect the visibility of the disc independently. In the first case, for example, in the clear but dark-coloured water of a bog lake, the light intensity is low at the point where the disc disappears. In the second case, for example in the milky water of a glacier lake, the water can be quite well lighted as if under a ground glass, even when there is a high level of opacity. The determination of the limit of visibility does little in this latter instance to elucidate the light conditions in the lake. Nevertheless, when properly employed, this simple and portable apparatus serves as a useful means of describing waters, and the Secchi disc will long remain a limnologist's tool (see also Vollenweider, 1960).

The natural differences in transparency found in the waters of different lakes are very considerable. The highest values, 50 metres and more, are found in tropical and subtropical seas and in crater lakes. However, in clear mountain lakes (Lago di Garda, Walchensee) the visibility is not infrequently 20 to 25 metres, and in Lake Tahoe it is 33 metres (109 ft.). In general, the white disc disappears at 10 to 15 metres in our alpine lakes, and in the lowland lakes at depths from a few decimetres to 10 metres. In the alpine lakes the transparency is generally greatest in the winter at the time of the snow cover of the drainage area and of the plankton minimum. In the Baltic lakes, because of the turbidity resulting from the winter rains, it is least at this time.

People have also tried to measure the decrease in light intensity with increasing depth *by lowering photographic plates* and sensitized paper. Thus, for example, in Lake Geneva 200 to 240 metres is the limit at which a day's exposure will produce a recognizable darkening. However, these methods give unsatisfactory results for various reasons (for example, because of differences in the sensitivity of the plates).

Birge and Juday (1929–32), working on North American lakes in the decades following the First World War, should be credited as the first actually to employ exact methods to investigate the penetration of light. These investigators used sensitive *thermopiles* suitably arranged so they could be lowered to various depths by a cable. A galvanometer was used to measure the thermal current produced. Thermopiles have the very appreciable advantage of being equally sensitive to *all* wavelengths and are therefore an unsurpassed instrument for measuring *total radiation.* (For a description of the method consult Birge, 1922.)

In Figure 9, p. 27, the results of the American investigators are shown. The transmission curve for a lake of moderate transparency (7 metres) is given as percentages of the incident light. Birge's and Juday's numerous measurements show the wide differences that are to be found among different lakes and between these and distilled water. While 47 per cent of the total radiation penetrates through 1 metre of distilled water, only 40 per cent penetrates a like distance in the most transparent of the lakes investigated, and in many cases the value drops to 5 per cent (in one, to 2 per cent). Below 1 metre the total transmission is greater, since the infra-red rays are no longer present (see p. 13). At greater depths there is no further substantial change in the total transmission. The table below, taken from the measurements of Birge and Juday, shows these relations clearly.[6]

Total radiation transmitted, % per m.

Stratum, m.:	0–1	1–2	2–3	3–4	4–5	5–6	6–7	7–8	8–9	9–10
Blue Lake	30	72	73	74	75	78	80	79	79	75
Crystal Lake	38	77	79	80	81	82	86	83	84	82
Green Lake	27	64	67	69	70	70	69	70	72	72
Seneca Lake	32	70	72	71	70	70	71	70	72	77

Since the vertical percentile transmission in water depends on the length of the light path, which changes with the sun's height above the horizon, measurements made at different times of the day and at different seasons and measurements under cloud are recalculated to correspond to the effect with the sun at an angle of 45° in a clear sky.

The light that penetrates various lakes is not only differently affected as regards the total amount passed but also in its *spectral composition.* The thorough work of Birge and Juday also illustrates this. Their data were obtained by placing filters over the thermopiles. In our consideration here we shall, however, use more recent measurements made on the alpine lakes and we shall begin with a few words on the method now customarily used.

With the discovery of the photoelectric cell (which is also used, among other things, as a light meter for photography), measurement of radiation in water received a new stimulus. These photocells are more sensitive than thermopiles and can be used to greater depths, but they have the disadvantage of not covering the whole spectral range indiscriminately. Their range of sensitivity is similar to that of the human eye; therefore they give only a partial measure of the infra-red and ultraviolet. In spite of this and other short-comings, the investigations of recent years have shown that, with due regard for all sources of error, very useful results can be obtained with photocells, and light investigations both in the sea and in fresh water are almost exclusively carried out with this type of apparatus nowadays.

[6]An apparent reduction of transparency below the surface occurs at all wavelengths because of the so-called "surface effect." This, however, is not real but simply the result of the reflection where the water is in contact with the window of the apparatus. See Mahringer (1958) and Berger (1958).

Figure 4 shows such an instrument. The photocell is housed in a watertight casing supplied with a cable. The whole is suspended from the cable, which transmits the photoelectric current to a measuring instrument in the boat.

Sauberer (1961) gives recommendations for the measurement of radiation under water which are accepted by the majority of physicists, limnologists, and oceanographers.

FIGURE 4. Apparatus for measuring radiation under water. *Ph*, photocell; *Gl*, glass cover; *Fi*, colour filter. From Eckel (1935), diagrammatic.

Because of their sensitivity, photocells are especially suitable for measuring the *spectral composition* of light at various depths. Filters of known transmissivity, either coloured filters for relatively broad regions of the spectrum or interference filters for narrow ones, are used above the photocell. For convenience, several can be arranged on a turntable. Results from Lunzer Untersee are given below as an example of the change in spectral composition of light on passing through lake water.

Figure 5 shows conditions in this lake on a clear summer day.[7] The abscissae are again wave-lengths in mμ and the ordinates the per cent of the surface intensity reaching the depth in question. The surface intensity

[7]In cloudy weather (thus when the light is diffused) the distribution of wavelengths is not essentially different. Indeed, measurements when the sky is overcast are very useful for exact observations and were therefore especially used by Sauberer.

is thus always taken as 100 over the whole wave band without reference to the energy curve of solar radiation, a simplification that does not, however, essentially alter the picture. We see that the transparency of Lunzer Untersee is greatest in the green (in which it shows an important difference in transmission as compared with distilled water, Figure 3, *DW*), and it rapidly decreases *not only towards the longer wave-*

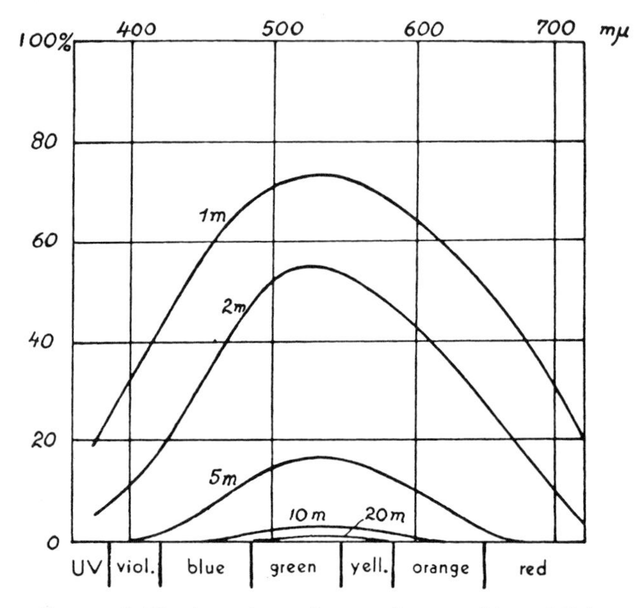

FIGURE 5. The intensity and spectral composition of light at various depths in Lunzer Untersee (transparency 8–12 m.) expressed as percentage of radiation reaching the surface (summer averages at the mean elevation of the sun). From Sauberer (1939).

lengths but also towards the shorter ones. At 20 metres there is nothing but green and some yellow and only a small percentage of these; all the rest of the spectrum has disappeared. At the greatest depths of the lake there exists a green twilight, which at 30 metres amounts in intensity (but not in spectral composition) to about full moonlight (about 1/500,000 of sunlight).

In other lakes the peak of the transmission curve (cf. Figure 3, *A*, *U*, *O*, *S*, *L*) is sometimes shoved towards the blue (Achensee) and sometimes towards the orange (Lunzer Obersee, Skärshultsjön, Lammen). Necessarily (because of the greater absorption of the longer wave-lengths even in pure water) the total intensity decreases more rapidly in "brown" lakes than in "blue" ones. This is clearly shown in Figure 3. With these differences in transparency the light at greater depths of different lakes varies from blue-green to brown. It can be seen, therefore, that *the transmission curve is very suitable for the demonstration of the optical properties of a lake.*

When the curves in Figure 3, together with the results of numerous measurements in other European and North American lakes, are compared with one another, it will be seen that the greatest variation is found in the region of short wave-lengths while the curves approach each other closely at long wave-lengths. Thus the transmission values for the shorter wave-lengths (approximately the blue region) are especially useful to characterize lakes. In all cases it is sufficient to give the transmission for three wave-lengths (400, 500, and 600 mμ) in order to provide an indication of the optical properties of a lake. As a further simplification it might be proposed that the percentage transmission for these three wave-lengths rounded off to tens might be grouped as a code number. Thus, for example, the code number 376 would describe Lunzer Untersee in which the transmissions are: 400 mμ, 32 per cent; 500 mμ, 70 per cent; and 600 mμ, 63 per cent.

We now come to that light phenomenon already studied precisely by Aufsess (1903), which is also one of the most striking aspects of a lake or river to the layman, the *colour of waters.* The rich variety of colour, which embraces shades ranging from the deep blue of the Mediterranean or one of the upper Italian lakes through the emerald green of many alpine waters to the dark brown of moorland tarns, has numerous causes, of which the selective transparency of the water in question is one of the most important but by no means the only one. The coloration is also influenced by the living and inanimate suspended matter. It is further influenced to a lesser degree by the quality of the incident light, that is by the colour of the surrounding country (for example, green forest or bare rock) and by the colour of the sky above it. Finally—in shallow bodies of water—the colour of the bottom influences the colour of the water.

A lake completely devoid of material in suspension must appear almost black when viewed from directly overhead if it is deep enough, since the light striking the surface will all be absorbed without hindrance. Such optically pure water, however, never occurs in nature. More recent investigations have shown that there will be a certain dispersion of light even in water containing no particulate matter at all in suspension. This is the result of variations in density caused by irregularities in the motions of the molecules (the fluctuation theory of Smoluchowski-Einstein). This dispersion is nearly inversely proportional to the fourth power of the wave-length; hence the short waves are dispersed to a much greater degree than are the long ones. Moreover, such dispersed light undergoes selective absorption towards the blue in the water before it reaches the eye, as Figure 3, *DW* indicates. When there are suspended materials present, as is always the case, the dispersion effect is tremendously increased by reflection from them. The clearer the water the greater will be the average depth from which the dispersed light will reach the eye, and as a consequence of the greater extent of the light-filtering effect of the water, the more saturated and darker will the resulting colour appear. Bright colours are found in turbid waters, because of the small average depth of reflection. Thus, for example, we can see why the dark green colour of a clear alpine lake changes to a bright luminous "sea green" after a flood that has discharged a fine argillaceous silt into it.

The particles suspended in the water are, however, not all colourless (clay, silica, carbonate); to a large degree they are pigmented, particularly in the case of *phytoplankton,* which have green, yellow, and also occasionally red pigments. *Coloured* light is reflected from these particles, and combined with the filtering effect of the water produces a mixed colour in which now this and now that component predominates. We shall speak further on the express subject of the relation of *vegetation* to colour in water in the chapter on plankton; it is sufficient here merely to recognize the influence of this factor.

The actual filter effect, that is, the *selective transparency* of water, which, as indicated above, can be seen as colour in deep lakes only as the result of the dispersion of light, is influenced by the *dissolved substances* and especially by those that are coloured. *Pure water* in thick strata is *blue,* as has been pointed out. Brown humus materials are always brought into our waters from the soil, from decomposing leaves, and from surrounding bogs, and these bring about, according to their proportion, a shift of the original colour towards *green, yellow,* and *brown.* Bog waters, which as a rule have the greatest admixture of these materials, appear appreciably yellow even in samples, and in thicker strata occasionally dark brown.

In many cases lake water in a glass vessel displays no appreciable colour when viewed by the eye, and samples can scarcely be distinguished from distilled water. However, the transmission curves will be greatly affected by the slightest trace of humic substances, which strongly absorb the short wave-lengths of light. The differences in the courses of the curves (Figure 3) and in the code numbers spoken of above are largely due to differences in humus content. On the other hand colourless mineral components in solution have no appreciable effect on the position of the transmission curve.

The so-called *Forel-Ule colour scale*, which has standards in 21 steps from blue to brown made from solutions of cuprammonium sulphate, potassium chromate, and cobalt ammonium sulphate, is used for describing the *colour of lakes*. The colours seen in deep waters within the reflection of a dark screen, or simply in the shadow of a boat, can be compared with these.

In the case of water samples, the colour under conditions of transmitted light is usually determined as outlined in *Standard Methods* (1960) by comparison with dilutions of the following standard solution: 1.246 g. potassium chloroplatinate (equivalent to 0.5 g. Pt) and 1 g. crystalline cobaltous chloride with 100 ml. HCl in 1 litre of distilled water. A content of 1 mg. platinum per litre is taken as the unit of colour. Ohle (1934) employed methyl orange as a standard. These determinations are especially important in the investigation of waters with higher humus contents.

Geographic names often express the dark brown coloration of bog waters; thus, for example, we have the Rio Negro, the tributary of the Amazon which drains a vast humus region. The waters of the Rio Negro can still be distinguished far downstream from where they join the main river, a circumstance first noted by Alexander von Humbolt and recently the subject of researches by Sioli (1954–57) and Gessner (1960).

From what has been said it can be seen that a blue colour can arise only in the absence of appreciable amounts of humic materials and of coloured materials in suspension, such as phytoplankton. Therefore, only a water poor in organic productivity can be blue: "Blau ist die Wüstenfarbe des Meeres" (Schütt, 1893). Just as the tropical seas with a low production of plankton (for all their multiplicity of species) are deep blue, so inland waters are the bluer the smaller the amount of free-floating organisms they contain. On the other hand, waters with a high plankton content are always yellow-green to yellow in colour. Observation of the colour of a water thus offers a certain criterion for estimating its productivity provided that the influence of humic material can be excluded.[8]

The true or, perhaps more properly, *specific* colour of a water must be sharply distinguished from its *apparent* colour, which is derived not only from the inherent properties of the water itself but is also affected by the surroundings. Thus in shallow waters, white sand below particularly brightens the colour tone; and a coloured bottom (for example green or brown plants) will, together with the selective transparency,

[8]Waters of particularly high calcium content (for example, brooks in chalk hills) are frequently characterized by a striking blue-green colour. This phenomenon is caused by the dissolved lime coagulating the humus colloids and in this way bringing out the true (blue) colour of water.

give rise to a mixed colour depending on the depth. The surrounding terrain will also have a further effect through its wooded heights or coloured rocks, which can change the spectral composition of the incident light. Finally, the varying composition of the sky reflection, which changes with the time of day and with the degree of cloud cover, should be mentioned.

The phenomenon of surface reflection is not the least factor in influencing the colour of a body of water. Even the darkest tarn can appear blue in the distance when it mirrors a cloudless sky.

Under certain circumstances it can be important to investigate the transparency of rather restricted strata of water at any given depth of a lake independent of the incident sunlight. For this reason Willer (1936) measured the turbidity of water samples by means of a photoelectric cell. For measurement in a lake itself the transparency meter originally constructed by Petterson (1936) and remodelled by Sauberer (1938) for limnological investigation is useful. An electric light of constant luminosity housed in a water-tight container together with an appropriate lens provides an almost parallel beam of light which falls on a photo-cell placed one metre away. The resulting photoelectric current, which is proportional to the transparency of water traversed by the light path, is in turn led to a galvanometer in the boat. Such measurements, which can now also be carried out by day with a type of apparatus recently developed by Åberg and Rodhe (1942), are especially useful for the observation of locally limited turbidity. Thus, Figure 6 shows the effect of a sharp stratification of phytoplankton on the transparency of water.

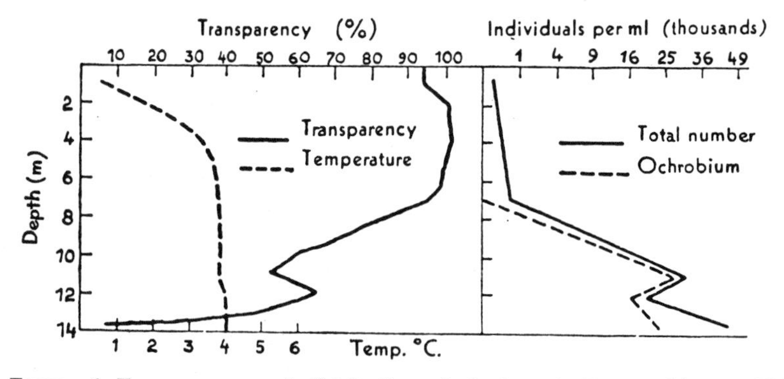

FIGURE 6. Transparency and distribution of plankton in Lunzer Obersee, February 25, 1938. The plankton maximum at 11 m. was chiefly due to the iron bacterium (*Ochrobium tectum*).

A part of the incident light is *scattered* in all directions by the material in suspension. The proportion of this diffuse radiation, the horizontal light and the light from below, increases with depth or with turbidity. Thus the *light from below* at one metre in the clear Lunzersee amounts to 1 per cent of the light reaching a horizontal surface from above, while in the turbid Leopoldsteinersee (near Eisenerz in Styria) it amounts to 9 per cent.

The transparency of *ice* is also of interest and importance to life in water. Recent measurements for various wave-lengths of the visible spectrum, made by Sauberer (1950), of the transparency of the ice covering Lunzer Untersee are given below as the percentile transmission per metre, togther with corresponding values for distilled water and water from the *lake itself.*

Wave-length, mμ:	400	500	600	700	800
Ice (Lunzer Untersee)	96.0	92.0	81.5	55.0	17.0
Distilled water	98.4	99.2	81.0	55.0	11.1
Lake water	33	68	63	31	(10)

It can be seen that the transparency of clear ice is very great and is much more like that of distilled water than like that of the lake water from which it was frozen. The difference in transparency between lake water and lake ice is particularly great in the ultra-violet (values for this are not included in the table). Thus Merker (1931) measured the transmission of Lunzer Untersee water as only 2 per cent for light having a wave-length of 313 mμ while Sauberer (1950) found the transmission through the ice of that lake to be 90 per cent for the same wave-length. Sauberer found the transparency of ice for the infra-red to be the same as that of distilled water.

A sheet of clear ice has no effect, therefore, on the utilization of light by aquatic life. However, it is quite otherwise if the ice is full of air bubbles or is covered by a layer of snow. Twenty centimetres of dry new snow reduces the radiation that penetrates it to 6.7 per cent before the water is reached. Dry old snow of the same depth would reduce the light to 1 per cent. The transparency of wet snow is still less (because of its high water content). It should also be borne in mind that on an average about 75 per cent of the light reaching the surface of the snow is reflected.

Those interested will find an extensive discussion of the radiation climate of inland waters in Sauberer and Ruttner (1941) and also in the "Recommendations" (Sauberer, 1961) mentioned above.

(*b*) *The distribution of heat*

Up to now we have considered only that portion of radiation that decreases constantly in intensity and changes in quality as it *penetrates* waters. Of just as great importance for life are the effects of the radiation that is *absorbed* by the water. The former provides the *light climate* at a given depth and hence determines the conditions for the assimilation of carbon dioxide by green plants. The fraction absorbed in the water produces *heat*, the most important regulator of living processes. The temperature relations also control the stratification of the water mass and the currents of a lake basin indirectly through changes in the density of the water. *We can, therefore, designate the thermal relations as the pivotal point of every limnological investigation.*

Two aspects of the thermal properties of water influence the aquatic environment: (1) its specific heat; (2) its temperature.

The *specific heat*—by which is meant the amount of heat that must be transmitted to a unit weight of a substance in order to increase its temperature 1° C.—is one of the most independent environmental factors and we can therefore dispose of it quickly. It is sufficient to say that water has a specific heat that is surpassed by only a few substances, for example, liquid hydrogen (3.4) or liquid ammonia (1.23). The specific heats of the common constituents of the earth's crust, for example, most rocks, amount to only 0.2 as compared with 1.0 for water. However, water has such a high specific heat only when in the liquid state; the specific heat of ice is only 0.5.

The special position that water holds in this respect contributes not the least to its outstanding suitability as an environment. The climate in water is much more equable than on land. Variations in temperature take place only very gradually, and the extremes between day and night and the changes in the course of the seasons are relatively small.

Water *temperatures* are usually measured with a mercury thermometer. Simple thermometers are obviously suitable only for determining surface temperatures; for observations at greater depths special apparatus must be employed, of which two types are in use: the *sampler thermometer* and the *reversing thermometer.*

The method using the sampler thermometer depends on the high specific heat of water, which allows a moderate amount of water to be brought up through strata of different temperatures without undergoing any appreciable change. A simple form of the sampler thermometer is the so-called *Meyer sampler bottle*, which anyone can make for himself (Figure 7). It is weighted with lead or stones, lowered with the stopper in place, and opened at the desired depth by a jerk on the cord. When it is brought up a thermometer is inserted through the neck and the temperature taken. In a modification of this sampler the thermometer is placed inside. Because of the increase in hydrostatic pressure with depth, such a sampler can be used only down to about 50 m.

Apparatus in which the sampling chamber remains open while it is being lowered so that the water can pass without hindrance, and which can be

FIGURE 7. Meyer sampler.

closed top and bottom by movable caps, has a wider range of usefulness. The sampler is closed at the desired depth by dropping a messenger. An apparatus of this type, which is made from a glass cylinder to permit the viewing of a thermometer inside, and which has been used for many years at the Biological Station at Lunz, is shown in Figure 8. The thermometer, graduated to 0.1 °C., is protected by the reservoir *R*, a precaution which even at high external temperatures permits an accuracy of 0.1 °C. This sampler has been used (during the German Sunda limnological expedition) down to 430 m. and is also suitable for taking water samples for chemical and plankton investigations. It therefore has the advantage that temperature distribution, chemical analysis, and an appreciable part of the biological investigations can all be carried out with the same sample. This is especially important in the determination of causal relationships where a sharp stratification exists.

The *reversing thermometer* is a special mercury-in-glass thermometer mounted in a metal case that can turn over. After being left about five minutes at the required depth it is released by a messenger, which allows the case to rotate through 180 degrees. The rotation breaks the mercury column at an ingeniously fashioned constriction and it travels to the other, now lower, end of the thermometer, which bears the graduations. Temperatures encountered later can have but little effect on the length of the mercury column, and even this source of error can be completely removed through the "column correction." The reversing thermometer is almost universally used in oceanography; limnologists, however, often prefer a sampler thermometer.

For certain investigations, especially when a continuous temperature record is desired or the temperature is required at points every few centimetres, a *thermocouple* or a *resistance thermometer* can be employed to advantage; the construction of these will not be gone into here.

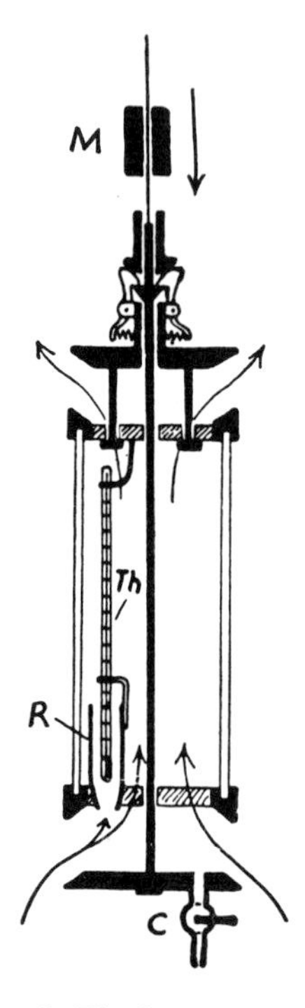

FIGURE 8. The Lunz water sampler for taking water samples and for temperature measurement. *M*, messenger; *Th*, thermometer; *R*, reservoir; *C*, petcock.

In recent years a great advance in the use of resistance thermometers has come with the use of "thermistors" (see p. 259), which are thermosensitive resistances. Type 2311/300 of Standard Telephones and Cables Ltd. is a specially suitable element for limnology. The reader will find a review of the usual methods of using thermistors to measure lake temperatures in Mortimer and Moore (1953).

Turning now to a consideration of *thermal conditions* in waters and especially in lakes we must first seek to establish the decisive factors. The largest source of heat is *solar radiation.* Heat is taken up to by far the greatest degree directly through *absorption* by the water and to a lesser degree through the *transfer* of heat from the air or from the bottom. This indirect heating plays an important role only in the case of ground water and springs. The *terrestrial heat* of our earth is the sole factor determining the temperature of hot springs and certain volcanic lakes. Finally, under certain circumstances the *condensation of water vapour* at the water surface can furnish important quantities of heat.

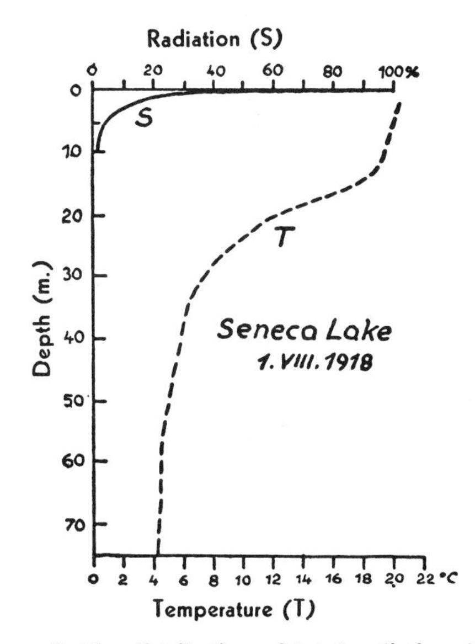

FIGURE 9. The distribution of total radiation (S) and temperature (T). From Birge and Juday (1919).

The *loss of heat* from waters takes place to the greatest degree through *radiation*, with further losses through *evaporation* and through *conduction* to the air and the bottom. Finally, the volume of water flowing through the lake, which in general carries away the topmost water strata and therefore brings about a considerable loss of heat, is of great influence on the thermal economy.

If, as stated above, the heating of a lake is chiefly the result of solar radiation, we must expect that the distribution of heat in a water basin completely undisturbed by other influences would correspond to the absorption of these radiations given on pages 15f.; that is, the temperature of this hypothetical lake must decrease very sharply from the surface downward. Such a state in no way approaches the normal summer condition of a lake, as the temperature curve (T) of Seneca Lake in North America in Figure 9 shows. This lake is used as an example because simultaneous measurements of the absorption of radiation (*S*) were available, so that the results can be shown in the same figure. We can see immediately that within the upper 10 metres, in which 99 per cent of the total radiation was absorbed, there was no appreciable difference in temperature. A rapid drop in temperature begins rather suddenly just below 12 metres and then the gradient (that is, the temperature difference per metre) decreases again below 20 metres and is almost inappreciable at the greatest depths.

This form of temperature curve, by which the water mass is partitioned into three parts differing in their thermal conditions, is characteristic of the midsummer state of almost all lakes, although considerable differences will be shown from one example to another in the sharpness of expression and the extent of the individual strata. Following Richter (1891), who made thorough observations and descriptions of these phenomena in Wörthersee (Carinthia), the zone of the sharp drop in temperature is designated as the *Sprungschicht* in Germany. It is termed the *thermocline* in America, and the *discontinuity layer* in England. More recently the term *metalimnion* has been adopted for this zone. The water mass overlying the metalimnion and showing slight or no thermal stratification is called the *epilimnion*. The layer that extends from the metalimnion to the bottom of the lake is called the *hypolimnion*.

How does this remarkable distribution of heat, which differs so greatly from the absorption curve, come about? We know that during the winter the water mass of a lake in temperate latitudes possesses a temperature of no more than 4°C. *Thus it is evident that even very deep strata, to which solar radiation can penetrate only to a minimal degree, gain an appreciable increment of heat in summer.* It is self-evident that wherever radiation is absorbed there is a heating of the water. Where heat cannot be given off to the air, as in the case at greater depths, it may be accumulated and, as Ricker (1937) has shown, lead to a sensible rise in temperature, even if the amount of energy absorbed is very small.

However, the distribution of heat cannot arise from the simple absorption of radiant energy, as is made clear by the temperature curve given above. There must therefore be some force at work which sends the water warmed at the surface to the lower depths. As a matter of fact, Birge and Juday have established by careful calculations for the example of Seneca Lake given above that the heating of the individual water strata through the *direct absorption* of solar energy amounted only to 10 to 12 per cent of the total distribution of heat.

The energy that distributes heat in a lake is derived chiefly from the wind, as has been shown by the researches of Wedderburn (1910), Birge (1916), Schmidt (1915), and others. The wind pushes water particles lying on the surface and generates a current, the speed of which is dependent on the strength of the wind. When it reaches the shore, the moving water mass is deflected by the resistance of the colder, and therefore heavier, quiet, deeper strata. The result is a current in the opposite direction in the stratum just below the surface. Such currents produce turbulent eddies as we shall see later, which lead to a vertical interchange of water particles, and so to a partial or complete intermixing. The interchange (known as eddy diffusion) is the greater and carried on to greater depths the greater the speed of the currents and thus the stronger the wind and the smaller the differences in water density in the lake. Accordingly, the transport of heat is most effective at times when the temperature difference between surface and bottom is still not great. It is therefore especially effective in the spring. With the progressive accumulation of heat and increasing differences in density, mixing can take place only to lesser depths. A boundary is thus formed between the turbulent surface stratum and the quiet water masses underlying it, so that with further heating of the former the temperature drop between the two must continue to become more and more marked. *The metalimnion thus indicates the limit of the mixing currents which are derived from the surface.*

These then are the processes of prime importance in determining the form of the temperature curve. An observer can easily be persuaded of the influence of wind-generated currents if he makes temperature measurements on two sunny days, one calm and the other windy, or on the same day if wind follows a period of some hours of calm. Owing to the manner in which the radiation is absorbed, a marked warming of the surface with a sharp drop directly below will be found when it is calm. When it is windy, a more or less uniform heating of the whole epilimnion will go on throughout the day.

The equalization of the temperature of the epilimnion is not brought about *solely* by wind-generated currents. The cooling of the water surface by evaporation, and also by radiation and conduction during the night and in cold weather, plays an important part. Since heat waves are absorbed by water to a very much greater extent than they are by air, the surface temperature of our lakes is often considerably above the mean air temperature in summer and fall. A lake, therefore, imparts—at night and on cloudy days—important amounts of heat to its surroundings. The cooled and hence heavier water particles sink down from the surface until they reach a stratum of the same temperature or density. In this way *convection currents* are set up, which likewise bring about an intermixing of greater or less extent. Obviously these vertical cooling currents can affect the form of the temperature curve only through the breakdown of stratification (paring it off, as it were, from the top downwards). They can never bring about a transfer of heat to the deeper strata. This vertical circulation is especially effective in late summer in completely equalizing the temperature of the epilimnion, and in conjunction with it there is usually set up an especially steep temperature gradient at its lower boundary (gradients of 3° C. per metre are not unusual; cf. Figure 11).

The metalimnion is of special importance not only in the mechanics of currents and mixing but also in connection with the *biota* in a lake. It divides the water mass into two regions characterized by fundamental differences. The epilimnion, remaining under the influence of the atmosphere through the turbulent currents that are created by every wind, is kept in motion, and any stratification set up in it can be only transitory. The individual water particles travel between the surface and the metalimnion through the many levels, as do also the passively floating forms, of which the phytoplankton are of particular interest. On the other hand, movements in the metalimnion and the hypolimnion—if they occur at all—occur predominantly within the one level; permanent vertical translocations of water particles take place only to a minor degree. *Above all, the metalimnion acts as a barrier between the upper and the lower regions* and prevents contact of the hypolimnial water and of the organisms suspended in it (in so far as they are not capable of independent movement) with the surface, that is, with the air, and (when the depth is great enough) bars them from the use of light.

(*c*) *The energy content*

A consideration of the *energy content* imparted by the temperature and density stratification cannot be avoided in view of the transcendent importance of temperature in almost all hydrographic and biological problems in a lake. It is often only by such a consideration that an understanding can be gained of the phenomena that are observed. Therefore, we shall briefly touch on the most important views in this field of investigation.

Each body of water possesses a certain *heat content*; thus it has a store of heat which it can impart to its surroundings on cooling to 0°C. This increment of heat at a temperature of t°C. is equal to the amount of heat in calories which must have been transferred to the water mass to heat it from 0°C. to t°C.

Since the specific heat of water is unity, the calculation is extremely simple. We first take a column of water from the surface to the bottom and by multiplying its volume in $m.^3$, $dm.^3$, or $cm.^3$ by its mean temperature in degrees centigrade obtain the heat content in ton, kilogram, or gram calories (if we use 0°C. as the base temperature). If our water column is so taken as to have a cross section of 1 $m.^2$, the product of the depth in metres by the mean temperature in centigrade gives the heat content directly in ton calories, since the volume of each metre of depth is one cubic metre. The mean temperature can be calculated as the arithmetic mean of the temperatures of the individual metre strata or it can be determined graphically (by planimeter from a temperature curve plotted on rectangular coordinates).

If we wish to determine the heat content of a whole lake the calculation is somewhat more troublesome. We must then take into account the diminution of volume with depth, because the lower strata make a smaller contribution to the total heat content than do the upper ones. From a bottom-contour map of a lake obtained by sounding, we can get the area of whatever contours are desired and find the volume of each metre stratum. These fractional volumes are multiplied by the corresponding mean temperature for each metre and the products summed to obtain the total heat content of the lake. There is a graphical method, introduced by Schmidt (1915), whereby this calculation can be performed rapidly and conveniently.

If we divide the heat content thus obtained by the volume of the lake, we obtain the true *mean temperature.* If we divide by the surface area, on the the other hand, we obtain a number that is of importance for the heat balance. It gives the *average heat content in the water below one square metre of the surface.* Changes in this value during the course of the year show what amounts of heat—either gains or losses—pass through a unit area of the surface in a given length of time.

The winter temperature of lakes in the Alps is, in general, if we exclude the uppermost layer, not far from 4°C.[9] (the temperature of maximum density). It is much better, therefore, in such cases, to take the winter temperature (instead of 0°C.) as the base temperature in the determination of the heat content, and in this way obtain the amount of heat the lake has accumulated, or the gain in heat, since the winter.

The caloric energy represented by the heat content of a lake is very considerable. Thus in the relatively small Lunzer Untersee (surface area 0.68 $km.^2$, volume 13.6×10^6 $m.^3$) the minimum heat content in winter is 40×10^6 ton calories and the summer maximum is about 160×10^6 ton calories. The heat gain, and correspondingly the amount of heat given up during the cooling from summer to winter, is therefore about 120×10^6 ton calories or equivalent to that released by the combustion of 15,000 tons of high-grade coal. From this example, it can be understood that large lakes (for example, Lake Constance) are able to influence the local climate appreciably while cooling.

The heat content is not the only form of energy of importance to us which enters a lake stratified with respect to temperature and water

[9]In the Great Lakes of North America, on the other hand, the winter temperature may drop to almost 0°C.

density. In many respects the *content of mechanical energy* is still more important. We shall first disregard the kinetic energy that is incorporated in water masses in motion and concern ourselves solely with the *potential energy* present in a lake in which lighter water lies over denser. In such a lake, as Schmidt (1915, 1928) has pointed out, a stable system is present which is not able to perform work and can only be upset by the introduction of new energy. We are thus speaking of "potential energy" in a sense which is opposite to the way the term is most generally used, for the characteristic of stability is just the loss of potential energy in contrast to other possible states (Schmidt).

Birge (1916) was the first to determine with some exactitude the amount of energy that must be expended to produce a given stratification in a lake. It is the sum of the work performed by the wind in driving the water from the surface to the various strata at summer temperatures against the resistance of an equal column of water at 4°C. (the spring temperature).

From other assumptions, Schmidt (1915, 1928) arrived at the following conclusions independent of Birge. While Birge calculated the work that was necessary to bring about the summer distribution of temperature starting from a homothermal condition at 4°C., Schmidt determined the expenditure of energy necessary to upset an existing stratification or to bring it to a state where the whole water mass would have taken on the mean temperature by mixing. Schmidt termed this expenditure of energy the *stability of stratification.* The idea of stability is of special importance to the limnologist. It gives a value for the resistance that a given state of stratification is able to oppose to the stirring effect of the wind and thus also a value for the degree to which the hypolimnion of a lake is shut off.

Since the centre of gravity of a stratified body of water lies lower than that of an unstratified one (because denser layers are below) we can define stability as the *work* required to raise the centre of gravity an amount corresponding to its displacement downward from its original position. This is equivalent to lifting the weight of the whole lake by a distance equal to the difference between the two centres of gravity. This elevation then offers a suitable means of calculating the stability. If the centre of gravity of a lake in a stratified state lies at a depth of s metres, and in an unstratified state at s_0 metres, and if we put the weight equal to the volume, V (since the specific gravity of water is always very near to unity), then the stability is

$$S \text{ (in kg-m.)} = (s - s_0) \cdot V.$$

Let us now derive from this simple proposition a formula useful under all conditions. This extension of the formula is dealt with by Schmidt, who also gives instructions for determining the stability graphically. The method

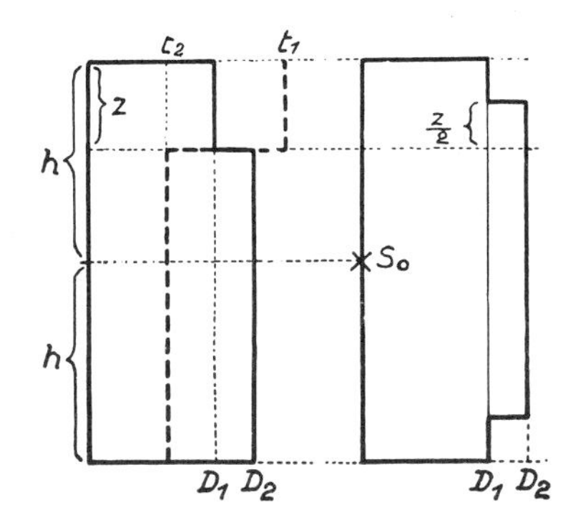

FIGURE 10. Diagram for calculating stability. For explanation see text.

is also to be found in Eckel (1950). For our purpose it will be sufficient if we, by making an extremely simple assumption (which in nature would scarcely be realized), establish a formula that will permit us to appreciate the fundamental laws of stability. Let us assume that our lake basin has perfectly perpendicular walls and that the temperature stratification in this column of water is such that at a given depth z there is a sudden change in temperature with completely uniform temperatures t_1 and t_2 with densities D_1 and D_2 existing above and below (Figure 10). The centre of gravity s_0 (in the mixed state) lies at depth h, the total depth of the basin being $2h$. It is clear that in the establishment of equilibrium by mixing, the excess weight in the part below the temperature change must be lifted to the level of the centre of gravity s_0. This excess is $(D_2 - D_1)$ multiplied by the height[10] of the lower colder stratum, thus multiplied by $(2h - z)$, and the elevation amounts to $z/2$. We therefore obtain the stability as

$$S = (D_2 - D_1) \cdot (2h - z) \cdot z/2.$$

From this formula it can be seen that the stability is dependent on the density differences in a given stratification. Since the density difference $D_2 - D_1$ is appreciably greater at higher temperatures than it is at lower ones (cf. p. 11) it is understandable that there is only a small degree of stability at the commencement of the spring warming and that at this time important amounts of heat are easily transported to the depths by wind-generated currents. Further it is clear that tropical lakes in which only small differences in temperature between surface and bottom occur (e.g., surface 28–30°C.; bottom, 24–26°C.) can nevertheless attain very large stability values.[11]

[10]In the example here the height is equivalent to the volume.

[11]As an illustration of these relations let us take the stability values for Altausseersee (the east Alps) and for Ranu Klindungan (east Java). The observations for these lakes give very similar temperature curves except for the absolute levels of temperature—in Altausseersee: epilimnion 12.2°C., hypolimnion 5.2°C. (difference 7.0°C.); in Ranu Klindungan: 31.4°C. and 26.2°C. (difference

We can further see from our formula that under otherwise equal conditions the stability first increases with the descent of the metalimnion (thus with increasing z), reaches the highest values when the metalimnion is at the level of the centre of gravity s_0 ($z = h$), and as the metalimnion proceeds to greater depths decreases again towards zero ($z = 2h$). This is in harmony with the phenomenon that when the metalimnion lies high, as frequently occurs with marked warming of the surface in spring, it does so only for a brief time and quickly disappears or is driven deeper, whereas the deep-lying metalimnia of summer often remain unchanged in position for month-long periods. In autumn, on the other hand the breakdown of stratification occurs the faster the deeper the metalimnion has been driven, as soon as it has passed below the centre of gravity. If it is desired to calculate the stability of stratification of an entire lake by Schmidt's method the volume in relation to depth must be considered as well as the heat content. As an example of the order of magnitude of such total stabilities the conditions in tiny Lunzersee may be taken again: values of 30×10^6 kg-m. are frequently obtained for it in midsummer; this is the work required to raise thirty 10-ton cars 100 m.

It is perhaps not superfluous to add that in inland waters the stratification of layers of different density and the resulting stability is not *always* due to temperature differences. *A difference in salt content* can also produce a similar phenomenon. This is frequently the case in saline lakes, and in these the stability due to the salt content can supersede that due to the stratification of temperature. Hutchinson (1937) has investigated such cases in his studies on limnology of land-locked lakes in desert regions and has given them a thorough mathematical treatment. However, even in the freshest water the universally occurring differences in concentration of dissolved substances can by no means be disregarded (Berger, 1955) (meromictic lakes, cf. p. 40).

(d) The annual temperature cycle

After this fundamental orientation, we can now come to a consideration of the *annual cycle of temperature in a lake.* It has already been mentioned several times that a lake in the temperate zone is, in general, uniformly at the temperature of maximum density (4°C.) from top to bottom at a certain time in the spring. In this state the stability is equal to zero. Wind-generated currents can proceed through it without hindrance and set the whole water mass of the lake in motion with consequent thorough intermixing. This is the time of the *spring turnover.*

5.2°C.). The sharp thermocline in both cases lay between 5 and 10 m. The stability between 0 and 40 m. depth amounted to 60 kg-m. per square metre of surface in Altausseersee. In Ranu Klindungan it was (in spite of the smaller temperature difference) 190 kg-m.

The duration of this period depends on weather conditions. Usually it is short, since surface warming generally sets in rapidly after the removal of the ice cover. However, as is pointed out above, the stability is slight at the beginning of the warming period because of the low prevailing water temperatures and because the thermal gradient is confined to the uppermost strata. Any stratification that arises is destroyed by mixing with the onset of even a relatively weak wind.

These circumstances lead to a gradual, more or less uniform, warming of the water mass to a considerable depth, occasionally right to the bottom. This is especially the case in lakes exposed to the wind, as for example those in the lowlands of north Germany or in Finland where total circulation by no means stops at the beginning of the warming period and a considerable rise in the temperature of the lower strata is the general rule. In sheltered situations, on the other hand, or when the break-up is immediately followed by a calm warm period, the stability brought about by the rapid warming of the upper strata quickly becomes so great that a thorough mixing by the wind is no longer possible. The lake then almost without change enters the state of *summer stagnation.* Under certain circumstances the vernal mixing can be so incomplete that conditions in the depths can remain as they were throughout the winter.

Meteorological conditions immediately after the break-up of a lake are therefore of the greatest importance in determining the *temperature of the deeper strata.* If the surface warming is rapid with no extended period of circulation, then the bottom temperature of the lake will be in the neighbourhood of 4°C. throughout the summer. However, if changeable windy weather prevails in the spring the whole water mass of the lake will be gradually warmed to the bottom before a resistant stratification can be formed. In consequence the bottom temperature of a particular lake will not be at all the same in different years. In Lunzer Untersee the variations observed up to the present have been about two centigrade degrees.

At the beginning of the summer stagnation, the thermocline does not by any means have the sharply expressed characteristic form shown in Figure 9. The active transfer of heat at this time generally produces a gentle gradient near the surface; only later does the temperature curve assume the typical midsummer form under the influence of variations in heating, cooling, and mixing by the wind. The concomitant approach of the metalimnion to the centre of gravity also brings about an increase in stability (see p. 34). Consequently, stronger and stronger winds are required in order to displace the metalimnion still farther downward.

Finally a state of equilibrium is reached in which the position of the metalimnion remains almost constant, and changes in thermal conditions in the epilimnion no longer take place.

With the autumn cooling, or in other words when the *loss in heat* is greater than the intake, the breakdown of thermal stratification begins. The position of the metalimnion remains unchanged at the beginning of this process. However, with the reduction of the temperature difference the stability of the stratification is reduced, and the time soon comes when a wind of normal strength (actually the current generated by it) can overcome the resistance of the metalimnion, and a progressive incorporation of its water mass and later of those masses underlying it can follow. This process is called the *autumnal partial circulation.* While the metalimnion sinks down farther and farther during this process and the temperature difference becomes less and less, the steepness of the gradient becomes greater and greater—especially at the beginning of the mixing process.

This partial circulation leads gradually to a uniform temperature in the whole water mass (homothermy) and with it to the end of the stability (cf. the temperature curves in Figure 15). The currents can pass unhindered to all depths and the lake enters the state of the *fall turnover.* The level of the uniform temperature displayed by all strata at the beginning of this stage depends on various circumstances but in particular on the depth of the lake. It should be remembered that a vertical transfer of heat takes place in the mixing processes brought about by the wind. We can therefore expect a sudden rise in the bottom temperature in late autumn; in one particular case (Lunzer Untersee) it was from 4.5° to 6°C. If the lake is very deep, however, this transfer of heat cannot raise the temperature of the extensive hypolimnion to any appreciable degree above its springtime level.

The *duration* of the fall turnover depends on meteorological conditions as well as on the depth of the lake. The greater the mass of water to be cooled, the longer the cooling process will continue. In the large and deep lakes of the Alps, therefore, the state of circulation usually continues through the whole winter and hence these lakes seldom freeze. In shallower basins, however, the whole volume of water from top to bottom is sooner or later cooled to 4°C., the temperature of maximum density. A further loss of heat is limited to the uppermost strata since the cooled water particles are then lighter than those that lie under them. As soon as the temperature reaches 4°C. a single cold calm night can lead to the formation of a sheet of ice.

Since, however, the difference in density of water between 0°C. and

4°C. is slight (see Figure 1) and the gradient lies just below the surface, only a minimum of stability is set up by this winter stratification and a very gentle wind is sufficient to break it down. For that reason it can often be observed that the cooling of the whole water mass proceeds further, to about 3°C. and less, before—with the intervention of calm conditions—the ice seal finally comes. The temperature prevailing at that time will be maintained throughout the winter at least in the deeper strata. The winter temperature of a lake is thus governed, as is the summer temperature of the hypolimnion, by the meteorological conditions that prevail immediately before stratification becomes stabilized.

Frequently there is a small rise in temperature under the ice during the winter, which will be noted also at greater depths in lakes where the water has been cooled appreciably below 4°C. in the turnover. Since clear ice is so transparent (see p. 23) such warming is essentially the result of the absorption of solar radiation. However, when there is a snow cover, which is the usual case, the effect of radiation is almost entirely removed. In addition, heat stored in the lake bottom during the summer can raise the temperature of the water in winter.

The ice cover protects the lake from forces impinging on its surface, in particular from the effect of the wind. After its formation a permanent stratification can be set up which begins with 0°C. immediately under the ice, and displaying a sharp rise, reaches the temperature of the deeper strata in a short distance. This state, which exists only under conditions of ice cover, is termed the *winter stagnation* (cf. the temperature curves in Figures 6 and 42c).

(*e*) *Types of stratification*

Very frequently the temperature stratification during the stagnation periods does not offer the simple picture sketched above. Variations from the norm occur which are caused both by meteorological conditions and by the individual characteristics of the lake.

In summer it is often found that not only is a metalimnion developed but also there are two or more regions where the temperature drops steeply. Such conditions are brought about through alternations of periods of warming with periods of extensive mixing. When a period of fine spring weather rapidly increases the temperature of the upper strata and then stormy weather with strong winds and cold sets in, the shallow metalimnion is driven down to a substantial depth—just as it is in the autumnal partial circulation. A new warming period can then lead to the formation of a second metalimnion overlying the first. This process can obviously be repeated and it is possible to infer the thermal history of a lake since the last total circulation from an observation of the temperature curve.

Special forms of stratification can also be brought about by the extent of inflow-outflow relations and by the form, size, and location of the lake basin. Large flows of water through a lake generate turbulence in it, especially in the stratum where the inflowing and generally colder water levels out—that is, in the region of the metalimnion. As a result the metalimnion increases greatly in extent and does not have as steep a temperature gradient. This phenomenon is particularly observable in lakes which are expansions of large rivers. (In Figure 11

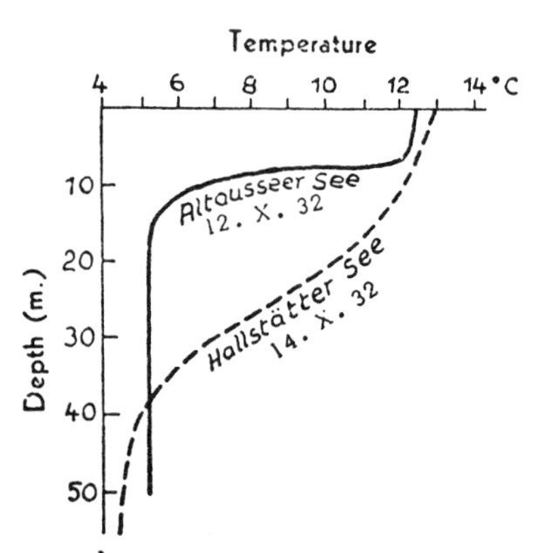

FIGURE 11. Influence of inflow-outflow relations on the extent of the thermocline in two adjacent lakes.

the temperature curve for Hallstättersee, an expansion of the Traun, is compared with that for Altausseersee, which has but little flow through it and in consequence exhibits an especially sharp thermocline.) The influence of the *surface area* is such that, in general, large lakes with their greater turbulence resulting from the extensive surface have deeper lying metalimnia than do small lakes. The data of Ruttner (1931) for various-sized lakes in Sunda offer a good example of the effect of area. Berger (1955) found that the "maximum depth of mixing" in (meromictic) lakes is proportional to the fourth root of the surface area. However, the influence of size can be cancelled by the location of the lake or by its orientation in relation to the direction of the prevailing winds. In small sheltered ponds very extreme temperature stratifications can be observed which in their form and divisions are identical with those of lakes, but in which the depth relations can be completely expressed in a few decimetres.

In general each lake—when the average summer stagnation is considered—has its own individual temperature curve, which is the resultant of the effects of the climate, of the lake's position, form, size, and the volume of flowthrough.

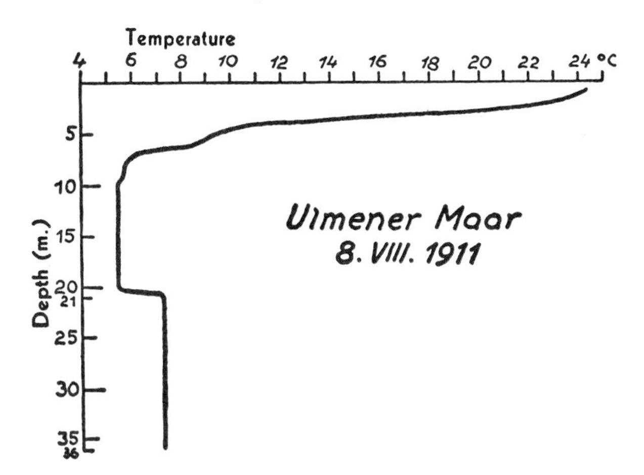

FIGURE 12. Temperature stratification in a meromictic lake. From Thienemann (1915).

In many lakes peculiar stratification phenomena are found which are not understandable at first glance. It is not altogether rare to find a considerable *increase* in temperature at a certain depth in the hypolimnion below a perfectly normal temperature curve in the upper strata. A very good example, the temperature curve of Ulmener Maar taken from Thienemann's (1915) observations, is reproduced here (Figure 12). This apparent paradox is explained by an equally marked stratification of the salt content. The salinity amounted to 185 mg. per litre in the upper and to about 500 mg. per litre in the deeper strata. The increase in density brought about by this difference in concentration is several times greater than the decrease brought about by a temperature rise from 5.4° to 7.2°C. This density gradient brings about considerable stability, which prevents a complete turnover at the time when total circulation would normally occur. The salinity and the higher temperature in the deep waters of the volcanic Ulmener Maar are brought about by springs which empty into the lake. Similar and even more extreme cases are found in the volcanic regions of Indonesia where temperatures up to 91°C. have been measured in the depths of crater lakes while the surface strata display temperatures near that of the air.

The difference in solute concentration need not be quite as great as in the example of Ulmener Maar cited above in order to make stratification possible in spite of the temperature rise below. As was outlined on page 8 a salt content of 1g./l. increases the specific gravity by about 0.0008. The difference in density attendant on a change in temperature of from 4° to 5°C. amounts to 0.000008. To compensate this difference in density thus requires an increase in salt concentration of only 10 mg./l. It is not at all rare to find lakes—even in the Alps—in which the difference in concentration between the epilimnion and the hypolimnion is several times this amount.[12] If such a basin is sufficiently deep and sheltered, stabilities can be set up in this manner that can prevent a fall turnover. Circulation in such lakes extends only to a certain depth and for that reason they are termed *meromictic*, following Findenegg's (1935) terminology, in contrast to the normal *holomictic* lakes. The water mass that does not take part in the annual turnovers has been termed the *monimolimnion* by Findenegg and the expression *chemocline* has been employed by Hutchinson (1937) for the stratum containing the concentration gradient. The latter author has given special consideration to stability relations in meromictic lakes and Berger (1955) has shown how to determine the relative importance of "concentration stability," which is always present to a greater or lesser degree, in relation to "temperature stability."

The continuously stagnant (or renewed in the course of years by gradual interchange) hypolimnion of meromictic lakes usually shows a slight rise in temperature of a few tenths of a degree, which can be due to a variety of factors. In many cases the lake bottom, the temperature of which at the altitudes in question nearly equals the mean air temperature is the source of heat. In others again it is the seepage of ground water or perhaps also the heat of combustion of oxidized organic substances.

The meromictic state with a chemocline is not restricted to deep lakes but can occur temporarily in shallow bodies of water and has been found in a small spring-fed lake only 3 m. deep (Lunzer Mittersee, Ruttner, 1955).

In this connection should also be mentioned the well-known Transylvanian salt lakes in which a highly concentrated salt solution (24 per cent!) is

[12]An example may demonstrate this. On June 24, 1933, Krottensee (upper Austria) had a temperature of 4.6°C. and a salt content (as measured by its electric conductivity) of 218 mg./l. at a depth of 20 m., and a temperature of 5.0°C. and a salt content of 264 mg./l. at 40 m. On the basis of temperature, the density at 20 m. would have been 0.999997 and at 40 m., 0.999992; thus the stratification was apparently unstable. When, however, the salt content is taken into consideration, the actual density at 20 m. amounts to 1.000182 and at 40 m., to 1.000217; the stratification is therefore entirely stable.

overlain by a shallower sheet of fresh water. In these lakes at depths of from about 1 to 3 metres, temperatures up to 56°C. are reached. The radiant energy of the sun is stored within the saline layer since the release of the heat at the surface cannot take place because of the impossibility of circulation. Anderson (1958) found similar conditions in Hot Lake, a salt lake 3.5 m. deep in the State of Washington (U.S.A.). Here the temperature at a depth of 2.5 m. was 30°C. even when the lake was frozen. The July maximum was 50°C. at a depth of 2 m.

Our considerations of the temperature stratification of lakes up to this point have concerned only regions which display a regular alternation of summer and winter. How do those waters behave *which lie outside the temperate zones* of our earth? We have seen that our lakes essentially conserve the winter temperatures in their depths throughout the summer. Only the anomalous expansion of water (the density maximum being at 4°C.) prevents the temperature of the hypolimnion from being still lower. In a climate where the air temperature does not fall below 4°C. for any considerable length of time there can naturally be very little cooling of the waters below this temperature. In lakes in such latitudes there is no inverse stratification and the temperatures of the depths will show a higher value than do ours. For example, even Lago di Garda, in which the temperature of the depths is about 7°C., is such a lake.

The closer we approach the tropics the less will be the annual variation both of air and of water temperature and hence the smaller also the difference between surface and bottom. In the equable climates of tropical regions we must expect high bottom temperatures, for these can scarcely be less than the mean temperature of the coldest time of the year. Accordingly, the lowland lakes of Indonesia, for example, have temperatures of about 26°C. in the hypolimnion down to their greatest depths. The monthly mean of the surface temperature in a given instance varied between 26° and 29°C. and thus is only a little higher than the profundal temperature. At higher altitudes these relationships change only in that the temperatures of all the strata are lower. Thus in Lake Titicaca, which lies at an altitude of 3800 meters, surface temperatures between 10.1° and 12.5°C., and bottom temperatures between 9.4° and 11.4°C., have been found. The characteristic feature of tropical lakes is therefore not the high surface temperatures but the slight seasonal variation in temperature and the small difference in temperature between the surface and the bottom.

Otherwise the temperature curves for tropical lakes (Figure 18) display exactly the same principles of stratification that we have recognized for temperate lakes. In spite of the insignificant temperature differences, the stability is scarcely less than in lakes of our latitudes (see p. 33) because of the very rapid decrease in density at high temperatures. Another phenomenon also arises from the property of the density of water just mentioned. In the tropics a given degree of cooling releases much more active convection currents than it does in the lakes of the temperate latitudes at the time of the fall turnover. In the latter, with their commonly lower temperatures and therefore smaller differences in density, the speed of the descending currents can only be small—indeed in the neighbourhood of 4°C. it is almost infinitesimal. On the other hand, the slightest cooling in a tropical lake sets up convection currents, which soon affect the warm hypolimnion also if this state continues, and must quickly lead to a turnover in the lake without the necessity of any wind action.

Since the annual variation in temperature is very slight in the true tropics, turnovers in the lakes there are not confined to certain seasons as they are in our regions. Turnovers can occur repeatedly at short intervals of time when periods of cooling quickly follow each other. On the other hand, after an exceptional cooling period with its depression of the temperature of the lower strata, none may occur again perhaps for years. Only in those places where rainy seasons alternate with markedly dry ones characterized by intense radiation, as for example in the east monsoon region of Java, do turnovers occur at regular seasons. There they are associated with the dry season.

In the *Arctic* and in *high mountain* regions there is another type of lake, which contrasts with the tropical lakes. This type is frozen throughout almost the entire year and does not reach a stable condition of direct stratification by warming in the short time it is ice free. The state of these lakes thus varies between winter stagnation and total circulation: their surface temperatures rise only a little, if any, above 4°C.

Forel (1901) was the first to propose a generally accepted classification of the lakes of the world according to their thermal conditions. He distinguished three types: (1) *temperate lakes*, which displayed a regular alternation of periods of stagnation accompanied by direct or inverse stratification with intervening periods of total circulation; (2) *tropical lakes*, which have a continuous direct stratification and whose temperature therefore never sinks below 4°C.; (3) *polar lakes*, with temperatures below 4°C. and inverse stratification.

Forel's tropical type, in the light of the results of limnological investigations in recent decades, is probably somewhat too broad. According to the definition given, numerous lakes in the Alps belong to it, such as Lago di Garda and other upper Italian lakes, which in spite of a lack of inverse stratification in winter are nearer in every respect to the temperate type than to the true tropical one. It is probably more appropriate to designate as tropical lakes only those that lie within the tropics and whose temperatures vary but slightly throughout the year, so that only relatively slight differences between the epilimnion and the hypolimnion occur. All other lakes that are never inversely stratified, and which are similar in their annual cycle of temperature to the temperate type, might be grouped together as (4) *subtropical lakes* (Yoshimura, 1936).

Hutchinson and Löffler (1956) provided another classification to replace Forel's, which is inadequate in many respects. They based their classification on the number of periods of complete circulation in the year, as follows:

Cold monomictic: lakes with only a summer period of circulation (polar and sub-polar lakes).
Dimictic: lakes with spring and fall turnovers (temperate lakes).
Warm monomictic: lakes with a winter circulation period (sub-tropical lakes).
Oligomictic: tropical lakes with rare periods of turnover at irregular periods.
Warm polymictic: tropical lakes with frequent periods of turnover.
Cold polymictic: tropical mountain lakes that have an almost continuous circulation.

The importance of these lake types, which are based on climate, to the consideration of questions of animal geography is discussed thoroughly by Löffler (1958).

3. *Water Movements*

Since the movements within the water mass of a lake which are of importance to limnologists depend largely on temperature or density stratification for their effect, a brief discussion of them might have been included in the thermal section. We can distinguish between *rhythmic* and *arrhythmic* water movements. To the first belong the travelling and standing waves; to the latter, currents of various kinds.

(*a*) *Waves*

The best known of all water movements, *travelling surface waves*, are of only subordinate interest to us in our considerations, for their effect is limited to the uppermost strata and they therefore take no part in the displacement of the more extensive masses. If we observe an object floating on the water surface which is being moved by a wave, a cork for instance, we see that it rises and falls in a nearly circular path (an orbit) so that after the completion of the wave cycle it is returned to its original position again. Like the cork the water particles also have only an oscillatory motion without undergoing any permanent translocation. The impulse for this oscillation arises from some force, perhaps the impact of an object or the force of the wind, which makes the water particles depart from their state of equilibrium. Acted on by the force of gravity, they then seek to regain their original position. The movement must be such that neighbouring particles will move in the same manner one after the other. This results in a travelling wave the cross section of which is termed a *cycloid*, that is, the path described by a point on the rim of a wheel rolling over a plane surface. We shall go no further into the fundamental theory of wave motion that has been particularly developed in oceanography but will refer the reader to the works of Forel (1901), Aufsess (1905), Thorade (1930), and Defant (1929).

We are interested here solely in the question of the depth to which the surface waves in our lakes will have an influence. Deeper particles will obviously be affected by the oscillation, although the diameter of the circles the particles will describe decreases rapidly with depth. A simple relation has been established for this decrease. For every increase in depth of one-ninth the wave-length (the distance between two peaks) the amplitude decreases by one-half. At a depth that is equal to the wave-length the amplitude is, therefore, less than 1/500 of that observed on the surface. The height of waves (surface amplitude) does not bear any fixed relation to their length. However, on the average, according to Forel, it is about 1/20 of the wave-length. At a wave-length of 18 m. (which approaches the maximum observed in Lake Geneva) and with a height of about 1 m., the amplitude of oscillation of a water particle at a depth of 4 m. (2/9 of the wave-length) is 25 cm. according to the relation given above, at 8 m. (4/9) it is 6.25 cm., and at 18 m. not quite 2 mm. Waves of such a size occur only on exceptional occasions except in very large lakes. The much smaller waves that are usual in basins of small and moderate size will not be appreciable even at minor depths. These considerations, however, hold only for deep water. In shallows, if the depth is much less than the wave-length, the extent of motion of the water particles is approximately the same throughout from the surface to the bottom and is rather like a to and fro movement of the whole water mass (Defant, 1929).

The average depth to which wave action is effective is important in the littoral zone of lakes since sedimentation of the beaches is prevented above this depth. Thus the marl banks on the steep sides of the alpine lakes do not increase in height above a certain point. Their almost level surface lies at a depth that is dependent on the extent of the wave action. In Lunzer Untersee, for example, this is about one metre.

Standing waves. Standing waves cause much greater displacements than do the travelling ones and are therefore much more important in limnology. In the *latter*, the water particles all move through the same distance, but *differ in phase*, while in the oscillation of a standing wave the *phase* of all particles is the *same* but the *extent* of the movements is *different*. At the *nodal points* no vertical movement takes place at all, while at the *antinodes* it reaches a maximum.

The processes entering into play in these oscillations can be easily demonstrated by a model in which we set water containing suspended particles of some sort to swinging from end to end in a trough. If the experiment is correctly performed, the water surface will oscillate about a horizontal line lying at the middle, the node line. At this place, no vertical movements occur, while at the two ends, the loops, the water rises and falls at a rate dependent on the length and depth of the vessel. Simultaneously the floating particles below the node line will be seen to move to and fro horizontally with the same rhythm, since of course the water that is drawn to one side by the sloping surface must flow over to it from the other. If the vessel is very long in relation to its depth, movements of considerable extent take place under the node line even with small amplitudes at the ends. The speed of the horizontal movements is greatest at the nodal point and gradually decreases to zero at the ends of the vessel.

The standing wave in our experiment was *uninodal*. We can, however, generate *binodal* standing waves if we exert a rhythmically varying pressure on the surface in the middle of the vessel. A complete antinode will arise in the middle, there being also two half antinodes at the ends as in the first experiment. *Plurinodal* waves can also be observed. Finally we distinguish between longitudinal and transverse standing waves according to whether the oscillation is about the transverse or longitudinal axis of the trough.

In a rectangular vessel with a plane bottom, the formula first calculated by Merian (1828) for the period t (the duration of the to and fro movement in a uninodal oscillation) can be reduced to: $t = 2l/\sqrt{gh}$, in which l is the length of the vessel, h the depth of water, and g the acceleration due to gravity. If the vessel is not rectangular in cross section and the bottom is not plane, quite complicated equations, which we shall not enter into here, take the place of this formula (see Forel, Aufsess, Halbfass, Defant). However, since in larger lakes the average depth is very small in proportion to the length, the simple formula gives a good approximation of the correct result.

These standing waves are a phenomenon occurring in all lakes, but only in large basins is the periodic rise and fall of the water surface extensive enough to be evident without the employment of measuring instruments. The fishermen on our largest lakes have known of them since antiquity and the scientific name for them, "seiches," originated at Lake Geneva (according to Forel it comes perhaps from the French *sèche* since shallows at the margin are left dry with the subsidence of the water). The most extensive seiche observed in Europe up to now, with an amplitude of 1.87 m., was also recorded in Lake Geneva, on

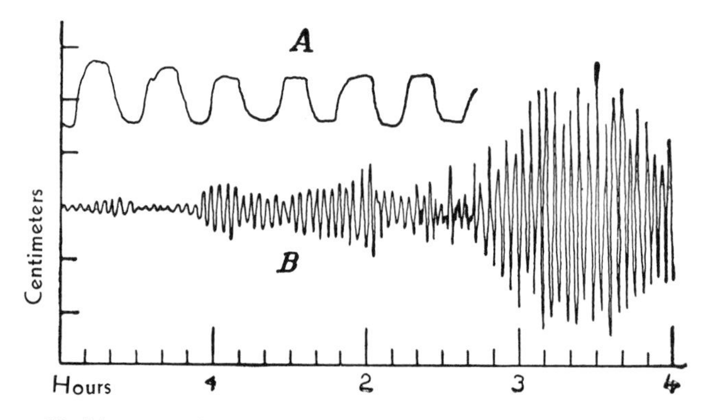

FIGURE 13. Limnograph records of standing waves (seiches). *A*, in Starnberger See (period of oscillation 25 min., from Aufsess, 1905); *B*, in Lunzer Untersee (period somewhat over 3 min.).

October 3, 1841. As a rule, however, the extent of these oscillations of the surface of this lake is far below this maximum. As an example may be cited the series of seiches (March 1891) given by Forel (1901) for the same lake, which lasted for 7½ days and consisted of 147 oscillations with a period of 73 min. and an amplitude which sank gradually from 20 to 7 cm. In order to make exact observations on such seiches in smaller lakes a recording level gauge (limnograph) is used. Figure 13 shows limnograph records for Starnberger See (Würmsee) and for Lunzer Untersee. Since in addition to the uninodal longitudinal oscillation in lakes, bi- and pluri-nodal as well as transverse oscillations, which are superimposed on the fundamental, almost always occur, the curves sketched by a limnograph are at times quite complicated.

The period of a seiche has a characteristic magnitude for each lake as follows from the formula given above. For example, the longitudinal fundamental is 73 min. in Lake Geneva; in Lake Constance, 56; Lago di Garda, 43; Attersee, 32; Grundlsee, 9.5; Altausseersee, 5.3. The dependence of the period of oscillation on water depth is displayed especially well in Lake Balaton, which is about as long as Lake Geneva but has a mean depth of only 3 m. and has a period of oscillation of about 10 to 12 hours. Complicated lake basins divided into arms and bays display numerous irregularities in their seiches. Thus an individual arm can have oscillations that are independent of the main basin. Bryson and Stearns (1959) have described such a system of changes in level with its associated currents for a bay of Lake Huron.

The force that generates standing waves is excess pressure on part of the water surface. This pressure can be exerted by local barometric variations, by wind, or by a rain shower which is limited to part of a lake. The energy required is relatively small. For example, in Würmsee (which has a period of oscillation of 25 min.) a rain limited to half the lake with a precipitation of 2 mm. in an hour, according to the calculations of Emden, set up an oscillation with an amplitude of 25 mm. (cited from Halbfass, 1923). Sauberer's (1942) observations on Lunzer Untersee showed that an ice cover does not prevent the occurrence of a surface seiche.

Standing waves are not limited to the surface, that is, to the boundary between water and air only; they occur also within the mass of fluid (without being in evidence as surface movements) if there are two or more overlying strata of different densities. To distinguish these from the surface seiches discussed above, they are called *internal seiches.*

We can again begin by studying the phenomenon experimentally. A coloured salt solution containing some suspended particles is introduced by a tube into the bottom of a glass trough, for instance, an aquarium not quite half full of water. As long as it is in a state of rest, the colourless water lies on the approximately equal amount of salt water in a more or less sharply marked plane. Standing waves are then generated by tilting the vessel or by applying pressure at one end. Both the surface and the bottom layer are set in oscillation, each, however, independently of the other in amplitude and period. While the rapid to and fro motion of the surface dies away in a short time, the slower oscillation of the bottom stratum continues long after the water surface has come to rest. If we introduce a weaker salt solution, thus one of lesser density, into a second vessel as the bottom stratum we shall find that its oscillation is still slower than that of the solution in the first. Thus the period of oscillation is evidently dependent on differences in density of the two strata in association.

The formula established by Watson (1904) for the period t of uninodal internal waves in vessels of rectangular cross section is:

$$t = 2l \Big/ \sqrt{\frac{g(d_1 - d_2)}{(d_1/h_1) + (d_2/h_2)}}$$

where l is again the length of the basin, h_1 the height of the lower heavier stratum, h_2 that of the upper one, d_1 and d_2 the corresponding densities, g the acceleration due to gravity. The dependence on the difference in density between the two fluid strata as well as on their heights can be found from this formula. We see further that the formula given on page 46 is only a special case in which the density of the upper medium d_2 (in that case the air) is very small in comparison with that of the lower (water, $d_1 = 1$) and can, therefore, be neglected.

From what is said above, we must expect internal seiches in our lakes in the boundary layer between the lighter water of the epilimnion and the heavier water of the hypolimnion. When such oscillations appear, the metalimnion swings up and down about one or more nodal lines with a period established by the conditions. This phenomenon can therefore be conveniently followed by temperature measurements taken at short intervals of time, simultaneously where possible, at two or more diametrically opposite places in the lake. It is evident that the metalimnion is scarcely ever completely still but rather that the same temperature can be found now deeper and now shallower. For this reason these standing waves were first called "temperature seiches."

The graph (Figure 14) taken from Halbfass (1923), which shows the varying position of the point of 12° C. at the north and south ends of Madüsee during a three-day period, is given as an example. The total period of oscillation here is nearly 24 hours, the maximum amplitude about 10 metres.

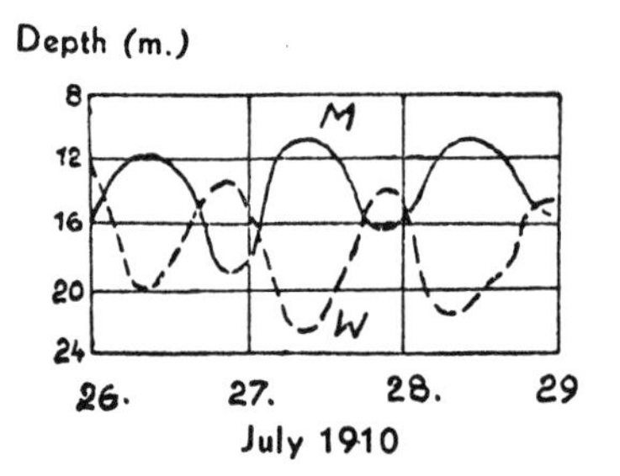

FIGURE 14. Vertical oscillation of the 12°C. stratum at two locations (*M* and *W*) in Madüsee; period of oscillation about 24 hours; from Halbfass (1923). This figure illustrates an internal seiche.

Mortimer (1950), who went back to models of the type described above, recommends a three-layer system (water, phenol, and dilute glycerol) for the study of internal seiches, since at least this degree of complexity is necessary to interpret these processes in many lakes. The reader may find the theory of displacements in such a three-layered system in Hutchinson (I, pp. 341 ff.), where there is likewise a description of its application by Mortimer (1952) to conditions in Lake Windermere.

In general the wind is the source of energy of internal seiches. In consequence the greatest amplitudes are found after severe storms. The drive of the wind leads to a piling-up of the epilimnial water at one end of the lake and a local depression of the metalimnion. When the wind drops, the water mass is set in oscillation by the equalization that follows, as in the experiments described above. However, such disturbances of the equilibrium can also be brought about by other forces, for example, by the inflow of flood waters. According to Mortimer (1955), the earth's rotation causes the oscillation of a seiche to rotate about a nodal point in large lakes rather than merely to swing about a nodal line.

The importance of internal seiches to limnology lies both in the considerable vertical displacement of the water masses at the loops and in the active horizontal movements in the region of the nodes. The rise and fall of the water masses leads not only to quite different results being obtained in consecutive temperature series, thereby making difficult the determination of the true position of the metalimnion in a state of rest, but it also alters, as Demoll (1922) has shown, the picture of chemical stratification and the distribution of plankton. Indeed all dissolved and suspended materials are raised and lowered rhythmically. The horizontal movements at the nodes lead to currents that are of much greater importance than those brought about by surface seiches because of the much greater amplitude. Two rhythmic but opposite current systems arise above and below the thermocline. Their speed—like that of a pendulum at the midpoint—is greatest when the metalimnion is horizontal.

Since the internal seiches are not dependent on the morphology of a lake alone but in particular also on the density difference and the depth of the superimposed water strata, their periods do not have any fixed length in a given basin. The oscillations are most rapid at the height of the summer stagnation, become slower and slower in the autumn circulation, and finally disappear entirely with the equalization of all density differences. However, the shortest periods in relatively small lakes are at least a few hours. Thus Exner (1928) observed, under normal wind conditions, a period of about 4 hours and an amplitude of approximately 1 metre in Lunzer Untersee, which is 1.6 km. long. The maximum speed of the horizontal current at the nodes was calculated to be about 1 cm./sec. under these conditions.

It is perhaps not superfluous to point out that internal travelling waves occur as well at the boundary between the epilimnion and the hypolimnion and that these can also make the location of the metalimnion difficult to determine.

(*b*) *Currents and eddy diffusion*

When we now turn to the arrhythmic water movements, the actual currents, we enter a field of science that has had a major development only in oceanography. In limnology, currents have been investigated only to a slight degree because the bodies of water usually studied are small with slow currents that are hard to measure. There are only observations (generalizations from which were somewhat premature) for a few large basins such as Lake Constance and some Scottish lakes. We must, therefore, limit ourselves to considerations which can be taken from physical laws and laboratory experiments and which, so far as they concern the course of the phenomena in any particular case, are not yet in any respect established by measurements in lakes themselves.

Currents can be generated internally (that is, they can originate in the lake itself and in stratification conditions in it) as well as externally. They first draw their energy from changes in density, which upset the stable equilibrium in a lake. We have already seen movements of this type in the *vertical convection currents* which are brought about by surface cooling. Among external influences the wind, as has already been pointed out several times, is the most important. It has in part a direct effect in that the uppermost water particles are carried along with the air current and a surface stream is set up in the direction of the wind. In addition the wind has an indirect effect, which is the result of the piling-up of the surface water driven onto the shore and of the pressure thus exerted on the strata immediately below. This induces a flow there in a direction opposite to that of the surface current.

If we allow water to flow slowly through a glass tube with a small crystal of dye (fuchsine, for example) adhering to its inner surface, we can observe the dye carried along in the form of a long uniform thread parallel to the longitudinal axis of the tube. The streamlines are parallel in this case. If the speed of the flow is now increased gradually, it will be seen that the path of the dye will suddenly be disturbed; vortices begin to be set up in it and a uniform pigmentation of the whole contents of the tube appears. The original *laminar* or *ordered* current has changed to a *turbulent* or *disordered* current, to a motion in which the individual streamlines intermingle in a tangled course with the result that the coloured and clear water are mixed together. Other experiments have shown that laminar flow can exist only below a certain velocity. This critical velocity depends on the viscosity of the fluid and on its density, and is inversely proportional to the diameter of the tube. For water it is about 18 cm. per second for a tube with a diameter of 1 cm. With a diameter of 1 m. it is less than 2 mm. per second. Under these circumstances it is clear that in natural waters laminar flow is scarcely to be expected in view of the magnitude of their channels and currents. Every brook or river offers us the opportunity of observing turbulence phenomena. The turbulences, which arise mostly from tiny irregularities in the bed, are carried along by the stream as helices with horizontal axes. These increase in size as they rise and finally end at the water surface in flat upwelling elevations. In places where neither the water surface nor the bottom dissipates the kinetic energy of these vortices, as in the open water of a lake, they spread at right angles to the original current and mix the water they embrace. These intermixing phenomena or, to use a better term, *interchanges* of adjacent water masses, with all their attributes—for example, temperature, dissolved substances, and suspended bodies (plankton)—are of far more importance in the economy of a body of water than the mere translocation of water by laminar flow. In a lake in which there is any current at all, whether it is due to the wind circulation mentioned on page 29 or to the displacements of mass which occur with internal seiches, turbulence is propagated perpendicular to the direction of the current. Since currents usually have a horizontal course and their speeds vary in different strata, the turbulence is generally vertical but with an intensity which changes gradually from stratum to stratum. These circumstances are often extremely striking in their effects on the intermixing of materials as well as on the flotation of the plankton organisms. We must for that reason devote some words to the laws of turbulence, which have been specially investigated by Schmidt (1925).

Since it is practically impossible to measure the path of an individual stream of turbulence, we shall follow Schmidt's line of reasoning and seek to draw general relations for mixing from the effects of turbulence. This is possible because the effects of the mixing processes are very similar to those of another physical phenomenon, the conduction of heat, so that we can use for our purpose formulae derived for the latter process. If a body has a temperature difference of s'°C. (that is, a temperature drop of s') between two surfaces 1 cm. apart, the flux of heat per second through 1 cm.^{2} perpendicular to the direction of the gradient is $S = k \cdot s'$, where k is the coefficient of conductivity of heat for the body in question.

In a quiet water mass that has a heat gradient from the top downwards, an equalization of the temperature that is not aided by external influences can only take place by the conduction of heat. If, however, a current with a speed above the critical velocity occurs in this water mass, then a turbulent transport of warm water down and cold water up takes place, which similarly equalizes the temperature but in an appreciably shorter time. This "apparent conduction" likewise transports heat through a surface at right angles to the gradient, and we can express the amount passing through a square centimetre per second in a fashion similar to that used in the case of simple conduction by the formula:

$$S = A \cdot s'.$$

Here s' again is the gradient per centimetre. A is a constant dependent on the degree of intermixing and thus on the intensity of the turbulent currents.

It is called the *coefficient of eddy diffusion* (Schmidt). A is not, as is k, a constant peculiar to a given substance, in this case water, but varies with the state of motion of the material.

It can be seen from our simple formula that the coefficient of eddy diffusion can be calculated directly when values for the gradient s' and the amount of heat transported per second through a given cross section are known. The latter is easily found from the increase in the heat content which takes place in a given period of time down to the depth under consideration.

The coefficient of eddy diffusion which Schmidt has given for four depths of Lunzer Untersee at various periods of the year is offered as an example of such calculations:

Depth (m.)	9.IV *to* 15.IV	*to* 8.V	*to* 20.VI	*to* 30.VII
5	A = 2.02	0.73	0.57	0.18
10	2.41	0.55	0.38	0.06
15	2.24	0.32	0.21	0.04
20		0.20	0.04	0.01

Consideration of these values is fruitful in several respects. First it can be seen that the mean coefficient of eddy diffusion decreases with time and with depth. Only in the first interval, which coincides with the spring turnover, is A uniformly high at all depths investigated. After the establishment of thermal stratification the values show a general decrease and also a great difference with depth, which is especially marked when summer stagnation has set in completely. It can be seen how mixing is greatly restricted by a fully developed metalimnion; but in spite of this there is still a slight degree of turbulence in the hypolimnion. The values given above do not give an indication of the degree of mixing in the epilimnion itself in Lunzer Untersee, for there the metalimnion generally begins higher than 5 m. The mixing in the epilimnion is doubtless greater than the value to 5 m. would indicate. In larger lakes more exposed to the effect of the wind and particularly in the sea, A reaches very high values. Thus Schmidt has calculated a value of $A = 50$ at a depth of 28 m. from observations in the Mediterranean. Mortimer (1941) has been able to determine an increase of A in the hypolimnion with increasing size and depth of the lake. Ricker (1937) refers to the importance of the daily vertical migrations of plankton for mixing and especially for the downward transport of heat.

Since not only the content of heat but also that of dissolved substances (for example, oxygen) is dependent on mixing, the mixing processes are of supreme importance in the supply and distribution of substances necessary for life in a lake, an importance that is being emphasized to an increasing degree in limnological investigation (cf. Grote, 1934). They are of greatest importance in the vertical distribution of plankton and other materials in suspension.

Up to this point, we have ignored temporal changes in the conditions influenced by mixing. In nature, however, such variations are the rule. Thus, for example, the heat content of the surface waters of a lake varies periodically in the course of the day and throughout the year. How are these diurnal and annual cycles of warming transmitted to the depths by mixing? In the first place it is clear that their phase shows a lag in time so that the temperature maximum appears later at a given depth than it does at the surface, just as is the case with simple conduction. Further, the amplitude of the variation decreases with depth. It is beyond our scope to derive the formulae that permit the calculation of these relationships; for these, reference should be made to Schmidt (1925). We shall be content with comparing only the retardation of the cycle and the decrease in the extent of the variation with increasing depth as related to the simple conduction of heat, on the one hand, and to the apparent conduction due to eddy diffusion on the other. With *simple conduction* alone, annual variations in temperature would extend down to a depth of 7.7 m. after one year, and the variation would decrease to one-tenth at a depth of 2.8 m. Yet, with a value of *unity* for A the same annual cycle would, aided by the *apparent conduction due to turbulence*, reach down to 199 m. and the reduction of the range of variation to one-tenth would take place at 73 m. With $A = 0.1$ (a usual value to find below a well-developed thermocline) the corresponding figures would be 63 and 23 m. The effect of these relationships on the annual temperature cycle at various depths is shown in Figure 15, which indicates the retardation of the maxima and the decrease in the annual variation at depths of 10, 15, and 20 m. (Schmidt, 1934). The coincidence of the curves in the autumn cooling period is attributable to the much greater mixing by convection, which is superimposed on the normal distribution of temperature.

An objection to the calculation of the eddy diffusion coefficient from temperature measurements is that temperature increases found at greater depths can be brought about by the absorption of radiation that penetrates to the depth in question (pp. 28f.) so that the penetration of an eddy diffusion current through the thermocline cannot be precisely determined by this means. However, the mixing process can be observed through changes in the salt content as well as through changes in temperature. Thus Mortimer (1941) calculated almost identical values of A on the basis of the distribution of various dissolved constituents and also from the temperature for the hypolimnion of a lake in northern England 13 metres deep. Further, in Traunsee, a large alpine lake nearly 200 metres deep, rendered meromictic through the introduction of saline industrial wastes, the annual fluctuations in salt concentration in the upper strata can be followed with a certain lag down to the greatest depths. In this case it was possible (by suitable changes

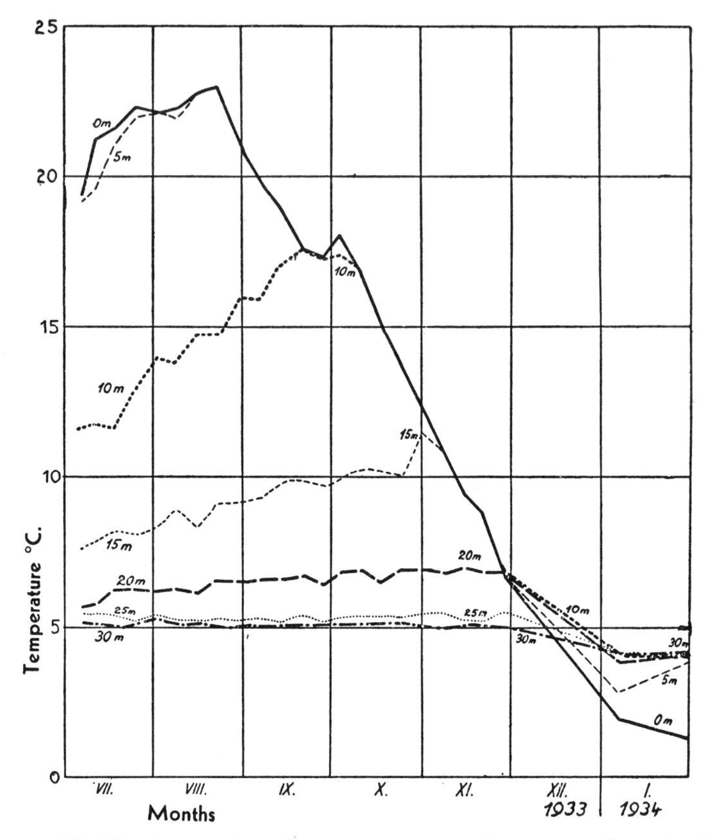

FIGURE 15. The temperature from summer to winter at various depths in Klopeiner See (Carinthia). After Schmidt (1934).

in the introduction of the effluent) so to improve the diffusion conditions that the decrease in oxygen content, which had already begun in the hypolimnion, was halted, and a gradual increase in the oxygen content followed (Ruttner, 1937).

As has been stated above, knowledge of the direction and position of *horizontal currents*, as opposed to their mixing effects, is not of the same importance for limnologists as it is for oceanographers. In comparison with the great ocean currents, which encircle the world and influence the climate of continents by transporting prodigious volumes at considerable speeds, the corresponding phenomena in inland waters are only feeble and confined processes, frequently of no more than local interest. We can only say in concluding this short discussion that but few exact measurements are available. The investigations of

Wedderburn (1910) on the wind-drift in Loch Garry (Scotland), where the speeds were great enough to be determined by the Ekman current meter used in oceanography, are an example of these. Figure 16 shows a diagram of the distribution of current (according to direction and strength) at various depths. The reversal of direction took place at about 10 metres and the return current reached a marked maximum at 18 metres, dropping to zero again immediately above the thermocline. These wind-produced currents completely express the scheme set forth on page 29 in their course and in their dependence on the temperature distribution.

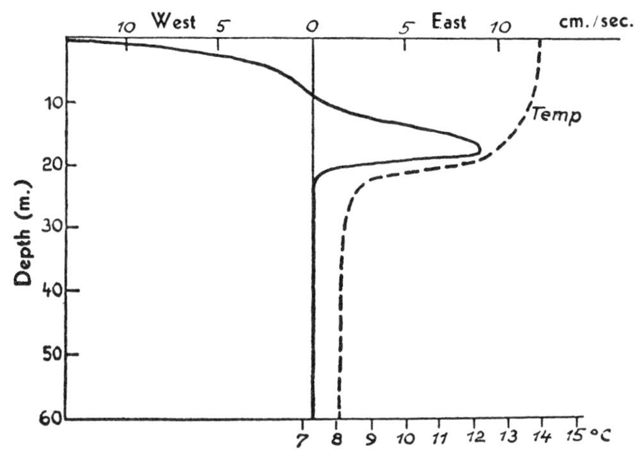

FIGURE 16. Direction and intensity of horizontal currents at different depths in Loch Garry (Scotland), August 26, 1909, with a west wind. After Wedderburn (1910).

It is clear without further consideration that both the formation of the basin and the direction of the wind have an important influence on such currents and that a complicated system of movements with various and changing directions can be set up in a given partial circulation. In large lakes, as in the sea, the rotation of the earth has an influence on the currents and the phenomena associated with them (the Coriolis effect). Observations indicating this effect were made and analysed by Elster (1939) for Lake Constance.

Currents of another type are brought about by the *inflow and outflow* of water and are of special importance in lakes on larger rivers. These first received precise study from Auerbach (1926) and his collaborators. They were further studied by Wasmund (1927, 1928) and later again by Elster (1939) in Lake Constance where the Rhine water flowing along the north shore can often be followed over wide expanses because of its different colour. The path this inflowing current takes is fixed by the primary direction of the water course, by the formation of the basin, and by the direction of rotation of the earth. The depth at which it travels is, in general, dependent on the density relations of the water masses. In the main (during summer) the inflowing water is colder and thus heavier than the surface water of the lake, and it therefore sinks. The submerged waterfall at the discharge of the Rhine into Lake Constance (the *Rheinbrech*) is well known. As soon as the inflowing water impinges on strata of greater density, its descent is checked and the current becomes horizontal. In these cases too, as with a return current in wind-drift, a metalimnion has a pronounced effect on the course of the current. From this results the fact, so important to the economy of the lake, that the flow through it in the summer and thus the renewal of water concerns only the strata lying above the metalimnion, and the hypolimnion is left undisturbed. The evidence for this was obtained through observations in Lunzersee (Ruttner, 1914). Of course the depth at which the introduced layer flows changes at different times of the year, and in winter the inflowing water of less that 4°C. flows on the surface if a higher salt content or load of silt does not impart an excess density to it. The outflow normally draws off water from the surface and from the strata immediately below it and thus brings about a considerable loss of heat in summer. This withdrawal of heat, and thus also the surface temperature of the lake, is influenced by the form and profile of the outlet (Brückner, 1909). If the outlet is wide and shallow the loss is greater than when it is narrow and deep. Lakes with either natural or artificial underground outlets often have particularly high surface temperatures.

II. DISSOLVED SUBSTANCES AND THEIR TRANSFORMATIONS

Water is never absolutely pure in Nature. There are always various substances dissolved in it which come from the soil, the air, and the metabolism of organisms. Even rain water contains very small, but by no means unimportant, amounts of nutrient substances as well as being saturated with the gases of the air at atmospheric pressure. Natural waters have all degrees of salt concentration from the minimum content of rain water to the saturated solutions of salt lakes whose shores and surfaces are encrusted with a crystalline coating. Organisms inhabit all of them—even the most extreme—if life is not made impossible by the presence of poisonous materials or by lethally high temperatures. The composition of the organic community is greatly dependent, of course, on the salt content of the environment. In general a limitation of the number of species begins at concentrations exceeding that of the sea (3 to 4 per cent). Beyond this level it becomes increasingly difficult for plants and animals to maintain their osmotic balance and the number of species decreases. Good examples of this are afforded by Thienemann's (1913) investigations on the saline waters of Westphalia. At salt contents up to 3 per cent, 64 species of animals live; from 3 per cent to 10 per cent, 38; from 10 per cent to 16 per cent, 12; and from 16 to 20 per cent, only a single species, but this one in tremendous numbers.

Every body of water—among all the myriads of brooks, rivers, and lakes of the land areas—whether a very dilute solution of a few fractions of a per cent or one of high concentration, must serve as a *nutrient medium* for the plants living in it which are not rooted in the bottom. They must be able to obtain in a usable form those elements they need for the formation of their tissues. Wherever plants grow, *changes in the chemical state of the environment* are caused by the withdrawal of nutrients on the one hand and by the release of the products of metabolism on the other. The life processes in the upper, well-lighted strata result in the removal of the dissolved nutrients; thus their effect is opposite to that of the processes at those depths of the lakes where the lack of light makes productive plant life impossible[13] and where bacterial decomposition of the sinking organic substances predominates. There the substances built up in the surface strata by means of the energy of the light are broken down into their inorganic constituents, which again go into solution and increase the nutrient content of the bottom waters.

Thus from the point of view of biological productivity the waters of a lake consist of two fundamentally different regions, one below the other, in which opposing processes take place. These are the region of photosynthetic production (the *trophogenic zone*) and the region of breakdown (the *tropholytic zone*) below. It is an especially important fact that during the summer stagnation the first of these is essentially the epilimnion (often the metalimnion is included) while the second region coincides with the stagnant hypolimnion. These differences in the distribution of chemicals brought about by living organisms we term biogenic chemical stratification. *It is quite impossible to understand the chemistry of an aquatic biotope without taking into consideration the causal relationships in the metabolism of its community of organisms.*

[13]See pages 200f. regarding chemosynthesis, which proceeds independently of light.

Only those dissolved substances that either are of no importance for life or are present in amounts in excess of those necessary do not undergo a marked change in their amounts and states.

In normal fresh waters the total content of dissolved solids (a determination that can easily be made by evaporating about a litre and weighing the residue, or by various other methods, for example, simply by measuring the electrical conductivity) consists largely of only a few salts: the carbonates, sulphates, and chlorides of calcium, magnesium, sodium, and potassium, silicic acid, and small amounts of nitrogen and phosphorus compounds. In addition there are compounds of iron and manganese but these reach noteworthy concentrations only under special circumstances. There are also dissolved organic substances in varying, but seldom important, amounts. Besides these "solids" all water has a certain amount of gas in solution which, when there is contact with the air, is approximately in a state of equilbrium with the atmosphere. It can safely be assumed that beyond the substances named many other elements occur in all waters, although perhaps in such great dilution that they cannot be detected by ordinary chemical methods. However, the importance of these "trace elements" for life must not be underestimated on that account. Investigations in plant physiology in recent years have shown that besides the ten known elements essential for the growth of plants—C, H, O, N, P, S, Ca, Mg, K, Fe—a considerable number of others not only stimulate growth but are also absolutely necessary although in inconceivably small amounts.

As examples of the order of magnitude of the residue on evaporation in normal freshwater lakes the following figures are quoted: the majority of alpine lakes have 100 to 200 mg. per litre, Baltic lakes have 200 to 400 mg., and finally there are the salt-poor lakes occurring particularly in Scandinavia which have less than 50 mg.

As do the chlorides in the sea, the carbonates greatly exceed all other salts in fresh water, and of course in most cases the carbonate is calcium bicarbonate.

There is a simple procedure for the determination of the carbonate in relation to the total salt content. The carbonate content is usually found by titrating 100 ml. of water with tenth normal hydrochloric acid,[14] using methyl orange as an indicator.

---

[14]The method mentioned is the German method and the unit of alkalinity obtained by it is one equivalent part per million, which is the same as 50 p.p.m. $CaCO_3$. The standard method in America is to titrate with 0.02 normal acid and express the results in p.p.m.

The electrical conductivity of calcium bicarbonate solutions is known and, for concentrations up to an alkalinity of 5, can be found from a graph plotted from the following data:

Alkalinity (milliequivalents per litre)	*Conductivity* ($K_{18} \times 10^6$ = reciprocal megohms)
1	85
2	166
3	244
4	320
5	394

If the conductivity for carbonate thus estimated from the alkalinity is compared with a measurement of the total conductivity, the proportion of carbonate to the total salts can be closely approximated,[15] for in dilute solutions the electrical conductivity is directly proportional to the concentration of dissolved substances and not very different for the various salts under consideration. The proportion of carbonate in the total salts in our lakes amounts to about 90 per cent in numerous cases.[16] It is customary to state the carbonate content of waters as *carbonate hardness*.[17] The carbonate hardness is to be distinguished from *total hardness*, which is the total amount of alkaline earths present without reference to the particular anions to which they are bound. In total hardness are also included the sulphates and chlorides of calcium and magnesium. The alkaline earths mainly present as chlorides and sulphates are generally termed the *permanent* hardness (i.e., the hardness not destroyed by boiling).

In most natural waters then, the predominant anions are bicarbonates, associated mainly with calcium, to a lesser degree with magnesium, and still

---

[15]This will only be exact in waters in which the carbonate fraction is $Ca(HCO_3)_2$. However, such waters occur widely and the conductivities of other carbonates that may be concerned do not differ greatly from that of the calcium salt so that the error they introduce can be tolerated.

[16]In regions where the water has a very low salt content because of geological conditions, the carbonate fraction is usually smaller. Thus investigation of 530 lakes in northeastern Wisconsin (Juday and Birge, 1933), whose values for $K_{18} \times 10^6$ lay between 5 and 124, gave a carbonate proportion up to 50 per cent in 131 lakes, from 50 to 75 per cent in 186, and more than 90 per cent in only 159 lakes (30 per cent of the total). Similar conditions probably exist in Scandinavia. Åberg and Rodhe (1942) found carbonate fractions of only 16 per cent (at a total conductivity of 50×10^{-6}) in a few Swedish lakes. Highly saline waters that are rich in sulphates or chlorides for any reason will obviously display a low carbonate fraction.

[17]In America, hardness is expressed as parts per million $CaCO_3$. Equivalent hardness units for various countries are as follows:

Unit	*Equivalents in p.p.m.* $CaCO_3$
1 p.p.m. $CaCO_3$	1.0
1 grain $CaCO_3$ per U.S. gallon	17.1
1 grain $CaCO_3$ per Imp. gallon (England)	14.3
10 p.p.m. $CaCO_3$ (France)	10.0
10 p.p.m. CaO (Germany)	17.9

less with sodium and potassium. Sulphates and chlorides predominate under special geological conditions (the occurrence of gypsum and salt beds in the watershed). Rodhe (1949), making use of the wealth of material that Lohammar (1938) gathered for Swedish lakes and Clarke's (1908–24) compilation of data for the chemical composition of water in all parts of the world, established the important fact that in waters in which carbonates prevailed the proportions of the various ions were almost constant regardless of the total concentration. These proportions (in milliequivalents %) approximate those in the table below:

Cations		*Anions*	
Ca	64	Cl	10
Mg	17	SO_4	16
Na	16	HCO_3	74
K	3		

Of course the precise proportions as well as the total concentrations depend on the particular conditions in the watershed. For example, the content of calcium and magnesium bicarbonates makes up 90 per cent and more of the total in the lakes of the eastern Alps.

Saline waters

The remarks above apply only to those inland waters, both surface and subterranean, that have ultimate connections with the sea. Conditions in lakes that have water flowing into them but have no exits and in which the inflowing water thus evaporates (conditions we find in deserts and on the steppes), are quite different, as you would expect. The proportion of the salts in solution changes as increasing concentration causes the less soluble ones to precipitate. Thus if you evaporate water from an aqueous solution containing carbonates, sulphates, and chlorides, you will find that first $CaCO_3$ and then $CaSO_4$ will crystallize out while the soluble remainder can be concentrated appreciably. This process is much modified through the leaching of salts from the ground, so that springs may even discharge brines, and through the blowing away by the wind to other places during periods of drought of the salts that have crystallized out, where they may go into solution again. Three types (with many intermediates between them) of saline waters can be distinguished: chloride, sulphate, and carbonate waters. Sodium chloride predominates in chloride waters, Na_2SO_4 and $MgSO_4$ in sulphate waters, and Na_2CO_3 (with K_2CO_3 approaching it) in carbonate waters. Sodium carbonate is the chief constituent of the mineral in the "soda" lakes, whose alkaline carbonate is probably derived from the weathering of crystalline rocks and in part from the decomposition of vegetable matter. —The borax lakes, which occur in a few regions (e.g. Tibet and California) and which for ages have been a source of borax, occupy a special place (as is well known, sea-water also contains a small amount of boron). Temperature relations in saline lakes are considered on page 40.

The saline waters we know as *brackish* waters have another origin. These are the result of the dilution of sea-water and are confined to the seacoast. Because of their origin, brackish waters are therefore the concern of oceanography, not limnology. This point of view was adopted at the I.U.B.S. symposium on brackish waters held at Venice in April 1958. Thus Caspers (1959) stated that brackish water is hydrographically and biologically related to the sea, being a mixohaline border region quite different from saline inland waters. The opposite point of view is taken by Remane and Schlieper (1958), who, in agreement with Redeke (1933), consider all waters of intermediate salt content as brackish. Since the life in such inland waters is almost always the same as that found in brackish waters, they hold that biologists cannot make any distinction between them. However, the difference in origin and composition of salts that distinguishes inland saline waters from true brackish waters puts more and more weight on Caspers' interpretation.

In many respects the various steps of concentration to be found in brackish waters form a bridge between the sea and fresh water. You only need to recall the intermingling of marine and freshwater forms to be found in the Baltic fjords that so impressed those who took part in the XIIIth International Congress of Limnology at the Biological Station at Twärminne.

Conditions in the tidal estuaries of large rivers are especially interesting, such as Caspers (1959) has recently shown for the Elbe. In such estuaries the brackish zone is *poikilohaline*, having a periodic change in its salt content. However, this circumstance affects only the sessile organisms. The plankton remains in its appropriate water mass and moves back and forth with the tide.

1. *The Carbon Dioxide Cycle*

Because of the almost universal prevalence of carbonates and also since carbon is the primary element in organic synthesis, we shall begin our consideration of the transformations of matter in waters with the carbon dioxide cycle. Rain falling on the earth contains small amounts of carbon dioxide in solution which it takes up from the air. Since the absorption coefficient of water for CO_2 at 15° C. is 1 and the CO_2 content of air is about 0.03 volume per cent,[18] rain water contains about

[18]In recent decades the CO_2 content of air has risen to 0.044 vol. % as a result of increasing industrialization.

0.3 cc. or 0.6 mg. per litre. This CO_2, however, enters into combination with the water and forms H_2CO_3, carbonic acid, which displays its acid character through its dissociation into H^+ and HCO_3^- ions.[19] As it percolates through the soil, rain water comes in contact with air trapped in the humus layer. Since this air is appreciably richer in CO_2 than the atmosphere because of the respiration of the roots and micro-organisms, the water takes up more CO_2 as it trickles through this stratum of vegetation in proportion to the length of time it is trapped there. Next it generally encounters calcareous rock, especially in the deeper strata of mineral soil. Although quite insoluble of itself, the limestone readily goes into solution as $Ca(HCO_3)_2$ (calcium bicarbonate) in the presence of H_2CO_3. The *bicarbonate content* of the spring water when it finally emerges somewhere is thus in the main dependent on *two* factors: on the *calcium content of the earth* and on the *carbon dioxide content of the water.* Since calcium is also often present in substantial amounts outside the limestone regions in the weathering products of feldspar, etc., in places where igneous rocks form the substrate, the carbon dioxide content of the water is the most important factor in many instances. Thus, for example, the springs of the lime-poor but impermeable soils of the northern approaches to the Alps are actually richer in lime than those of the limestone mountains themselves, through whose permeable rock the rain passes rapidly with little opportunity to become charged with CO_2 in the upper layers of the soil.

However, *not all* the CO_2 content of waters is used for the solution of limestone, that is, not all goes into the combination $Ca(HCO_3)_2$. For calcium bicarbonate to be stable a certain *surplus* amount of CO_2 must remain free in solution. This amount increases very rapidly with increasing lime content (see the table below). We call this free CO_2, which is necessary to retain the calcium in solution, the *equilibrium carbon*

$Ca(HCO_3)_2$ (milliequivalents)	*Equilibrium* CO_2 (mg./l.)	$Ca(HCO_3)_2$ (milliequivalents)	*Equilibrium* CO_2 (mg./l.)
0.5	0.15	3.5	10.1
1.0	0.6	4.0	15.9
1.5	1.2	4.5	24.3
2.0	2.5	5.0	35.0
2.5	4.0	5.5	48.3
3.0	6.5	6.0	64.1

[19]This dissociation takes place in two steps, which have markedly different dissociation constants:

$$K_1 = \frac{[H^+][HCO_3^-]}{H_2CO_3} = 13 \times 10^{-4} \text{ at } 25°C.,$$

$$K_2 = \frac{[H^+][CO_3^=]}{[HCO_3^-]} = 41 \times 10^{-11} \text{ at } 25°C.$$

dioxide. It is obviously incapable of dissolving further calcium carbonate since this requires an amount in excess of the equilibrium value, the so-called *aggressive carbon dioxide.* If a solution of calcium bicarbonate loses the equilibrium CO_2, a reaction takes place according to the following formula:

$$Ca(HCO_3)_2 = CaCO_3 + H_2O + CO_2.$$

This dissociation proceeds with the attendant precipitation of $CaCO_3$ until there is enough free CO_2 in the solution to bring about an equilibrium. The equilibrium CO_2 begins to escape into the air as soon as the spring comes out of the earth. If the water is very rich in lime, we find that at a short distance from the source the stream bed, and the stones, leaves, moss, etc., lying in it, are encrusted with a dense precipitate of calcium carbonate. When there is less lime present, the precipitation is very slow, and hence the water of many streams and lakes is supersaturated, that is, it contains more calcium carbonate than it should for its CO_2 content.

Bäckström (1921) was the first to remark, concerning the time course of the precipitation of calcium when the equilibrium CO_2 is removed, that the equilibrium had a rather broad region over which the rate of reaction was practically zero (see Pia, 1933, and Ohle, 1934, 1952). This slowness of precipitation is one of the important factors influencing the salt content of waters, for, as is indicated above, a lake at sea level in equilibrium with the atmosphere would contain 0.6 mg./l. free CO_2 and one at an altitude of 500 metres, 0.53 mg./l. Such a content of equilibrium CO_2 corresponds to a $Ca(HCO_3)_2$ content of no more than 1 milliequivalent. The calcium content of standing waters is often markedly greater than this. Thus in the lakes of Schleswig-Holstein (Ohle, 1934) the average level of alkalinity is 2.41 milliequivalents with a maximum of 6.71. Lakes in the calcareous region of the northern Alps have an average alkalinity of 2.03 with a maximum of 3.07 (Ruttner, 1937). Hence, as Ruttner (1960) points out, the surface waters of many lakes and streams are supersaturated with calcium bicarbonate. Thus it is not important for limnologists to deal further with the well-known dependence of the solubility of CO_2 on its partial pressure since the dissolved CO_2 in the majority of natural waters is not in equilibrium with the atmosphere.

The compounds of calcium and magnesium behave quite differently in the carbonate–hydroxide system. While $CaCO_3$ is almost insoluble, $MgCO_3$ is quite soluble, and on the other hand $Ca(OH)_2$ is soluble but $Mg(OH)_2$ is quite insoluble. Because there is much less Mg than Ca in most waters its place in the metabolism of natural waters has received little study so far. Except in waters of dolomite regions, it rarely exceeds 20 per cent of the total carbonate (see p. 68). Thus the magnesium content cannot explain the supersaturation mentioned above, particularly since in general the magnesium fraction of the alkaline carbonates is much less than 20 per cent.

Birge and Juday (1911) give a beautiful example of the difference in solubility between calcium and magnesium carbonates in their investigations of Lake Mendota. Like many Wisconsin lakes, Lake Mendota water contains a striking amount of $Mg(HCO_3)_2$ in relation to its calcium content. The Ca:Mg ratio in the water is 0.6:1; but in the sediment (because of the greater solubility of $MgCO_3$) the ratio is 11:1, and in the lime encrustations on the *Potamogeton* leaves it is up to 36:1.

Calcium bicarbonate is almost completely dissociated in a very dilute solution:

$$Ca(HCO_3)_2 \rightleftharpoons Ca^{++} + 2\ HCO_3^-$$

The second step in the dissociation (K_2, p. 62),

$$HCO_3^- \leftrightharpoons H^+ + CO_3^=,$$

takes place to any extent only at the higher pH values (about 30% at pH 10, 85% at pH 11).[20]

Further, since carbonic acid is a weak acid, hydrolysis takes place:

$$HCO_3^- + H_2O \leftrightharpoons H_2CO_3 + OH^-.$$

Thus in a solution of calcium bicarbonate there is a quite complicated system: free carbon dioxide (CO_2); partly undissociated, partly dissociated, carbonic acid (H_2CO_3, HCO_3^-, H^+); dissociated calcium bicarbonate (Ca^{++}, HCO_3^-, $CO_3^=$); and finally hydroxyl ions resulting from the hydrolysis (OH^-). Since there is an excess of hydroxyl ions over the hydrogen ions produced by the dissociation of the H_2CO_3, a solution of calcium bicarbonate is weakly *alkaline*.

The *hydrogen-ion concentration* of water is one of those environmental factors that are very strikingly linked to the species composition of communities and their life processes. Because of its effect on pH, the bicarbonate content is (quite apart from its value as the source of the most important nutrient) of prime importance in many problems in

[20]That is, if the value of $H+$ decreases and K_2 remains constant in the equation

$$K_2 = \frac{[H^+][CO_3^=]}{[HCO_3^-]},$$

then there must be a corresponding increase in $CO_3^=$ (or a decrease in undissociated HCO_3^- in the denominator) so that the result is a greater dissociation.

limnology. We must therefore go more thoroughly into these relationships and the part that the dissociation system of bicarbonate plays in them.

It is well known that even absolutely pure water itself, distilled with the greatest of precaution and protected from the air, is weakly dissociated. The product of the [H^+] and [OH^-] ions present at any given time is a constant (the dissociation constant of water, K_w); it is a very small value, approximately 10^{-14}. Completely pure water is neutral in reaction, that is, H^+ and OH^- are present in exactly equal amounts and, of course, since $H^+ \times OH^- = K_w = 10^{-14}$, each is present in the order of magnitude of 10^{-7}. A litre of neutral water, therefore, contains a ten-millionth of a gram-ion of H^+ or OH^-.

If a salt, an acid, or a base whose dissociation contributes H^+ or OH^- ions is present in solution, these ions are added to the corresponding ions of the pure water, increasing their concentration by a certain factor. However, since K_w is fixed under all circumstances according to the equation given above, the concentration of the opposite ion must simultaneously decrease by the same factor. Hence only the amount of one kind of ion is required to determine the reaction, and the hydrogen-ion concentration is universally chosen to describe the reaction of a solution. For convenience, simply the logarithm of the number of hydrogen ions is used (without the negative sign). This hydrogen exponent is designated pH. At neutrality [H^+] is accordingly 10^{-7} and the pH is 7; when the reaction is more acid [H^+] is larger than 10^{-7} (for example 10^{-5}, pH = 5); when the reaction is more alkaline [H^+] is less than 10^{-7} (for example 10^{-9}, pH = 9).

Distilled water that has been in contact with air or rain water in nature, is never neutral but has an acid reaction because of the dissolved carbonic acid, which contributes H^+ ions. The pH is generally 4–5. If the carbon dioxide is driven off, for example by boiling, the pH value rises to the neighbourhood of 7. The hydrogen-ion concentration is not affected by neutral salts (such as sodium chloride), which are the salts of strong acids and strong bases, but is determined entirely by the absorbed carbon dioxide. Any addition of even a trace of acid or base brings about large variations in the pH value in distilled water and in solutions of neutral salts.

Quite a different situation is found in water containing a *bicarbonate* in solution, for example, calcium bicarbonate. In this case, as was pointed out above, *the pH is determined by the relation between CO_2 and carbonate, or more precisely by the H^+ ions arising from the dissociation of H_2CO_3 and the OH^- ions arising from the hydrolysis of the bicarbonate*. If some acid is added to this solution, a part of the bicarbonate is split off, which simultaneously combines with the acid. The free CO_2 released in its place is only weakly dissociated and increases the number of hydrogen ions by but a slight amount according to the relation given above, and thus the hydrogen-ion concentration of the solution is very little altered. The decrease in the pH value thus brought about is very small in proportion to the amount of acid added. Not until further additions have exhausted the bicarbonate will the reaction suddenly become strongly acid. This process is observed immediately after the methyl orange endpoint is reached when determining the bicarbonate content by titration with hydrochloric acid. If on the other hand some alkali is added or the CO_2 driven off, the shift of the reaction is likewise retarded, for equilibrium CO_2 is continually split off from the remaining bicarbonate while the hydroxyl ions of the added alkali are locked up in the precipitation of $CaCO_3$:

$$Ca(HCO_3)_2 + KOH = CaCO_3 + KHCO_3 + H_2O.$$

The variations in reaction observed in lakes of normal calcium content seldom exceed the limits of pH 7–9. The pH can be less in springs of very high CO_2 content and higher in ponds crowded with plants.

Carbonate–carbonic acid mixtures share their remarkable and, for living organisms, important property of preventing major fluctuations in reaction with other acid-salt combinations. These mixtures of *weaker* acids with their salts are called *buffers*. The reaction of their solutions is not determined by the concentration but by the salt-acid relation and is therefore little altered by dilution. Further, their hydrogen-ion concentrations are less affected on contamination by acids or alkalis than are those of unbuffered solutions. The degree of buffering is of course diminished by dilution, as is clear from what has been said above. Whether a water is well or poorly buffered depends on its bicarbonate content. Buffers play an extremely important role in the life of organisms. All body fluids are more or less buffered; blood is a carbonate buffer, urine a phosphate buffer.

Accordingly it is clear without further discussion that the presence or absence of bicarbonates determines whether a water is alkaline or acid in its reaction. Since the point of neutrality brings about a sharp separation in the organic world in the sense that many (but by no means all) species thrive only in alkaline or only in acid conditions, the content of calcium bicarbonate is a predominant factor in the composition of the organic community. With samples of algae, a fleeting glance in the microscope is often sufficient to establish whether they came from acid or alkaline water. Hence the buffering capacity is of great importance, as on it depends the pH of the water and the range of its variations. Very poorly buffered water may be acid or alkaline according to the circumstances. The determination of the bicarbonate content (which is easily done by the methods described above) must thus supplement the measurement of pH in limnological investigations.[21] The measurement of pH in natural waters is most conveniently made by a suitably refined calorimetry method and often such a method is also the most accurate. The electrometric method of measuring pH has found widespread application in limnology since the development of the glass electrode and many suitable forms of apparatus are available.

The natural solutions of calcium bicarbonate, which comprise the majority of natural waters, are continually under the influence of plant and animal metabolism. The uptake and release of CO_2 alters the buffer mixture and disturbs the equilibrium. These interferences are profound and persistent, appearing to a greater or less degree in all habitable waters, and they are especially striking in the chemical stratification of lakes.

If we wish now to consider these events more thoroughly we must begin with the fundamental synthetic process of life, the *carbon dioxide assimilation* or *photosynthesis* of plants. This complicated process, of which only the initial and the end products are known for certain, proceeds by utilization of light energy according to the following basic formula:

$$6\ H_2O + 6\ CO_2 = C_6H_{12}O_6 + 6\ O_2.$$

Thus, in the course of synthesis of carbohydrate (sugar, starch) photosynthesis enters into the economy of waters in two ways: by the removal

[21]Because of the importance of the relation "bicarbonate—free carbon dioxide—pH" in bicarbonate solutions for many limnological questions, a reference table is presented from which one of the three factors may be taken if the other two are known. The relations are given by the formula

$$\text{mg. } CO_2/\text{l.} = \frac{44}{3.72 \times 10^{-7}} \cdot 10^{-\text{pH}} \cdot A$$

(44 = molecular weight of carbon dioxide; 3.72×10^{-7} = first dissociation constant for carbon dioxide at 15°C.; A = alkalinity expressed as equivalent parts per million).

pH	*Factor*	*pH*	*Factor*
6.0	118	7.0	12
6.1	94	7.1	9.4
6.2	75	7.2	7.5
6.3	59	7.3	5.9
6.4	47	7.4	4.7
6.5	37	7.5	3.7
6.6	30	7.6	3.0
6.7	24	7.7	2.4
6.8	19	7.8	1.9
6.9	15	7.9	1.5
7.0	12	8.0	1.2

Alkalinity × factor = mg. CO_2/1. at 15°C. Note that an increase of 1 pH unit decreases the factor to 1/10 its former value. The table may be extrapolated by use of this rule.

of carbon dioxide and by the release of an equal volume of oxygen. We shall concern ourselves with the effect of the latter presently and at the moment consider only the uptake of the CO_2. The free CO_2 in solution and, after this is exhausted, the supply stored in the bicarbonate are available to aquatic plants for photosynthesis. Since the bicarbonate decomposes with a precipitation of $CaCO_3$ on the withdrawal of CO_2, half the chemically bound CO_2 (the halfbound CO_2) is made available. The calcareous deposits covering the upper surfaces of the leaves of many submersed water plants, for example *Potamogeton* and *Elodea*, are an evident proof of the precipitation of calcium carbonate by the process of photosynthesis.

It was first assumed that the aquatic plants could take up only free CO_2 and that the precipitation of calcium carbonate resulted automatically from the disturbance of the equilibrium and thus that the assimiliation of submersed plants did not differ in any way from that of land plants. However, Ruttner (1921) showed that the relations are not nearly so simple as this. The release of calcium carbonate in water by assimilating plants proceeds much faster and certainly more completely than would be the case if only the free CO_2 were withdrawn. Further, the occurrence of a strong alkaline reaction (up to pH = 11) in the experimental vessels as well as large changes in the equivalent conductivity[22] indicated that other processes must also be entering in. Thus true aquatic plants take up HCO_3^- ions from the dissociation of calcium bicarbonate and substitute OH^- ions for these. Some water and the greater part of the $CaCO_3$ precipitated by the plants arise from this ion exchange. *Bicarbonate* is therefore *actively broken down* by plants and used in the *assimilation process*. However, the process does not stop when all the bicarbonate is broken down and nothing but the monocarbonate ($CaCO_3$) remains. This also yields HCO_3^- ions by hydrolysis that can be exchanged against OH^- ions ($CaCO_3 + H_2O = Ca^{++} + HCO_3^- + OH^-$). The result is the formation of calcium hydroxide, which explains the strongly alkaline endpoint of the experiment. The same phenomenon occurs when solutions of potassium or sodium bicarbonate are used; indeed the results are even more marked because in these cases there is no drop in concentration due to precipitation (see Figure 17, *Elodea*).

These relations can be at least partially demonstrated if one or two three-inch sprigs of *Elodea* are placed in two small flasks filled with tap water to which a few drops of phenolphthalein have been added, and one flask is placed in the light while the other is darkened. Red cloudlets will

[22]The equivalent (Λ) is the quotient of the electrical conductivity K_{18} of a solution divided by its equivalent concentration. Thus, for example, Λ for $NaHCO_3$ = 80, for Na_2CO_3 = 112, and for NaOH = 208 (in solutions of alkalinity = 1). Hence if the concentration of a solution is known, the per cent carbonate or hydroxide can be established by ascertaining the electrical conductivity (Ruttner, 1948). A nomograph giving the conductivity curves of bicarbonate–carbonate–hydroxide mixtures at various equivalent concentrations can be used for this purpose.

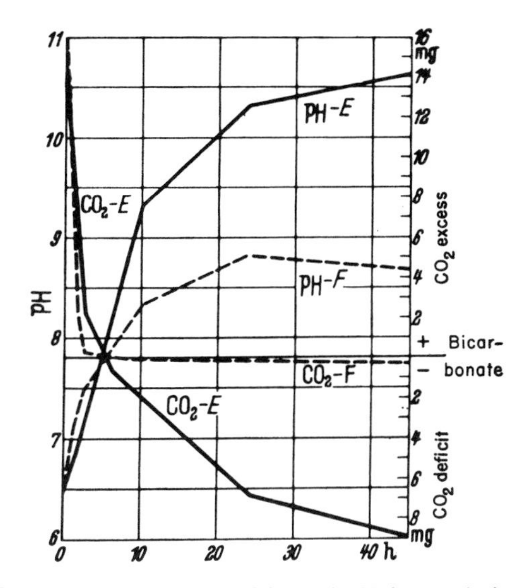

FIGURE 17. Changes in CO_2 and pH in a solution of *N*/1000 $KHCO_3$ enriched with CO_2 brought about by photosynthesis of *Elodea* (E) and *Fontinalis* (F). Free CO_2 is shown above the axis, bicarbonate deficit below. From Ruttner (1948).

soon appear on the upper sides[23] of the leaves in the lighted vessel and in a few hours the whole liquid will have a deep red colour because of the resulting high alkalinity, while the darkened flask will remain colourless. By bubbling expired air (CO_2) through the coloured water, the red coloration can be made to disappear. If the electrical conductivity is measured, or the alkalinity determined at the beginning and at the end of the experiment, a very marked decrease in salts will be found owing to the precipitation of calcium carbonate. In this way the amount of calcium carbonate precipitated and CO_2 assimilated can be measured.

However, not all aquatic plants possess the ability to decompose bicarbonate actively, as has been established for the submersed phanerogams and many but not all the algae. It is lacking, for example, in the aquatic mosses. These are only able to utilize free carbon dioxide, and correspondingly in the experiment the pH remains relatively low (about pH $= 9$) and there is no appreciable increase in the equivalent conductivity (Ruttner, 1947; Steeman-Nielsen, 1947). An experiment to illustrate the contrast in behaviour of *Elodea*, a phanerogam, and *Fontinalis antipyretica*, a moss, is illustrated in Figure 17. It can be seen that both plants utilize free CO_2 at approximately

[23]Correspondingly, the carbonate deposits found in nature on the leaves of *Potamogeton*, *Elodea*, and other submersed aquatics are only on the morphological upper side. Ruttner (1921) was the first to observe this and designated the phenomenon "physiological differentiation." It was later treated exhaustively by Arens (1936) as a "polarized mass action exchange" in significant experiments.

the same rate. However, when the free CO_2 is exhausted *Fontinalis* practically ceases photosynthesis while *Elodea* continues by utilizing much of the bicarbonate but at a decreasing rate, with the result that the solution becomes almost a mixture of K_2CO_3 and KOH at the end of the experiment. The pH curves of the two experiments show the same effect as do the CO_2 curves. This physiological behaviour corresponds also with the distribution of the aquatic mosses in nature. They are found only where free carbon dioxide is present in the water: for example, in springs, mountain streams, moorland pools, and in lakes only in the eulittoral zone where there is exchange with the atmosphere and at greater depths where the CO_2 content again increases. On the other hand, the submersed phanerogams and the algae, which can break down bicarbonate, develop luxuriantly also in the alkaline and usually carbon-dioxide-free littoral region. The freshwater Rhodophyceae, *Batrachospermum*, behaves like the aquatic mosses and its distribution is correspondingly the same as that of the latter (Ruttner, 1960).

The consideration of all these processes connected with carbon assimilation, as well as the opposing processes of animal and vegetable respiration, is unavoidable if we wish to understand the changes in carbon dioxide and bicarbonate content and the hydrogen-ion concentration of water. We can recognize these processes most clearly in the distribution of dissolved substances in lakes. At the time of the spring turnover after the break-up of the ice, the CO_2 and bicarbonate content and the pH are nearly uniform for all depths because of the thorough mixing. In many cases the state is that of a bicarbonate solution in equilibrium with free CO_2, such as is found as a rule in mountain streams and tap water. In carbonate-rich lakes, to be sure, a CO_2 deficit will frequently be found even at this time if the water (perhaps in the inflow) has an opportunity to come in contact with air. This loss obviously goes on at the surface of the ice-free lake with the tendency to bring about a solution in equilibrium with the CO_2 content of the air, and there is as a result a certain loss of calcium carbonate. However, these changes brought about by purely *physical* conditions are slight, particularly in the case of deep lakes, and are completely insignificant in comparison with the *biogenic* effects. Within the phototrophic stratum both the rooted aquatics in the littoral zone and the microscopic algae of the open water plankton use up CO_2 or HCO_3^- during the day as soon as they begin to be active in the spring. The following developments result from this:

1. Reduction of free CO_2.
2. Reduction of calcium bicarbonate (through the precipitation of $CaCO_3$).
3. An increase in the pH.
4. In special cases, such as in sunlit pools with luxuriant submersed

vegetation, the decomposition of bicarbonate can proceed up to the formation of hydroxide and the pH can be raised up to about 11.[24] In larger bodies of water, the breakdown does not go so far because the water mass is too great in relation to the assimilating plant material.

Since the rate of assimilation depends on the intensity of light it is to be expected from the data on absorption of light (pp. 13ff.) that these changes would be greatest in the uppermost, best-lighted strata, and that a clearly expressed stratification would be produced there. This is not possible, however, because of the turbulent mixing that goes on in the epilimnion. The differences that arise are continually equalized and we ordinarily find a more or less uniform distribution of CO_2, bicarbonate, and pH down to the upper limit of the metalimnion. At night, photosynthesis stops and the respiration of the animals and plants returns CO_2 to the water. However, as the amount returned is usually less that that removed during the day, the daily balance of assimilation remains positive and leads to a gradual summation of effect. Thus at the height of the summer stagnation we ordinarily find that the epilimnion is impoverished with respect to CO_2. The content often drops to zero and indeed may have a "negative" value; that is, the removal of HCO_3^- is more rapid than the precipitation of calcium carbonate and a supersaturated bicarbonate-monocarbonate solution results. The pH is high, as to be expected when there are major withdrawals of CO_2. While a pH of 7.6 is usually found in the spring or tap water of average calcium content, the values in the upper strata of lakes with the same concentration may be from 8 to 9, and in shallows overgrown with plants even higher values are the rule. Lakes that have a pH of less than 8.5 (the endpoint for phenolphthalein) in offshore waters may, however, have waters that turn red on the addition of the indicator in the littoral zone because of the alkaline reaction there. The bicarbonate content of the upper strata is always less than that of the lower ones because of the precipitations of calcium carbonate.[25] Measure-

[24]A pH of 11 is reached with a $N/1000$ hydroxide solution (alkalinity $= 1$). Higher pH values cannot be reached under normal conditions even when there is a complete breakdown of the $CaCO_3$ solutions and they are originally supersaturated, for the calcium carbonate is always precipitated before the OH concentration reaches a high enough level. On the other hand, even if higher pH values are possible in alkaline carbonate solutions a pH of 11 would not be greatly exceeded in such an experiment. This is probably because photosynthesis is checked by the high pH.

[25]This phenomenon is not always due to the biological causes stated above. In lakes in which the formation of a metalimnion begins during the high water of spring, which is frequently the case in the mountains, the water of the epilimnion will, as indicated on page 56, be replaced by the salt-poor snow water and in this way a marked chemical stratification will be produced.

ments on *Elodea canadensis* made at Lunz show the magnitude of calcium precipitation: 100 kg. of fresh *Elodea* precipitated 2 kg. of $CaCO_3$ in a day with 10 hours of sunlight. It is easy to realize the amounts of calcium that are precipitated every year in a large lake if we stop to consider the masses of submersed aquatics that cover the slopes of our lake basins, and further that some hundred kilograms of plankton algae (which because of their great surface area are especially active in taking up carbon dioxide) are suspended in the open water below each hectare of surface. The effect of *biogenic calcium carbonate precipitation*, an appropriate term originated by Minder (1922), is well shown in the gleaming white marl banks of alpine lakes. These great deposits of almost pure calcium carbonate, often many metres thick, owe their origin to the deposition of suspended particles of chalk behind points of land where the water is sheltered from the prevailing currents (see also p. 192).

Two examples of the stratification of carbon dioxide, bicarbonate, and pH compared with the calculated values for equilibrium are shown in Figures 18 and 19. One is from a eutrophic tropical lake in Indonesia, the other from an oligotrophic temperate lake in the limestone region of the northern Alps. Both show essentially the same relations in their stratification.

The water in the epilimnion is uniformly markedly *undersaturated* with CO_2. In the metalimnion and thus below the zone of intensive eddy diffusion the gradient usually is quite evident: the CO_2 content increases, the pH decreases. At a certain depth these values reach a level that approximates the state of equilibrium expressed by the bicarbonate content in question (about 10 m. in Figure 18, 15–20 m. in Figure 19). If no disturbance is brought about by mixing processes this depth can be taken as the lower limit of effective CO_2 assimiliation, that is as the depth at which the uptake of CO_2 by photosynthesis and its output by respiration and decomposition balance each other in the daily cycle. This is the boundary between the trophogenic and the tropholytic zones. Below this depth, lakes show differences in behaviour. In those with a considerable calcium content (see the examples given in Figures 18 and 19), the state of equilibrium is maintained over a considerable depth. The free CO_2 does not increase further at greater depths, except in the strata immediately adjacent to the bottom, nor, because of the increasing bicarbonate content, does the pH. The explanation of this is easily found: the calcium carbonate precipitated in the trophogenic region falls gently to the bottom and, meeting the aggressive CO_2 in the tropholytic zone, goes into solution and increases the concentra-

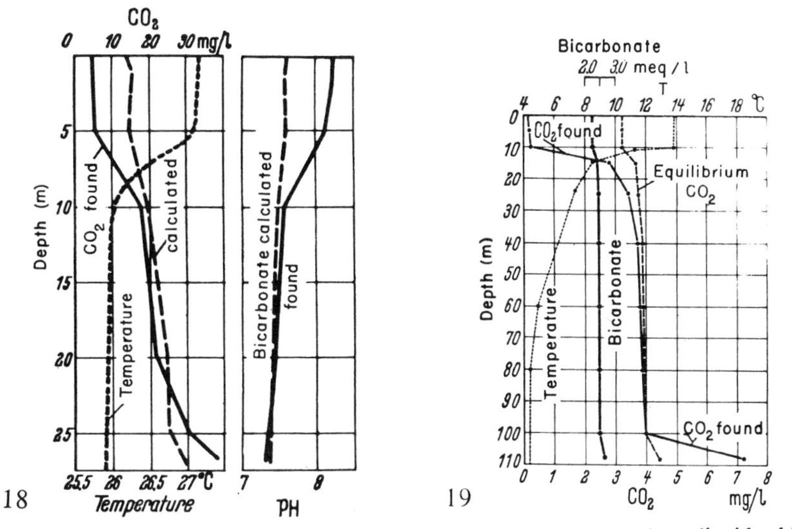

FIGURES 18 and 19. Observed stratification of temperature, carbon dioxide, bicarbonate, and pH in comparison with calculated equilibrium values: 18, Ranu Lamongan, East Java, from Ruttner (1931); 19, Wolfgangsee, Salzkammergut, from Ruttner (1937).

tion in the deep water. On the other hand, in lakes low in calcium, the calcium carbonate precipitated is not sufficient to fix the CO_2, the equilibrium value is quickly overshot beneath the thermocline, and the result is a considerable excess of (aggressive) CO_2, and a lowering of the pH. In the examples in Figures 18 and 19 such a state is found only in the strata adjacent to the lake bottom.

2. *Dissolved Oxygen*

Thanks to the introduction into limnological investigation of the simple and exact Winkler iodometric method of determination[26] we are well informed of the oxygen content of waters, which is closely linked with the CO_2 cycle. Oxygen and carbon dioxide are the two great complements in metabolism, as is well known. Wherever a chemical gradient of biogenic origin occurs in nature, the distribution of these two sub-

[26]Alsterberg's modification of the Winkler method in which the sample is prebrominated should always be used in limnological investigations. Otherwise, values that are too low will always be obtained in what is exactly the biologically important region where the oxygen is under 1 mg./l. because of the presence of reducing substances which combine with the iodine. For a rapid survey Todt's (1958) electrometric method is very convenient when strictly accurate results are not required; with this method the oxygen curve can be rapidly found (see also Ambühl, 1955).

stances is exactly opposite. We must therefore expect in lakes a stratification of oxygen opposite to that of carbon dioxide. When the CO_2 content of the trophogenic layer decreases owing to photosynthesis, the oxygen content increases proportionately. On the other hand, as the oxidation processes of breakdown decrease the oxygen in the lower tropholytic region, it is enriched with carbon dioxide (or its salts). However, while the CO_2 given up below the thermocline remains in the water either free or chemically bound—that is, it cannot suffer any appreciable loss either to the air or through physiological processes—the oxygen given off in the epilimnion is in contact with the atmosphere and may be lost to it, and is in part used in respiration. With a reasonably active rate of carbon assimilation, a supersaturation of the water with oxygen occurs, which is accompanied by a diffusion gradient at the surface, and an appreciable fraction of the excess oxygen escapes. Thus a lake behaves like an assimilating plant; it gives off oxygen in the light. On the other hand at night or in cloudy weather the oxygen loss due to decomposition and respiration can be replaced again through the surface.

As would already have been expected, the contact with air is important for the CO_2 economy as well. But in view of the great stores of bicarbonate which most waters contain, it does not play the same essential role as it does in the oxygen economy of the upper strata of a lake.

In round numbers a litre of air contains 210 ml. of oxygen and 790 ml. of nitrogen (including small amounts of the rare gases). Since the absorption coefficient (which depends on temperature) for the first is about 1/32 and for the latter about 1/65 at 20°C., a litre of water in equilibrium with air at 20°C. contains 6.4 ml. of oxygen and 12.3 ml. of nitrogen. Because the absorption coefficient increases with decreasing temperature the values at 5°C., for example, change to 8.9 and 16.8 ml.

Since the solubility of gases in water depends on pressure as well as temperature, waters at various altitudes contain different amounts of dissolved oxygen when at air saturation. In certain considerations it is important to know the saturation values for dissolved oxygen in relation to temperature and to height above sea level. Ricker's (1934) nomogram is convenient for determining the percentage saturation of a given oxygen content at a given temperature. The correction for the lake's altitude can be made by multiplying the percentage saturation so found by $760/p$, where p is the average atmospheric pressure at the altitude of the lake's surface. The atmospheric pressure at the lake's surface can be estimated if the elevation of the lake is known from the formula

$$\log p = \log 760 - \frac{273 \cdot h}{18{,}400 \cdot T},$$

where log 760 = log mean pressure at sea level, h is the altitude in metres, and T is the mean absolute temperature of the air column between sea level and the altitude in question.

Mortimer's (1956) extensive discussion of the literature and nomogram are helpful in the bothersome matter of calculating saturation values. Burkhard's (1955) " oxygen calculator," a circular slide rule based on the same principles, further simplifies the task.

Even at the surface, saturation values are not established rapidly and even there we may find respectively supersaturation or undersaturation following intensive photosynthesis or rapid consumption of oxygen.

During the autumnal and vernal circulation periods, not only are all differences in stratification equalized, but under normal conditions the whole water mass of the lake is repeatedly coming into contact with the atmosphere. In most cases an equilibrium between water and air is reached and the whole lake is then "saturated" with the atmospheric gases from top to bottom. If the water is at 4°C. at this time and if the barometer is also at 760 mm. the oxygen content will be 9.1 ml. or (using a unit of measurement more suitable for many purposes) 13.1 mg./l. If the lake does not freeze, the oxygen content will remain at essentially air saturation all winter. The same is approximately true for the deep, unproductive, steep-shored mountain lakes even when frozen over. Shallower lakes with higher organic productivity have a well-expressed oxygen stratification in winter when covered with ice. Such a stratification can display a variety of forms of oxygen loss depending on the nature of the relief of the basin or it can (more rarely) be a supersaturation at the surface under clear ice. If the warming is rapid enough after the breakup so that the vernal circulation is not completely to the bottom, the winter stratification can be maintained into the summer with no equilibration. However, when the upper waters warm after a normal period of spring turnover and the normal summer distribution of temperature takes place, then changes in the oxygen content are to be expected on physical grounds alone. In the warm epilimnion there will be a decrease even with continued saturation because of the decrease in solubility at higher temperatures. In the hypolimnion, however, the relations will be almost unchanged. If the temperature in the epilimnion is 20°C. and in the hypolimnion still 4°C., a curve in oxygen content which is the mirror image of the temperature curve and which will show a rise in oxygen content with depth from 9 to 13 mg./1. must result even under conditions of saturation. However, such a summer curve in oxygen content is only rarely encountered. In general the following types of stratification occur. Either the oxygen content remains almost unchanged in the whole column of water (*orthograde* O_2 curve of Åberg and Rodhe, 1942) or it decreases in the metalimnion and the hypolimnion in a manner almost parallel to the temperature curve (*klinograde* O_2 curve). In extreme cases it drops to zero in the lower strata. *In both types of summer stratification, even in that with uniform distribution, there must be a gradual decrease in the oxygen content of the hypolimnion* (there can only be transitory departures from a state of saturation in the epilimnion because of the continuous interchange with the atmosphere).

The decrease is brought about by oxidation processes taking place in the free water as well as at the boundaries of the lake in the contact zone between the ooze and the water (see further pp. 214f.). The extent of these processes, among which the most important are the bacterial and enzymatic breakdown of organic matter originating in the phototrophic zone and the respiration of organisms, depends on a variety of circumstances.

Without doubt the most important of these factors is *the amount of oxidizable substances.* As a rule these originate within the trophogenic zone of the lake itself. Under otherwise equal conditions a lake with a richer biota will have a greater oxygen deficiency in its deep water than will a poorer lake. *The productivity of a lake can thus be estimated in many cases from the nature of the oxygen curve.* The intensity of the decomposition within the tropholytic zone can be used as a measure of the production in the trophogenic zone. This circumstance was already surmised by the famous physiologist Hoppe-Seyler (1895). It was demonstrated in 1911 by Birge and Juday through numerous observations on American lakes and it has assumed great importance since 1915 in the classification of lakes from the point of view of fisheries biology, especially because of the significant investigations and valuable contributions of Thienemann (1918). Lakes are broadly classified into two opposing types: *eutrophic* (rich in nutrients) and *oligotrophic* (poor in nutrients). A glance at Figure 20 will immediately demonstrate the oxygen relations in these. The temperature curves given there clearly indicate once more the influence of density stratification on the chemical conditions.

Another factor that influences the consumption of oxygen in the hypolimnion is *temperature.* We know that respiration and the other oxidative processes that must be considered here are dependent on temperature and according to van t'Hoff's law the rate is doubled to trebled by a temperature rise of 10°C. In temperate latitudes where the profundal temperatures of lakes are not very different from one another this relation cannot be detected except by careful comparative investigations, if at all. In tropical lakes, however, it becomes of great importance in determining the course of the oxygen curve. In equatorial

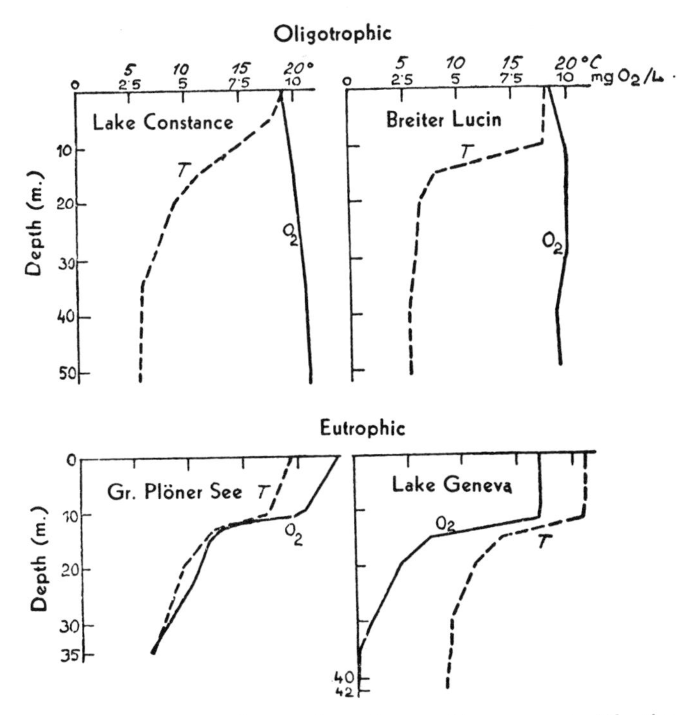

FIGURE 20. Examples of oligotrophic and eutrophic oxygen stratification: Orthograde—Lake Constance (Alpenrand), Breiter Lucin (Mecklenburg); Klinograde—Gr. Plöner See (Holstein), Lake Geneva (Wisconsin). From Auerbach, Maerker, and Schmalz (1926); Thienemann (1918); Birge and Juday (1911).

regions the profundal temperatures of the lakes are about 20°C. higher than they are in the temperate zone. Chemical reactions there are four to nine times more rapid, that is, other conditions being equal, four to nine times as much oxygen will be used per unit time in a tropical lake as will be consumed in a temperate one. In the latter the low profundal temperatures have a preservative effect like that of an ice-box, and a large proportion of the organic material that sinks down from the trophogenic zone is deposited undecomposed in the sediment. Conversely, the "incubator" temperature of a tropical lake brings about far more decomposition even while the material is still sinking down, a fact that is demonstrated by the much greater mineralization of the deposits in these waters as well as in other ways. We find, therefore, a more or less marked oxygen deficiency in the hypolimnion of all tropical lakes that have been investigated regardless of whether they are eutrophic or oligotrophic in productivity. *In tropical lakes temperature is the determinative factor and the oxygen curve loses its importance as an indicator of the magnitude of organic productivity.*

Further, it may be taken that the oxygen stratification is also greatly dependent on the velocity, governed by the size of particles, with which the organic remains sink; this speed controls their stay in the hypolimnion. Finally the *bottom profile* and particularly the *depth* of a lake (or the relation of the volume of the trophogenic zone to that of the tropholytic) play a decisive role, since in a very large hypolimnion the products of decomposition are diluted to a much greater extent than they are in a less extensive one.

For the assessment of biological productivity it is of great importance to measure exactly the extent of oxygen consumption that occurs in the hypolimnion during the summer stagnation (see p. 172). The determination of the *oxygen deficit*, the difference between the amount of oxygen present at the beginning and at the end of stagnation below a given depth (for example, the metalimnion), is used for this purpose. If the oxygen content during the spring turnover is not known it is usually assumed that the water was saturated at the temperature in question and the deficit is calculated from this saturation value. Such considerations are vitiated by a source of error which is difficult to assess, namely, the generally unknown and fluctuating coefficient of eddy diffusion (see p. 51) which determines the amount of oxygen transport from the trophogenic stratum to the hypolimnion. For this reason calculations on the basis of oxygen deficits are only permissible when made with the greatest of caution.

The use of an "absolute" oxygen deficit based on the difference between the oxygen content in the hyplimnion at the end of stagnation and the saturation value at 4°C. (Alsterberg 1927, 1929) has been proposed instead of the "actual" deficit described above, on the assumption that the oxygen content at the end of the vernal circulation corresponds to the 4°C. value. Since the latter is by no means always the case, Münster-Strøm (1931) recommends taking the oxygen content observed at the end of the spring turnover, or some later date, as the point of departure ("relative" deficit). See also Hutchinson (1938) and Elster (1955) in this connection.

The type of oxygen stratification outlined above, which may be found in all gradations between the extremes of oligotrophic and eutrophic lakes, frequently shows disturbances which are difficult to explain with certainty. The most frequent of these are oxygen *maxima* or *minima* in the metalimnion ("heterograde" curves, Figure 21). The restriction of the phenomena illustrated in our example to the metalimnion would make it appear obvious that they are related to the stable stratification existing there and to the absence of eddy diffusion. The explanation of the metalimnial oxygen maximum is generally simple. When photosynthesis takes place actively within a high-lying metalimnion there is an increase of oxygen, which can be given off only by a slow diffusion upwards

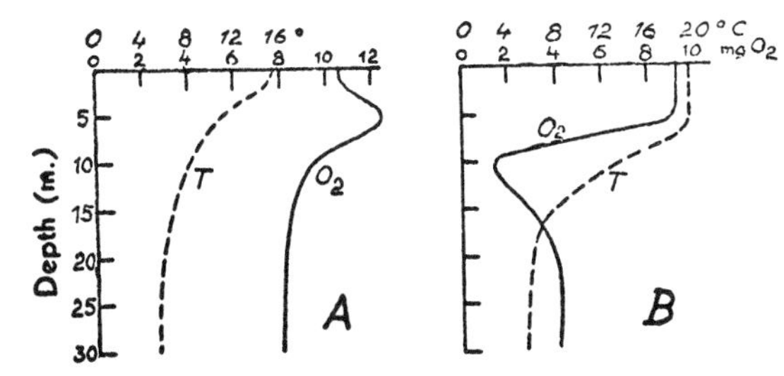

FIGURE 21. Metalimnial oxygen maximum (*A*, May 14, 1932) and minimum (*B*, September 25, 1932) in Ossiacher See (Carinthia). From Findenegg (unpublished).

since mixing by turbulence is more or less impossible. The oxygen is retained in the stratum in which it originated and apparent supersaturation of considerable magnitude appears. It is not a real supersaturation because the hydrostatic pressure allows a greater solubility.[27]

The occurrence of minima cannot always be explained so easily. They can be caused by *accumulations of oxidizable materials* in the metalimnion (which may consist of not only the respiring and decomposing plankton of the open water but also of extensive banks of ooze) or they may be brought about by the temperature gradient alone. In either case the assumption is that the metalimnion is below the depth for effective photosynthesis.

The influence of temperature can be considered as follows. Let us assume that we could eliminate contact with the air and the effect of carbon assimilation in a lake with normal temperature stratification and with a uniform content of oxygen and oxidizable substances at all depths. According to van t'Hoff's law, the oxygen will be consumed appreciably sooner in the upper warm strata than in the cold depths. After a certain time oxygen would first be encountered in the metalimnion and—corresponding to the drop in temperature—would increase with depth. If carbon assimilation is then permitted, production of oxygen will take place from the top down, the rate decreasing with depth as the light decreases, and will reach the null point in the metalimnion under certain conditions. The curve for the final distribution of oxygen, which results from the summation of production and consumption, must then of necessity show a more or less well expressed minimum in the metalimnion. If the lower limit of photosynthesis penetrates deeper, the oxygen minimum will gradually become a maximum, as can easily be seen by constructing the corresponding case (Ruttner, 1933).

[27]These maxima need not always be due to the photosynthetic activity of the phytoplankton. In small clear lakes with high thermoclines they can also appear when the steep banks are covered with a dense growth of aquatic plants, since the water enriched with O_2 at the shore spreads only in a horizontal direction within the stratum of equal density. Thus in Lunzer Obersee the metalimnial oxygen maximum first appears with the growth of the *Elodea*.

Another deviation is shown by the oxygen curve adjacent to the *ooze*. In the last metre or so above the bottom, the drop is steeper and steeper. From Alsterberg's (1927) investigations, we know that the ooze of lakes is free of oxygen even at very small depths. There must therefore be an oxygen gradient between the surface of the ooze and the well-oxygenated water stratum above it, since oxygen is taken up from the water by the sediment or is consumed by reducing substances diffusing out of it. These gradients will at first be confined to a *micro-stratification* in the shortest possible distance from the surface of the ooze but will gradually be imparted to the more extensive overlying strata by eddy diffusion.

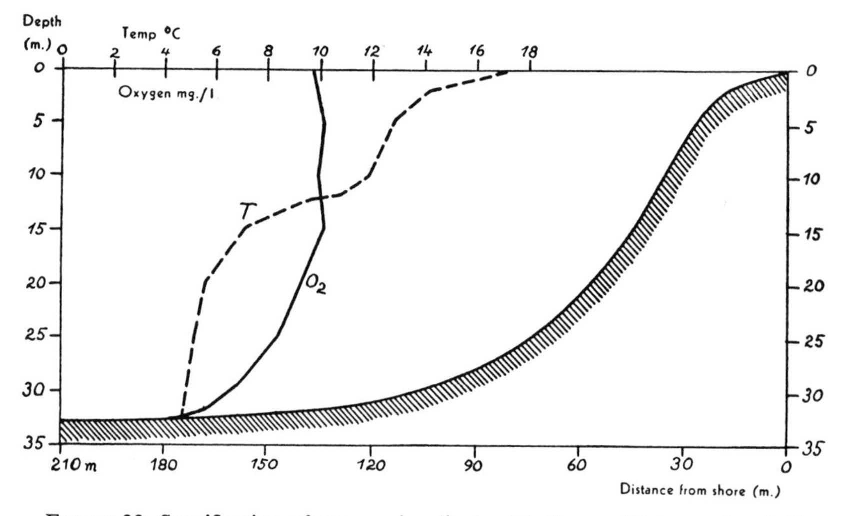

FIGURE 22. Stratification of oxygen in oligotrophic Lunzer Untersee in relation to the bottom profile (August 11, 1924).

The oxygen consumption at the lake bottom can also influence the oxygen content of the free water masses through horizontal currents in a fashion similar to the enrichment of the metalimnial oxygen content by the rooted aquatics, as has been seen, for example, in Lunzer Obersee. Thus, especially in smaller lakes if they are oligotrophic, we can also observe an oxygen curve in which the decrease closely parallels the profile of the lake bottom. For example, in oligotrophic Lunzer Untersee (surface area 0.67 km.^{2}) the decrease in oxygen content is first appreciable below the thermocline and the drop is the greater the flatter the profile of the lake bottom and hence the greater the relative surface of contact with the ooze in relation to the water mass below the depth in question (Figure 22). The difference of this oxygen curve, arising from circumstances of morphology, from that for a eutrophic lake is clear (Figure 20).

The *meromictic lakes* discussed on pages 39ff. present a special case with their stagnant profundal waters which do not circulate every year. These, like the tropical lakes mentioned above, display a marked and often complete loss of oxygen in the hypolimnion even when in their other characteristics they belong to the oligotrophic type. The reason for this is clear from what has been said above. It is the result of the continued oxygen consumption over the course of years owing to the absence of the fall turnover.

The oxygen curve in normal lakes in the temperate zone is determined chiefly by the amount of *oxidizable matter*, in tropical lakes by the *temperature*, and in meromictic lakes by the *duration of stagnation*. Thus, in these last, time is the most important factor.

It might further be mentioned that the organic matter, which by oxidation determines the character of the oxygen curve, *by no means always has its origin within the lake itself*. Often it comes from substances brought into the lake in solution or suspension from neighbouring or more distant regions. Frequently it is the result of man through agriculture or the effluents from sewage systems or industries. Under natural conditions it is the drainage from bogs in particular, which introduces reducing substances into otherwise poor lakes, and which causes a disappearance of oxygen from the hypolimnion.

3. *Elemental Nitrogen and Methane*

While oxygen and carbon dioxide owe their presence in the water to both absorption from the atmosphere and to organic activity in the limnetic biotope, the content of elemental nitrogen comes almost entirely from the atmosphere and that of methane exclusively from anaerobic decomposition, which takes place predominantly in the ooze.

Of course such an inert gas as elemental nitrogen enters into the metabolism of a body of water only to a limited extent. It is known that nitrogen fixation by bacteria takes place, especially in the ooze, and we also now know that certain Cyanophyceae are able to fix nitrogen (see pp. 87f.). We likewise know that nitrate reduction takes place in lakes with the release of elemental nitrogen. However, these processes take place on so minor a scale as scarcely to be able to influence the course of the nitrogen curve. Thus we see from the classical discoveries of Birge and Juday (1911), on which our knowledge of the distribution of N_2 in lakes is chiefly based, that the content of elemental nitrogen at all depths is approximately at the saturation value.[28]

Figure 23 shows the winter distribution of O_2, N_2, and CH_4 in a small eutrophic lake near Moscow, Lake Beloje. It will be seen that there is little difference in the amount of nitrogen except for a decrease just above the bottom brought about by the nitrogen-fixing bacteria. In contrast, the CH_4 curve shows a maximum at the surface of the ooze with a rapid decline in the free water (because of oxidation).

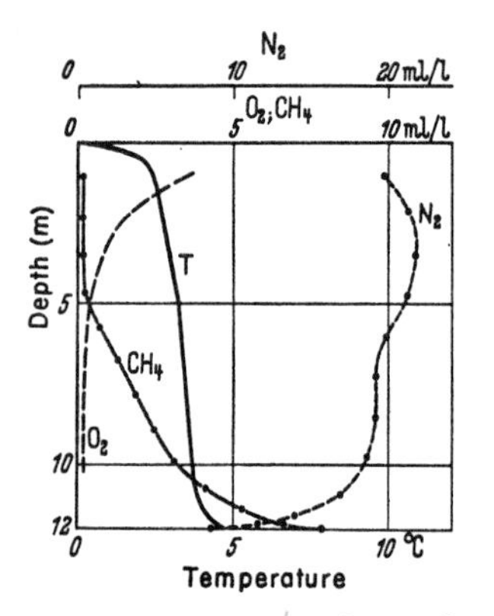

FIGURE 23. Vertical distribution of temperature, oxygen, nitrogen, and methane in Lake Beloje, March 7, 1938. From Kuznetsov (1959).

Methane arises from the anaerobic decomposition of organic matter especially in the ooze (see pp. 198ff.). Here saturation levels are frequently exceeded, and bubbles are formed, which are predominantly methane. These rise to the surface, especially when the barometer is falling. We must once more thank Birge and Juday (1911) for the first investigations of the methane content of water. There can be substantial amounts of methane in the oxygen-free water adjacent to the bottom ooze (for example 38.5 ml./l. in Garvin Lake). However, there is a rapid decrease in the strata above. In Lake Kivu (in the African Rift) there is such a high methane content that it may have commercial value.

[28]The metalimnion and hypolimnion are usually slightly supersaturated with N_2 in summer, a circumstance that arises because the gas is held in by the hydrostatic pressure while the water is warmed slightly.

4. *Iron and Manganese*

These two elements, which are allied in their chemical properties, are intimately connected in their distribution with the reciprocal relation of carbon dioxide and oxygen. We shall first consider the behaviour of iron and later take up the differences, generally unimportant, seen in manganese. Both elements are important in plant metabolism and are of particular interest in limnology as a prerequisite for the appearance of the interesting community of the "iron organisms."

Iron is one of the most widespread materials in the earth's crust but in spite of that it is only under special circumstances that it appears as a dissolved component in water. The reason for this lies in its peculiar solubility relations and in particular in the circumstance that it can be bi- or tri-valent as the ferrous or the ferric forms. The ferrous form can only exist in the absence of oxygen and the ferric form is almost completely insoluble (in the compounds that come under our consideration). The relationships in natural waters have been investigated by Ohle (1934) and particularly thoroughly by Einsele (1936, 1938, 1940). We shall follow the latter principally in our presentation.

The situation is relatively simple where water containing CO_2 comes in contact with iron in the ferrous form; in this case ferrous bicarbonate goes into solution. Such a solution, however, can exist only when the water is almost free of dissolved oxygen. If O_2 is present, oxidation to insoluble ferric hydroxide takes place:

$$4\,Fe(HCO_3)_2 + 2\,H_2O + O_2 = 4\,Fe(OH)_3 + 8\,CO_2.$$

The fact that the course of oxidation is influenced by other factors is important for an understanding of the relations of iron in natural waters. Among others, pH has a particularly strong effect. At neutrality ($pH = 7$) Einsele (1940) has shown that a solution of ferrous bicarbonate can only exist when the water contains no more than 0.5 mg./l. of oxygen. If the oxygen content is higher or the reaction more alkaline, $Fe(OH)_3$ is almost instantaneously precipitated. At lower pH values, however, the precipitation is much slower and we therefore usually find detectable amounts of iron in acid waters even when they are saturated with oxygen.

If, however, water containing carbon dioxide comes in contact with those upper strata of soil and rock that bear ferric compounds (which are much more common, e.g., limonite) almost no iron is dissolved even in the absence of oxygen. A reduction to the ferrous form must first take place and the presence of a reducing agent is necessary for this (for example, decomposing organic matter is very effective even in small amounts). The optimum for this reduction lies at pH 6.5, but organic matter is also effective in the alkaline range up to about pH 7.5.

We can thus conclude that water can dissolve greater amounts of iron as ferrous bicarbonate (1) when it is nearly free of dissolved oxygen, (2) when it contains adequate amounts of carbon dioxide, (3) when the pH is not above 7.5, and (4) when organic substances arising from decomposition are present which can reduce ferric hydroxide.

All the possibilities for the occurrence of iron in fresh waters are not exhausted in the discussion above. It can also occur in organic compounds (masked) and as a colloid. According to Sven Odén (1922), the humic acids, which occur so plentifully in the so-called brown waters of moorland regions, form colloidally dissolved humates with iron. These organic iron sols are much more stable than the inorganic bicarbonate solutions, and for that reason humus waters usually contain considerable amounts of iron even in the presence of oxygen. For example, Ohle (1934) found 0.48 mg./l. Fe in the surface water of Hertasee on the island of Rügen in the Baltic.

When ground water that contains iron comes in contact with air, $Fe(OH)_3$ is precipitated. This brings about the well-known deposits of *iron ochre* which are especially noticeable in bog or seepage springs. When such waters enter below the surface of a lake, precipitation likewise results from contact with the oxygenated and generally alkaline lake water. The "ochergyttja," which is the first product, hardens in the course of time to bog ore. We are especially indebted to Naumann (1930) for knowledge of the origin of this material and the information that until recently it was produced commercially and smelted in Scandinavia.

In the *epilimnion* (with the exception of acid bog waters), the conditions (a high oxygen content and an alkaline reaction) are exactly the opposite to those that are shown above to favour the solution of iron. Thus theoretically the surface water of an alkaline lake cannot contain any detectable amount of iron. If small amounts of the order of magnitude of 0.001 mg./l. are nevertheless detected, they must be present as colloidal iron or as a "supersaturated" solution (the result of incomplete precipitation).

The relations in the *hypolimnion* vary with the nature of the oxygen stratification. Iron does not go into solution in *oligotrophic* lakes, where there is a high level of oxygen even in the depths and generally an alkaline reaction. On the other hand, when the oxygen content in the bottom water of a *eutrophic* lake sinks to nearly zero, then all the conditions for the reduction of ferric hydroxide and the solution of ferrous bicarbonate are realized, since there are decomposing organic substances and aggressive CO_2 also present. Iron is reduced in the mineral fraction of suspended particles and in the bottom sediments, and dissolved as ferrous bicarbonate. Throughout the duration of the stagnation, the iron content of the bottom waters constantly increases and it reaches very high values especially in some of the meromictic lakes, for example, 18 mg./l. in Krottensee (Figure 27) and as much as 41 mg./l. in Zellersee (Einsele, 1944). These values are exceptional, however; in general only a few milligrams per litre are found in eutrophic lakes.

As soon as circulation begins in the autumn and oxygen is driven into the depths on the breakdown of stratification, ferric hydroxide must be precipitated. Since this takes place at even very low oxygen contents, as has been mentioned above, it is only necessary for the circulation currents to speed up the eddy diffusion in order to have the necessary amount of oxygen introduced into the bottom water and the precipitation initiated. The ferric hydroxide that is precipitated can settle to the bottom before it can be caught up by the circulation and transferred to the upper strata in suspension. In this way there need be no significant loss through the lake's outlet and the total iron content will be sedimented on the lake bottom. As soon as the next stagnation period begins and the oxygen disappears from the bottom waters, the conditions for the reduction to the ferrous form arise again at the boundary of the ooze, and at least a part of the precipitated iron goes into solution again. In spring-fed lakes, or those with only a limited influx of surface water, the whole exchange of iron, as Einsele (1936) has shown for Schleinsee, can be limited to this intralacustrine cycle of solution at the surface of the ooze and the subsequent reprecipitation, since scarcely any new iron comes into the lake from outside sources.

The interrelation of the precipitation and solution of iron with oxygen stratification is shown in Figure 24. The example is Lunzer Obersee, a

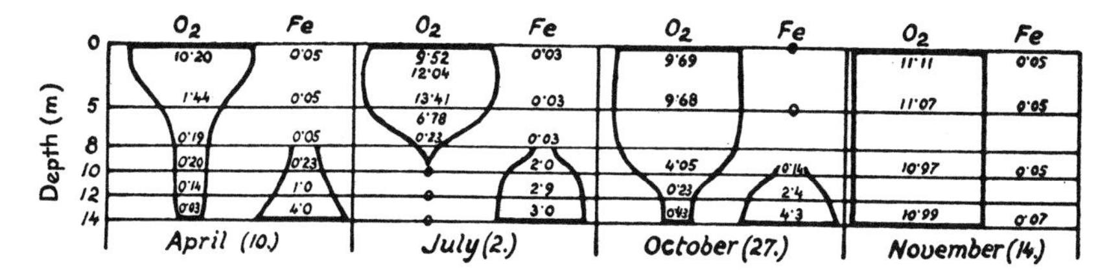

FIGURE 24. Stratification of oxygen and iron in Lunzer Obersee from spring to autumn. From Müller (1938). The method of plotting is the same as that used in Figure 27.

eutrophic lake of the Alps to which the watershed makes a comparatively rich contribution of iron. The distribution in April (when the winter ice cover still exists) shows clearly that the iron content of the water during the winter stagnation originates predominantly from the ooze boundary. In July it is evident that the equally high level of iron in the hypolimnion is chiefly brought about by the solution of sinking suspended materials. The October observation shows the precipitation beginning, and in November, with the onset of total circulation, the iron content of the whole water mass has disappeared to a trace because of the precipitation and sedimentation of $Fe(OH)_3$. F. Berger's unpublished observations on the extent and composition of the sedimentation in Obersee are in agreement with this picture.

MONTHLY DEPOSITION OF SEDIMENT
(in mg./dm.^{2} of lake bottom)

	Winter	*Spring*	*Summer*	*Autumn*
Total sediment	76	257	184	225
Inorganic	41	131	103	116
Iron (Fe) mg. and % inorganic	4.2 (10%)	21.2 (16%)	12.6 (12%)	24.2 (21%)

It can be seen that the greatest amount of iron (absolute and in percentage) is sedimented during the periods of circulation. However, an appreciable sedimentation takes place during the summer stagnation, which in the greatest part may be owing to the precipitation of iron by organisms at the lower boundary of oxygen.

The result of the processes is a progressive iron *enrichment* in the sediments of eutrophic lakes. This occurs because other mineral constituents, in particular calcium carbonate, are also dissolved out of the suspensions and the sediments by the aggressive carbon dioxide in the hypolimnia of such lakes. However, these other dissolved materials are *not* precipitated by the introduction of oxygen but remain in solution and, in contrast to iron and manganese, are gradually lost to the lake by outflow. *The hypolimnion of a eutrophic lake thus operates as a trap for iron.*

If the lower water contains hydrogen sulphide, a precipitation of ferrous sulphide can take place when there is an alkaline reaction. According to Einsele, 1 mg. Fe and 1 mg. H_2S per litre can exist together at pH 7; at higher pH values the amounts are smaller; at lower, greater. Since the content of hydrogen sulphide, even in the relatively few instances where this substance is detectable at all in open waters, only rises above a few milligrams per litre in exceptional cases, a *complete* precipitation of iron through the agency of sulphide can scarcely occur. However, a decrease in iron content can usually be found towards the bottom in lakes in which the hypolimnion contains H_2S.

In all essential ways *manganese* has a behaviour very similar to that of iron but it does display a certain difference in regard to stratification since it reduces more easily and is harder to oxidize than iron. Thus manganous bicarbonate is still stable in the presence of 1 mg./l. of oxygen at pH 7 and no oxidation takes place below pH 6. As a result of this property, dissolved manganese occurs at somewhat higher levels in the oxygen gradient than does iron (see Figure 27). A sharply expressed manganese maximum can occasionally be observed in tropical lakes in the upper hypolimnion. This may be interpreted as resulting from the process by which the manganic hydroxide precipitated by progressive eddy diffusion of oxygen from above is reduced again in the oxygen-poor stratum immediately below. This phenomenon is not seen in the case of iron because that element is more difficult to reduce. Manganous sulphide is not precipitated over the pH range normally found in natural waters.

The part played by organisms in the precipitation of iron and manganese is discussed on page 192.

The survey above by no means exhausts the manifold problems of iron and manganese cycles in waters. The tendency for insoluble ferric and manganic hydroxides to form sols, the absorptive properties of these colloids, their relation to humus material with which they partly form chemical compounds and partly act as protective colloids, and finally the special role the iron organisms play, often lead to complicated conditions which are extremely difficult to understand.

5. *Nitrogen and Phosphorus*

Nitrogen and phosphorus, together with carbon and hydrogen, are recognized as the most important constituents of living matter, the protoplasm of the cell, and are for that reason nutrients of outstanding value. These elements often escaped detection in the older analyses of uncontaminated spring or lake waters. The dilutions in which they occur in natural waters are so great that discovery of them was usually beyond analytical methods formerly in use. With the refinement of technique (especially in colorimetry) in the last decades, we are now able to determine these "traces" exactly and thus to elucidate the distribution of these two elements.

Inorganic *nitrogen* compounds are present in small amounts even in *rain-water*, where they are of course predominantly nitric acid and ammonia. These come from the *atmosphere*, in which they occur (if we exclude impurities from industrial smokes) as the products of electrical discharges, terrestrial decomposition, and volcanic eruptions. Rainwater at Lunz contains on the average about 0.36 mg. per litre of nitrate and ammonium nitrogen. This results in a contribution from the atmosphere of about 600 mg. per square metre or 6 kg. per hectare from a mean annual precipitation of 1650 mm.

A considerable amount of elemental nitrogen is fixed in the soil by the nitrogen-assimilating bacteria and becomes available for the metabolism of plants.

There can scarcely be any doubt that nitrogen-fixing organisms, bacteria of course and blue-green algae as well (cf. Fritsch and Demoll, 1938), are also active in the water. The fixation of nitrogen has been clearly recognized in ponds (cf. Demoll, 1925) and Einsele (1941) has traced back an increase of nitrogen in Schleinsee following the introduction of phosphorus to nitrogen fixation at the ooze surface in the littoral region. Hutchinson (1941) was able to establish with greater certainty the fixation of considerable amounts of nitrogen in Linsley pond by an *Anabaena.* More recent work has shown that nitrogen fixation by the blue-green algae is not limited to *Anabaena* but is also carried out by numerous other species. Kuznetsov (1958) was able to show that in "Black Lake" near Kossino, the activity of the blue-green algae in this respect exceeded that of the bacteria. For example, he calculated that the total summer assimilation of nitrogen for the whole lake by a planktonic species of *Anabaena* was 13.1 kg. N, while *Azotobacter* assimilated only 0.14 kg. N. Similar findings have been made by Hutchinson and Gessner.

Ammonia is the chief decomposition product from plant and animal proteins. In the presence of oxygen, however, this is immediately transformed by the nitrifying bacteria into nitrate[29] (see p. 200), which can be given up to the water percolating through the soil. For this reason, ground and spring waters not contaminated through human activities generally contain nitrogen only as nitrate and of course usually in amounts of the order of from a few tenths of a miligram to milligrams per litre.

Nitrate[30] is also the universal inorganic form of nitrogen from surface to bottom in oligotrophic lakes except for occasional traces of nitrite and ammonia. Nitrate can also occur to a level of milligrams per litre

[29]According to Karcher (1939) nitrification proceeds less rapidly in acid lakes than it does in alkaline ones.

[30]The method of Tillmans and Sutthoff (1911) which uses diphenylamine, is recommended for the quantitative determination of nitrate, an analysis which limnologists often have to make under difficulties. However, strict attention must be paid to all the details of the method, especially to the rapid cooling of the sample after the introduction of the reagents dissolved in concentrated H_2SO_4 (Müller, 1933).

in moderately eutrophic lakes.[31] During the summer stagnation there is, as was first recognized by Minder (1926), a clearly recognizable stratification, which is the natural result of a gradual utilization in the epilimnion, and this at times can lead to the complete disappearance of nitrate from the upper strata.

The distribution of nitrogen compounds in *markedly eutrophic* lakes and also in *meromictic lakes* presents quite another picture. In these the nitrate is also taken up by the phytoplankton as pointed out above and because of the more abundant flora the amounts found in the epilimnion in the summer months are usually smaller than in oligotrophic lakes; it is not unusual for nitrate to disappear completely in such cases. Below the lower limit of oxygen in the hypolimnion, where the oxygen amounts to only a few tenths of a milligram per litre, the nitrate suddenly disappears and with a further increase in depth there are gradually increasing amounts of dissolved *ammonia* (Figure 27, nitrate and ammonium N). There is occasionally, as Müller (1934) has shown, a limited occurrence of nitrite between the nitrate and ammonia; thus all three steps in the oxidation of nitrogen are linked with the course of the oxygen curve. The disappearance of nitrate at the limit of oxygen is very likely brought about by *denitrifying bacteria*, which reduce it through nitrite and nitrous oxide to elemental nitrogen. The ammonia that arises in the hypolimnion comes for the most part from the *breakdown of protein* in the water and the ooze.[32]

In many regions, especially on the plains, for example in the lakes of the Yahara basin near Madison (Domogalla and Fred, 1926) and in the Masurian forest lakes (Karcher, 1939), conditions differ from the state described above in that there is less nitrate in the epilimnion than ammonia. This apparent reversal of the usual condition is probably brought about by the morphology of the basin and by allochthonous influences. However, it is also favoured by the circumstance that the amounts of both nitrogen compounds are very small (the means for the Wisconsin lakes being 0.06 mg./l. nitrate N and 0.19 mg./l. ammonium N).

Phosphate is one of the nutrients which is least abundant in our

[31]See page 207 regarding the sudden appearance of ammonium and phosphate after the disappearance of the oxygen.

[32]It is not known whether reduction of nitrate to ammonia through nitrite, which occurs widely elsewhere in nature, also takes place to a significant degree in the water. Should this be true then part of the ammonia present in the hypolimnion could result from this process. Experiments in which nitrate was added to deep-water samples from Lunzer Obersee resulted in no formation of ammonia at all but only in the formation of nitrite and elemental nitrogen.

waters, being often present in amounts inconceivably small. The water in oligotrophic lakes often contains less than one thousandth of a milligram (1γ) of inorganic phosphorus per litre; indeed the content not infrequently drops below the limit of detectability of the extremely sensitive molybdate method. The ability of the plankton algae to utilize even these traces of an essential element must cause more wonder than our ability to detect them. Phosphorus is not taken up by the plankton algae solely in proportion to their demands for growth and reproduction. An excess over actual needs is stored if the supply allows this. This extremely important fact was established by Einsele (1941) during fertilization experiments in Schleinsee in which phosphate added to the water disappeared in a few days, having been taken up by the plants. From related experiments it was found that the plankton algae were able to store more than ten times as much phosphorus as they normally contained. This important fact has since been confirmed physiologically by Rodhe (1948) and more recently by Mackereth (1953) in careful culture experiments. Experiments with radio-phosphorus (P^{32}) have also given confirmation and further information on the subject (Hutchinson and Bowen, 1947; Coffin, Hayes, Jodrey, and Whiteaway, 1949; McCarter, Hayes, Jodrey, and Cameron, 1952). In all cases there was a rapid loss of phosphorus from the open water of the lake, the loss being due not only to storage by the phytoplankton but also by the zooplankton (*Diaptomus*) and especially by the littoral vegetation.

While, as is mentioned above, the nitrogen compounds in water are derived to an appreciable degree directly or indirectly from the atmosphere, the phosphorus comes exclusively either directly or indirectly (indirectly through organisms) from the weathering of phosphatic rocks (apatite) and from the soil, and is (in inorganic form) present as dissolved phosphate. In contrast to nitrate, phosphate is avidly held by the soil and for this reason is not as easily leached by rain water as is the former substance. This is likewise the reason why spring waters contain so much less phosphate than they do nitrate. Thus, for example, the waters of the springs around Lunz (a limestone region) contain about 0.1 to 0.8 mg./l. nitrate nitrogen and only 0 to 3 γ/l. phosphate phosphorus. In comparison rain-water at Lunz occasionally may contain up to 10 γ/l. phosphorus (presumably originating in dust), especially at the beginning of a shower. The phosphate stratification which can be seen in most lakes during the summer stagnation is naturally less strongly expressed in oligotrophic lakes than it is in the eutrophic type. Amounts of phosphate of the order of tenths of milligrams per litre often accumulate in the hypolimnion of the latter.

Einsele (1936) has found an interesting relation between the phosphorus and the iron cycle in eutrophic lakes. When ferrous iron and phosphate occur together in the hypolimnion of such lakes an insoluble ferric phosphate is precipitated at times when oxygen is introduced and the reaction made alkaline. The excess iron usually present is precipitated out as ferric hydroxide as mentioned above. Both compounds are deposited in the sediments of the lake. In this way almost the whole phosphorus content of a lake can be carried to the bottom at the fall turnover. When there is a lack of oxygen in the sediment the iron can be reduced from the ferric to the ferrous form and the phosphorus can be freed and go into solution. Thus, particularly in small lakes which are not too deep and in which the interaction between the bottom and the water mass proceeds to a considerable extent, the phosphate is subject to a persistent cycle once it reaches the hypolimnion (precipitation during the period of circulation and solution during stagnation). (See p. 217.) If H_2S is present FeS is formed, especially under alkaline conditions, and in that event phosphorus is set free (Ohle, 1954).

As the investigations of Steiner (1938) have shown, it is not only through the putrefactive activity of bacteria that phosphate is set free in the tropholytic zone; the sinking plankters are also partially broken down by autolysis in which the phosphatases split off the phosphate from the phosphorus-containing proteins (the nucleoproteins).

See page 96 concerning organic nitrogen and phosphorus as well as "total nitrogen" and "total phosphorus." An account of fertilization experiments is given on page 101.

6. *Other Mineral Substances*

A further element taking part in the constitution of protein is sulphur. It is present in the "nutrient solution" as sulphate, usually as gypsum ($CaSO_4$) or under certain circumstances as hydrogen sulphide (H_2S). Since the latter is easily oxidized (to sulphuric acid) it can only be stable in the absence of oxygen. Therefore it is never found in the epilimnion. It occurs only in the oxygen-free hypolimnion, in the mud, or in volcanic springs.

Sulphate is present in almost all waters in greater or smaller amounts. If we exclude lakes without exits (see pp. 60f.) and brackish waters, the upper limit of solubility for sulphate is in general limited by the solubility of gypsum which, for example, is found in the sedimentary rocks of the Triassic. The solubility of $CaSO_4 \cdot 2H_2O$ is about 2 g. per litre at 18° C. Since the uptake by the plants is ordinarily small in relation to the amount in solution, it is not to be expected that there would be a chemical stratification such as is spoken about below (Figure 25) and it can be assumed that the substantial differences in vertical distribution that have been encountered in the not too numerous observations (e.g. Birge and Juday, 1911; Ohle, 1934) do not have a biological cause

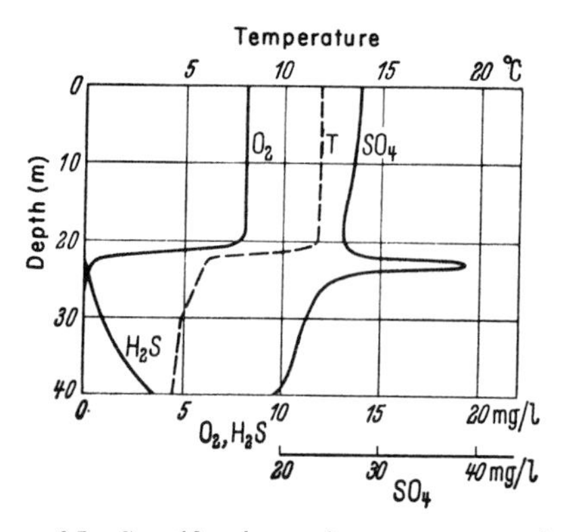

FIGURE 25. Stratification of temperature, hydrogen sulphide, oxygen, and sulphate in Grosser Plöner See, September 30, 1952. From Ohle (1954).

(however, see below). In another respect, however, the sulphate content is very important. As a rule only those eutrophic lakes that are also rich in sulphates have large amounts of hydrogen sulphide in the hypolimnion. There are a number of species of bacteria that are able to reduce sulphate to H_2S (see pp. 202f.). By far the greatest part of the hydrogen sulphide that appears, often to the extent of several milligrams per litre, in the oxygen-free depths of certain lakes originates from this *reduction of sulphate.* The amount of sulphur contained in the protein molecule is too small under ordinary conditions to bring about any considerable enrichment of the water with this gas when protein alone is decomposed. In agreement, Mortimer (1941) was able to establish a significant decrease in the sulphate content of the oxygen-free hypolimnion of a north English lake (Esthwaite Water). Because of their high sulphate content, sea-water and its dilutions (brackish water), together with saline and volcanic waters, are prone to the formation of hydrogen sulphide in the absence of oxygen. The high H_2S content of the Black Sea, in which the depths below 200 metres are made uninhabitable for higher animals because of this poisonous gas (cf. Caspers, 1957), is well known. In the depths of valley impoundments polluted by the sulphate-rich effluents from pulp mills, large amounts of H_2S, which can bring about catastrophic kills of fish at the turnover, can come from the reduction of the sulphate. A notorious example of this is given by Liebmann (1938) for the well-investigated Bleiloch impoundment in Thuringia. Enormous quantities of hydrogen sulphide have been reported in saline lakes. Thus Hutchinson (1937) found 780 mg./l. H_2S in the deep waters of Big Soda Lake.

That H_2S is not poisonous for all organisms but is indispensable for a highly interesting group of bacteria will be discussed in a later chapter (p. 201). Since these sulphur organisms can oxidize H_2S to H_2SO_4, a sharp discontinuity in the sulphate content can occur at the stratum where H_2S and O_2 are in contact, as is the case in Figure 25 (Ohle, 1954).

Under special circumstances sulphur can also occur as H_2SO_4 in natural waters. This is frequently the case in volcanic regions where the sulphur dioxide from the solfataras is oxidized to sulphuric acid. (See p. 201 regarding the oxidation of H_2S.) Lakes with sulphuric acid in their waters are known in Japan and Indonesia. Yoshimura (1933) found a pH of 1.4 in one of these lakes. Kawah Idjen in east Java is probably a similar case.

The "sulphuric acid pond" described by Ohle (1936) near Rheinbeck owes its free sulphuric acid (pH 3.6–3.1) to another process, the oxidation of the pyrites in which the clay is rich there. The free H_2SO_4 that arises from the oxidation of H_2S or S by sulphur bacteria (p. 224) can bring about low pH values in unbuffered waters.

While *silicic acid* is not a constituent of protoplasm it forms the basis of the skeletal structure of the most important group of algae in the water, the diatoms. It also plays an important part in the sclerification of organs, for instance in cyst integuments and the spicules of chrysomonads, in certain Heliozoa, siliceous sponges, and so on. It resembles carbon dioxide in its chemical properties. However, since it is much more weakly dissociated than the latter it is removed from its strongly hydrolyzed compounds, the silicates, in the presence of carbon dioxide or bicarbonates, and is then held in the water as free silicic acid in a dissolved (or colloidal) form. The amounts in which it is present vary greatly and in general are of the order of milligrams to centigrams per litre. Bog water generally has less than 1 mg./l. Its stratification is always clearly expressed since it is used to a considerable extent by the diatoms. A major decrease in dissolved SiO_2 is regularly found in the epilimnion after a bloom of diatoms. The investigations of Einsele and Vetter (1938), for example, in Schleinsee show this clearly (Figure 26). The solid line shows the course of the diatom population in μ^3/ml., the line with dashes and dots shows the annual change in silicate content in the epilimnion, and the broken line the change in the hypolimnion. The great rise in the silicate content in the hypolimnion after the diatom maximum of late summer and autumn is noteworthy for it indicates that silicates are released from the sediments.[33]

[33]However, there is also the possibility to be considered that SiO_2 is set free from an organic combination (ferri-silicic-humate complex) following the solution of the latter when the iron goes over into the ferrous form (Mortimer, 1941).

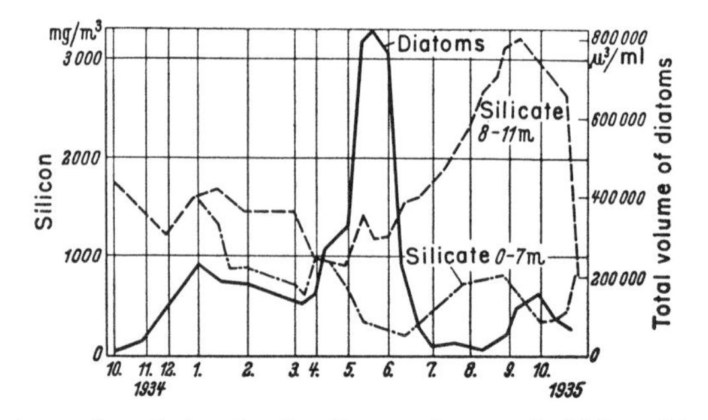

Figure 26. Annual variation in the diatom flora and silicic acid content at two depths in Schleinsee. From Einsele and Vetter (1938).

Only a little is known of the economy of *magnesium, potassium, sodium,* and *aluminium* but we have in flame photometry a sufficiently sensitive and rapid means of analysis to increase our knowledge of the place of these elements in limnology. We know now from the total analyses we have from predominantly surface waters that these elements occur in widely varying amounts. Aluminium occurs in the least amounts in normal waters (tenths of mg./l.) and magnesium in the greatest, especially in regions with dolomitic limestone. A biogenic chemical stratification is especially to be expected with potassium and magnesium since these belong to the elements that are necessary for life. Potassium is usually present to the extent of a few milligrams per litre according to Collet (1925), Järnefelt (1935), Kuisel (1935), the Wisconsin investigators (Lohuis *et al.,* 1938), Lohammar (1938), and Ohle (1940). Ohle established that there was a definite increase in the potassium content with depth in the Baltic lakes.

Magnesium is generally present in solution as bicarbonate, in which form the relations in water are similar to those of calcium bicarbonate. There is one difference: the monocarbonate of magnesium ($MgCO_3$) is much more soluble than is $CaCO_3$ and therefore when CO_2 is separated from the bicarbonate (for example through the photosynthetic activity of plants) it does not precipitate so easily as the latter. On the other hand magnesium hydroxide is practically insoluble in comparison with calcium hydroxide (cf. p. 80).

It is not particularly surprising that sodium should occur universally in various concentrations and is the predominant cation in brackish water.

Finally, chloride ions should be mentioned as a dissolved substance regularly present in waters. These are scarcely absent anywhere, so wide is the distribution of chlorine in nature, but they are present in major quantities only in salt and brackish waters. Since chlorine is not essential for vegetable life, differences in vertical stratification appear only when they are caused by hydrographic factors.

We should be going beyond the compass of this introduction if we pursued the possible limnological importance of other elements in the light of our still insufficient knowledge of them. It is enough at the moment to know that the "trace elements," which need only occur in extremely small amounts, can come to be of great significance in the productivity of waters. Often the unexplained differences in the composition of biocoenoses in neighbouring waters, which are apparently similar in all respects, are the result of the presence or absence of such obscure trace elements. Hutchinson has brought together in his great work our present knowledge of these elements. He has also assembled our knowledge on the occurrence of radioactive substances in water, a subject that is such a live issue nowadays.

7. *Organic Matter*

It was long assumed in limnological investigations that "pure" waters of lakes, streams, and rivers could be considered as inorganic solutions as long as they were not contaminated by human and animal excreta or by certain industrial wastes (for example, from sugar refineries, distilleries, rayon factories), and that the organic matter contained in them consisted predominantly of organisms and organic detritus, that is, of particulate material. It was assumed that the plants of this community were entirely *autotrophic*—supported by the assimilation of inorganic nutrients. A *heterotrophic* or rather *mixotrophic* habit involving the assimilation of organic as well as inorganic carbon appeared possible only in sewage and waters contaminated by it, where, as will be mentioned later, the process has been found in numerous organisms.

In the practical examination of waters the content of organic matter is expressed as an *oxygen demand* in terms of oxidation with potassium permanganate. However, it must not be forgotten that the uptake of $KMnO_4$ (which at best gives only a partial oxidation) gives no measure of the organic carbon since it depends on the state of oxidation of the matter in solution. Further the *loss on ignition* of weight from the residuum on evaporation cannot be taken as a measure of the content of organic matter either, without qualification. The *biochemical oxygen demand* (B.O.D.), which is the bacterial consumption of oxygen in a stated time under stated conditions, provides information concerning the easily assimilated organic fraction. Each of these methods has its importance with regard to special questions, particularly in the investigation of drinking water and sewage effluents.—The B.O.D. throws an interesting light on the nature of chemical stratification in lakes (e.g., Stangenberg, 1959).

Ohle (1934) has developed a method for the estimation of the organic content of water by a comparison of its electrical conductivity (which, being due to the electrolytes, is governed essentially by the content of inorganic salts) with the interferometer values (which depend on the total concentration of dissolved matter).

Investigations in the last decades, especially those under the direction of Birge and Juday (1934) in the course of the Wisconsin lake survey, have yielded the knowledge that "pure" lake waters contain considerable amounts of dissolved organic matter and, what is most important, that *the amount of dissolved organic matter exceeds by several times that in particulate form* (which is chiefly plankton). Thus for example, the eutrophic Lake Mendota contains 12.5 mg./l. of organic matter in solution and only 1.5 mg. in plankton. Even markedly oligotrophic lakes of that region had a content of about 5 mg./l. of dissolved organic matter and less than 1 mg. of plankton.

Because of the importance of this subject for the understanding of the metabolism of lakes, a table from the work of Birge and Juday is given which shows the results from a large series of lakes in the northeastern highlands of Wisconsin. In these investigations the lakes were grouped according to their content of organic carbon. The dissolved and particulate matter were separated by centrifugation and the ratio of dissolved to particulate matter calculated. The dissolved fraction was further separated into crude protein, ether-extractable material, and carbohydrate.

Grouping by carbon content (mg./l.)	*Organic matter* (mg./l.)		*Ratio dissolved: particulate*	*Crude protein* (%)	*Ether extract* (%)	*Carbohydrate* (%)
	Particulate	*Dissolved*				
1.0–1.9	0.62	3.09	5.0	24.3	2.3	73.4
5.0–5.9	1.27	10.33	8.1	19.4	1.3	79.0
10.0–10.9	1.89	20.48	10.8	14.4	0.4	85.2
15.0–15.9	2.32	31.30	16.5	12.9	0.2	86.9
20.0–23.8	2.22	48.12	21.7	9.9	0.2	89.9

It is a remarkable fact that the increase in total carbon occurred predominantly in the dissolved organic fraction. Further the proportional amounts of crude protein and ether extracts diminish with increasing total carbon. The causes of these phenomena lie in the difference in origin of the dissolved organic material. Part arises autochthonously in the lake in question, part is allochthonous material from the watershed. In lakes with the highest carbon content, the allochthonous fraction predominates, as is indicated by their humus-stained waters. The lakes with lower total carbon reflect conditions of autochthonous production (by plankton). Further investigations dealing with amino acids, vitamins, and antibiotic substances, concerning all of which our knowledge is still in a state of flux, are discussed in Hutchinson and Gessner. Shapiro's (1957) investigations of the chemistry and biological importance of the humic substances, which have such a great influence on the transparency of water and its buffering capacity, are of particular significance.

Total nitrogen and total phosphorus

Findings in lakes of North America, in central Europe, and in Scandinavia have all shown that organic forms of nitrogen and phosphorus are always present in larger amounts than are the inorganic forms mentioned on pages 87ff. This relation is particularly striking in the case of phosphorus. The total phosphorus is generally more than ten times that bound in phosphates (0.01–0.05 mg./l. in the epilimnion of the alpine lakes).

This total phosphorus, like total nitrogen, consists of three components: (1) the inorganic, (2) the dissolved organic, which is colloidal, and (3) the particulate, that is, the fraction bound up in the animate and inanimate particles in suspension. The proportions of these three fractions can display considerable variations. Thus the proportions of total phosphorus in Linsley pond are 8 per cent, 29 per cent, and 63 per cent in the order mentioned above, while in some of the Wisconsin lakes the relation of the colloidal to the particulate has an exactly opposite ratio (61 per cent to 28 per cent). The variable amounts of phosphorus stored by the plankton algae must have some bearing on these circumstances.

The dissolved organic fraction of the total nitrogen content displays a remarkable agreement in the various European and North American lakes as well as in the various depths of both oligotrophic and eutrophic types. In general it amounts to about 0.5 mg./l. This agreement suggests that the dissolved organic nitrogen is largely in a form which is very little used in metabolism under normal circumstances (cf. Ruttner, 1937; Naumann, 1941; Einsele, 1941) and that it is an end product which is very resistant to further decomposition. This conclusion is probably also true for colloidal phosphorus.

The experiments of Kuznetsov (1959), who showed that samples of water from various lakes lost only a fraction of their organic matter even after long standing, give support to this conclusion. The fraction of dissolved organic matter broken down (B.O.D. p. 94) can be taken to be the portion that is easily assimilable by bacteria. According to Kuznetsov (his p. 25) the easily assimilable fraction of dissolved organic matter depends on the level of productivity of the lake. In oligitrophic lakes it amounts to only 2–3 per cent while in eutrophic lakes it is 10–15 per cent of the total dissolved organic matter. A decrease in organic dissolved nitrogen by which the concentration may approach zero is found only in meromictic lakes. This can be taken to indicate that breakdown of this fraction can occur slowly under completely anaerobic conditions. However, the number of observations is too small for a final opinion to be formed yet. On the other hand the organic substances which continuously reach a lake from outside sources or are produced in it by metabolism, and which are usable as nutrients, are plainly quickly utilized and are not detected by analysis.

Biological water analysis

Even from these few remarks it can be seen that the content of dissolved organic matter in water can by no means be neglected in considerations of biological productivity. Moreover, because of this circumstance the whole problem is made more difficult. While green plants all utilize inorganic nutrients in a similar way, in the sense that they all assimilate carbon dioxide, for example, and they all need the elements mentioned on page 58 for growth (although they differ greatly in their quantitative requirements for these), many are specialized in respect to utilization of organic matter. There are algae that can supplement their photosynthesis by organic carbon only in the form of carbohydrate while others can only make use of organic acids for this purpose and still others do best when they are provided with peptones. These are only a few well-known examples of the many mixotrophic organisms whose metabolism is for the most part completely unknown. This knowledge is particularly lacking for the phytoplankton, which are especially difficult to cultivate in the laboratory. Thus we know little concerning whether various species may be able to assimilate organic substances and which of these they can utilize. Likewise little is known as to precisely what compounds make up the dissolved "organic substances" in water in a particular case.

In spite of the limited knowledge in this field the simple recognition of certain correlations has already been of practical value for some years past. From the observation that the appearance of certain species is related to certain levels of impurities, we have learned that the degree of pollution in water is indicated by the appearance of these organisms. Similarly in the inorganic field, the thriving of sulphur bacteria indicates the presence of H_2S and that of iron organisms indicates iron (biological assessment of water after Kolkwitz and Marsson, 1902). The inhabitants of pure water are distinguished as *katharobes*, those of polluted waters, as *saprobes*; the latter are sub-divided according to their affinity for organic matter into oligosaprobic, mesosaprobic, and polysaprobic organisms. Through the activity of all these organisms, bacteria, fungi, green algae, and animals, the organic content of the water is gradually broken down or oxidized—*mineralized*. This *biological self-purification* of waters, which can be accelerated by appropriate measures, plays an important role in industry.

With increasing settlement and industrialization, biological investigation of water supplies and effluents becomes a more and more important complement to bacteriological and chemical investigation. Biological assessment requires the most exact knowledge possible of the environmental requirements and the physiological capacities of the organisms in question. This knowledge must in the main be gained through culture experiments in the laboratory (e.g., Wuhrmann, 1945). Such investigations are continually on the increase at the present time and we can expect that this important branch of applied limnology will continue to be established on a more and more exact basis (cf. Liebmann's handbook, 1951).

8. *The Law of the Minimum*

Whenever natural waters have been called nutrient solutions in the preceding sections, the designation has meant only that the plants of this biotope grow by reason of the nutrients dissolved in the water. The term obviously was not used to indicate that the individual elements were present in an amount corresponding to that needed, as is approximately the case in an artificially prepared nutrient solution. In most waters there is a great disproportion between the composition of the salts present and the demands for organic growth. Single ions such as Ca^{++} and HCO_3^- are often present in amounts far in excess of the requirements, while others, like phosphate ions, generally are present only in disproportionately small quantities. Thus the plants cannot utilize an abundant ion because they suffer from lack of another. This *Law of the Minimum* (which states that productivity is limited by the nutrient present in the least amount at any given time) was first formulated by Justus von Liebig and later, when it was recognized that it held good not only in chemistry but also for the majority of living processes and environmental conditions, it was extended in the *Principle of Limiting Factors*. Just as in the case of carbon assimilation an increase in light intensity beyond a given level increases the photosynthetic activity only when the temperature is increased at the same time (see p. 138), so also a rich supply of nitrogen will increase productivity only when the amount of the minimally occurring phosphorus is increased at the same time (assuming that all other nutrients are present in excess). In the first instance temperature was the limiting factor, in the second the phosphorus content.

Obviously, other substances that are essential to life besides phosphorus and nitrogen can also act as limiting factors. Thus, although it is usually present in abundance, carbon can operate in this fashion under appropriate circumstances, since there is the greatest need for it as a predominant constituent of living matter. Carbon is most likely to be limiting in those waters that are low in bicarbonate. Such waters not only do not have the reservoir of bound carbon dioxide but they display also a particularly low level of equilibrium CO_2 (p. 62). Even those plants that can assimilate bicarbonate ions (submersed phanerogams and many algae) assimilate free CO_2 more rapidly so that the epilimnia of even relatively well buffered lakes are frequently almost devoid of free CO_2 (see Figures 18 and 19, p. 73). Under such circumstances carbon can easily become a limiting factor. It can be taken that the low productivity of many lakes in granitic basins arises from this cause. Examples of such lakes are: many in Scandinavia; the north Saskatchewan lakes of the Canadian Shield, recently investigated by Rawson (1960); and in particular the African mountain lakes, which are so extremely poor in electrolytes (Löffler, pers. comm.; Margalef, 1958).

Changing one of the factors will have the greater influence on production the more marked this factor is in the minimum with respect to the other factors. On the other hand the more the concentration of a substance exceeds this minimum value the less will be its relative effect as a factor (the *Law of Relativity* of Lundergårdh, 1925, or Mitscherlich, 1923).

Fry (1947) has made the principle of limiting factors more specific by subdividing the appropriate environmental factors into the three categories below:

Limiting factors. Limiting factors in the strict sense, which operate by being the sources of material and energy (light).

Controlling factors. Factors that influence the metabolic rate by conditioning the state of activation of the metabolites without themselves entering into the metabolic chain (e.g., temperature or the effect of Ca and Mg on cell permeability).

Lethal factors. Factors that destroy the organism through disturbance of the metabolic chain.

Further it should be recognized that the increase of an organism is a resultant of many physiological events all of which obey the law of the minimum in specific ways, and that the relations of these vary in the different species. Each species makes its own special demands on the environment.

When we consider the nutrient content of waters from this point of view, we must look for the limiting factor among the ions in solution that occur in the smallest amounts at the time (minimum constituents). Among the elements important for life these constituents are usually nitrogen or phosphorus, or, as is pointed out above, carbon. The plants take up the minimum nutrient to the greatest extent that they can. The great ability of living cells to take up phosphorus is shown by the fact that this substance is found in the epilimnia of our lakes in extremely small amounts, often below the limits of detectability. A minimum substance can thus frequently be recognized by the very fact of its being no longer detectable. Its introduction (perhaps through the affluents) results in its immediate use by organisms.

Schreiber (1927) has developed a basic approach by which the limiting nutrient may be recognized. His methods were applied, with some modifications, by Gerloff and Skoog (1957) to two Wisconsin lakes. If various nutrients are added to samples of lake water containing *Microcystis* there will be an increase in numbers of that alga only when the nutrient in question is limiting. In the two cases Gerloff investigated the limiting nutrient was nitrogen (not phosphorus).

The uptake of nutrients by organisms does not necessarily immediately lead to an increase in production. We have already seen from Einsele's (1941) findings that the phytoplankton is able to store many times its normal content of phosphorus. In a fertilization experiment in Schleinsee (see p. 174) no increase in the amount of phytoplankton was found until a month after the addition of phosphorus and an increase began only after nitrogen assimilation (as a result of the addition of phosphorus) first brought about a significant increase in the nitrogen content (which was the minimum substance). Rodhe (1948) found that cultures of algae were able to increase actively for periods of a week to a month after storing a supply of phosphorus, nitrogen, or iron, although the nutrient solutions in which they then were placed were completely free of these elements and were renewed repeatedly (see also Mackereth, 1953).

In this way a *dynamic equilibrium* arises between the amount of the influx of the limiting nutrient and the growth or reproduction of organisms. The nutrient introduced is immediately transformed into living matter or is bypassed into preliminary storage in the cells and the level of chemical in the water remains below the limit of detection. This state of affairs obtains in eutrophic as well as oligotrophic waters, that is for both lesser and greater influxes of nutrients. Hence the attempt to detect differences in productivity from differences in the detectable content of minimum substances will usually be fruitless.

The great differences in the environmental requirements of the various organisms make it necessary that we have a thorough knowledge of the physiology of races and species in order to understand the metabolism of waters. We are just entering this field through a wearisome task involving pure culture. However, important progress has already been attained (cf. Rodhe, 1948; Vollenweider, 1950).

The nutrient increment which thus determines organic productivity in open waters comes from three sources. Two of these, *tributary streams* and *precipitation from the atmosphere*, have already been mentioned a number of times. The third, and by no means the least important, is *interchange with the bottom sediments of the lake*. These sediments, as will be mentioned later (p. 190), consist largely of remains of organisms produced in the phototrophic zone which were not decomposed when they were free in the water. Such remains undergo a gradual breakdown in the ooze. The soluble products of this process are given off to the water in the strata adjacent to the bottom and are carried up again as nutrients in proportion to the magnitude of the process of eddy diffusion. When—at low bottom temperatures—the breakdown in the ooze proceeds more slowly than the deposition of organic material there is a gradual accumulation, and concomitantly there is also an increase in the contribution of the breakdown products to the free water. In this way, under especially favourable circumstances, an oligotrophic lake can become a eutrophic one as it ages.

If we now review once more at the conclusion of this section the total organic economy of a lake, we shall see that the changes in the distribution of dissolved substances and the stratification phenomena, which are frequently so striking, result from the interplay of synthesis and breakdown—the reduction process of *assimilation* and the oxidative process of *catabolism*. Both these processes go on together in the *phototrophic zone*, although catabolism is not in evidence, not because there is less breakdown in the epilimnion (in fact there is considerably more because of the high temperature) but because it is outpaced by the much more intensive assimilation. All the end products of catabolism, the carbon dioxide as well as the mineral substances, are immediately taken up again by assimilation. On the other hand, there is no photosynthesis in the lightless *tropholytic zone*; here the metabolic processes are predominantly processes of breakdown. The extent of these processes is the greater the higher the temperature of the hypolimnion, the larger the contribution of decomposable organic substances from the trophogenic zone, and the longer the duration of the period of stagnation. In the lakes of the temperate zone with their low hypolimnial temperatures, the two latter factors have the greatest effect. In *eutrophic* lakes (and also in *meromictic* ones with their extended stratification period) the oxygen of the lower waters, which was gained during the preceding turnover and by gradual intermixing with the water above, may be used up extensively through oxidation even to complete disappearance, and is replaced by corresponding oxidation products, which often accumulate in considerable amounts. The elevated hypolimnial temperatures of tropical lakes lead to a rapid disappearance of the oxygen store there regardless of the level of productivity of the lake in question. In *holomictic oligotrophic* lakes, on the other hand, the amount of oxidized substances is too small in relation to the water mass to bring about appreciable changes in the original distribution of materials.

Figure 27 shows extremes of these two types of stratification, the examples being two alpine lakes, the oligotrophic holomictic Grundlsee, and the meromictic Krottensee, which displays the same type of stratification as an extreme eutrophic lake although for different reasons (see p. 39). These graphs may be taken to represent the two extremes of a series, which has infinite gradations between them.

The nutrients which are gradually set free by decomposition in the depths and which accumulate there are locked in the hypolimnion and prevented from re-entering the production cycle as long as a stable stratification exists. The amount of such materials in the hypolimnion is very great, especially in eutrophic lakes. As an example, two tropical lakes may be cited for which such calculations have been made and which, moreover, are not essentially different from our own eutrophic lakes. Ranu Lamongan in east Java, a tiny volcanic lake only 750 m. in diameter and 28 m. deep, had 1,400 kg. of phosphorus and 12,000 kg. of ammonium nitrogen below the metalimnion. One of the larger Sumatran lakes, Danau Manindjau, which has a surface area of 98 km.^{2} and a depth of 170 m., contained about 1,500 metric tons of phosphorus and 7,000 tons of ammonium nitrogen below the 60 m. level. Knowledge of these relations has a special significance in the tropics since the lake waters there are used for irrigating rice paddies. Thus there promises to be an advantage in using water from the hypolimnion instead of from the surface for this purpose (Ruttner, 1931).

It is not only the difference in the biogenic chemical stratification that distinguishes a eutrophic from an oligotrophic lake. The optical properties of the waters are affected by the different content of plankton

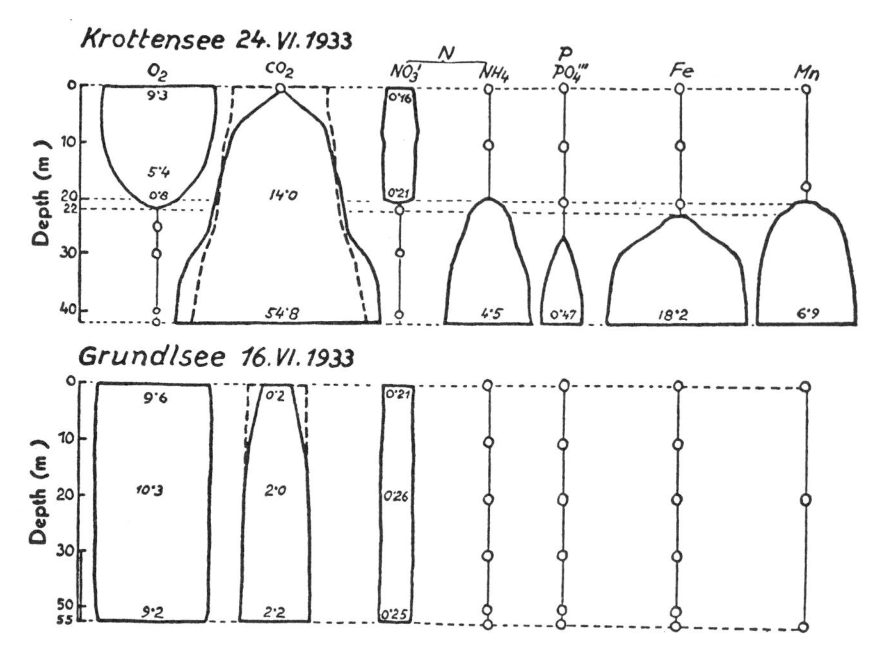

FIGURE 27. Extreme types of biogenic chemical stratification: Krottensee (Salzburg), meromictic eutrophic type; Grundlsee (Upper Austria), holomictic oligotrophic type. The abscissae are the cube roots of the content per litre. The diameter is thus a measure of linear concentration (Kohlrausch, 1916) and thus represents the edge of a cube. This type of illustration is a better representation in many respects than the curves that are generally used. The dotted outlines under "CO_2" are the bicarbonate concentrations calculated from the solution values for the equilibrium carbon dioxide. The values given are mg./l.

organisms, some of which are pigmented. A lake with very transparent and dark blue, blue-green, or green water is always oligotrophic. On the other hand, eutrophic lakes always have a relatively low transparency and are yellow-green to yellow-brown in colour; but the determination of these optical properties alone will not establish the productivity type, for the turbidity can have an inorganic origin and the colour can come from humic substances (cf. p. 21).

PART B

Biotic Communities

I. PLANKTON

LIMNOLOGY offers many fine examples of how greatly dependent the progress of science is on the current condition of the techniques of investigation. That the introduction of a very simple, even obvious, method of investigation can provide an unexpected stimulus is best shown in the history of plankton science. Until about the middle of the last century our knowledge of the various types of plankton, particularly marine plankton, was still very meagre. What was known was derived in large part from the investigation of the stomach contents of animals or from the accidental finding of forms that had wandered into the littoral region. Then it occurred to the distinguished physiologist Johannes Müller to sample the open sea with a net made of fine mesh silk of the type used in the milling industry for grading flour, and immediately there was disclosed to the astonished eye an extensive biotic community, previously almost unsuspected, possessing an almost inexhaustible richness of forms and problems. Investigators rushed with great enthusiasm into this new field: expeditions (Hensen, 1889) were organized with the sole purpose of ascertaining the composition, distribution, and quantity of plankton in the ocean. Soon men from Hensen's school, for example such pioneers as Apstein (1896), armed with plankton nets, were attracted to inland lakes to test the practical knowledge learned in the sea and to discover the problems of pelagic life in fresh water. For a while it seemed that hydrobiology would exhaust itself in plankton investigations. Hydrobiologists thought that in this community they had discovered the basic production of standing waters. And yet an error in method had been generally overlooked, or believed by Hensen to be insignificant: namely the fact that the meshes of the finest bolting cloth (no. 20) are about 70 μ (0.07 mm.) wide,[34] and that small organisms

[34]Recently nets made of Perlon or nylon, with meshes as small as 35 μ, have been used to good advantage.

of the plankton can slip through these openings. It was not until nearly the turn of the century that Kofoid (1897) and Lohmann (1908) became aware of this error. Lohmann, while investigating the unusually fine mesh branchial chambers of the Appendicularia (tunicates), found an abundance of very minute plankton that did *not* occur in net catches. He then attempted to centrifuge samples of sea water and found in their sediments the same small flagellates, diatoms, etc. Indeed these not only greatly outnumbered the "net plankton" organisms in the same volume of water but also often surpassed them in weight. The discovery of these *nanno* (dwarf) plankton provided the impetus for the development of a new procedure, which made it possible to recover the total plankton contained in a volume of water and thus to supplement the results obtained by straining water through a net. This procedure consisted of centrifuging the living organisms or of precipitating them in special chambers of from 0.1 to 100 ml. capacity by killing them with an iodine–potassium iodide solution (Kolkwitz, 1911, and Utermöhl, 1931). By this method of plankton investigation not only could the composition of the community be determined completely, but also the spatial and temporal distribution of individual species could be studied statistically.

Johannes Müller called this community "Auftrieb" since it was assumed that it occurred only at the surface of the sea. Hensen (1895) first used the term "plankton" to include all those organisms that float free in the water "involuntarily" independent of the shore and the bottom. Even though developmental stages (larvae) of bottom dwelling animals occur abundantly in the plankton, particularly in coastal regions of the ocean, nevertheless the plankton represents the community that is most completely and exclusively dependent on water as a place of life. No parallel situation occurs in the flora and fauna of the terrestrial environment; for although the air is often filled with seeds, spores, bacteria, and pollen grains, and although birds and insects move about in it temporarily, still there are no known organisms that can live floating free in the air, continuously independent of the ground.

1. *Flotation in Water*

The prerequisite for the existence of plankton, especially among those organisms possessing no power of movement of their own, is the ability to *remain suspended* in water. A continuing flotation depends on the density of the organism being the same as that of the surrounding medium, a condition that is scarcely ever completely realized. Since the density of plants and animals is somewhat greater than, and only in rare instances is less than, 1.0, this "floating" for the most part is a slow sinking. The slower the rate of sinking of an organism is, the better it is adapted for living in the plankton, unless it has the ability through its own movements to maintain its position. In marine plankton, on the other hand, true "floaters" commonly occur, because the high salt content of the environment makes it possible for the organisms, by changing the composition of their body fluids (while maintaining isotonicity), to equalize differences in density more or less completely (*Noctiluca*, radiolarians, pelagic fish eggs, etc.). But since we are concerned with freshwater plankton, the first question we must consider is what factors influence the *rate of sinking*. The distinguished Danish limnologist Wesenberg-Lund (1900) had already reached an understanding of this problem at the turn of the century. His "flotation theory" was later elaborated by Wo. Ostwald (1902) and expressed by the following formula:

$$\text{rate of sinking} = \frac{\text{excess weight}}{\text{form resistance} \times \text{viscosity}}$$

The factor *excess weight* in the numerator represents the difference between the density of the "floating" body and that of the surrounding medium, the water, multiplied by the volume of the organism. There are numerous examples of an increase in floating ability brought about by reduction of excess weight. Since the density of living protoplasm itself (approximately 1.05 g./ml. in most instances) can be changed only very slightly, such an increase in floating ability can be accomplished only by a reduction in weight of integuments and skeletal parts or by the secretion of light-weight material inside or outside the body.

Hence we find, for example, that many plankton diatoms have particularly delicate siliceous shells, and that the chitinous integument of pelagic crustaceans is as a rule distinctly thinner than that of littoral forms. In a great many forms, particularly among the blue-green algae, desmids, green algae, and some diatoms, but also among animals (for example, the rotifers *Conochilus* [Figure 28] and *Collotheca*, and the crustacean *Holopedium*), well-developed gelatinous sheaths of almost the same specific gravity as water reduce the density of the whole. Occasionally water is accumulated inside the body, the best-known example being the rotifer *Asplanchna* (Figure 29), which resembles a large bladder. In many plankton organisms, food reserves of oil reduce the excess weight to a considerable degree. Thus oil droplets occur regularly in many flagellates, diatoms, heterokonts, and rotifiers, and are especially abundant in the entomostraca. In the colonial alga *Botryococcus* the oil, which is not secreted within the cells but rather in the sheaths, often reduces the density to such an extent that the colonies rise to the surface instead of sinking.

The most effective means of reducing excess weight are gas bubbles contained within the body. Included here are the commonly occurring and long-disputed *gas vacuoles* of many blue-green algae (Figure 30)—those inclusions of a greatly different refractive index in the plasma which reduce the density of the algae below that of the water and cause them to rise to the surface ("water bloom"; cf. p. 131). In isolated instances, as in the phantom midge larva *Chaoborus* (*Corethra*) or in fish equipped with swim bladders, such gas inclusions provide the means of adjusting the floating ability to prevailing conditions through voluntary changes in volume.

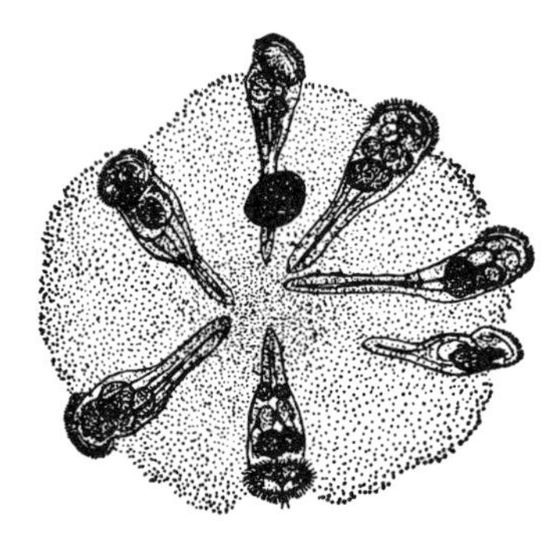

FIGURE 28. *Conochilus unicornis*, a free swimming, colonial rotifer living in a gelatinous envelope, × 60.

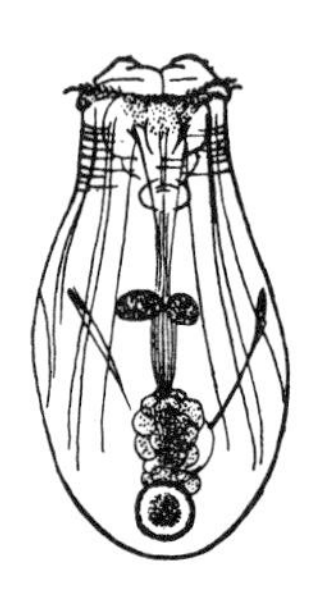

FIGURE 29. *Asplanchna priodonta*, a rotifer of the "bladder type," × 50.

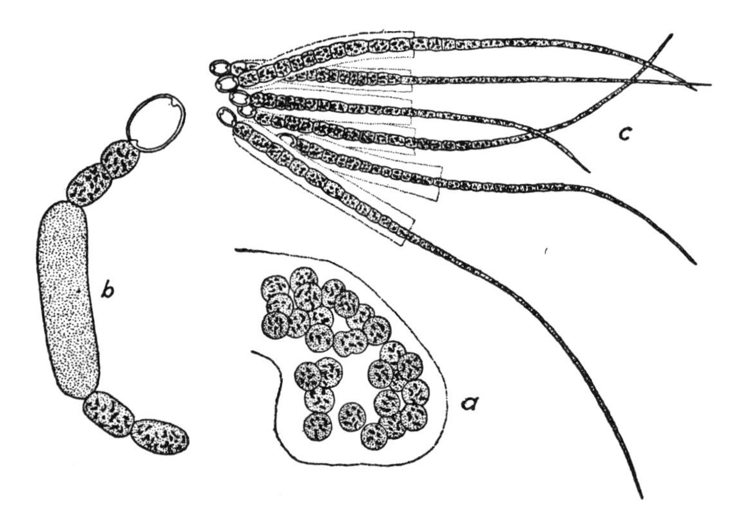

FIGURE 30. Planktonic blue-green algae with gas vacuoles (coloured black): *a*, *Microcystis aeruginosa*, part of a colony, × 825; *b*, *Anabaena flos aquae* with heterocyst and resting spore (akinete), × 825; *c*, *Gloeotrichia echinulata*, part of a spherical colony (from Smith), × 400.

The factors in the denominator of the formula—the form resistance of the body and the viscosity of the water—are closely related to one another. Their product represents the resistance that sinking bodies have to overcome. The *form resistance* is determined mainly by two characteristics: the *specific surface area* and the *area of an orthogonal projection of the body* onto a plane at right angles to the direction of movement.

The *specific surface area* is the ratio of the total surface area to the volume of the body. The larger it is the greater will be the relative friction with the water, and hence the slower will be the rate of sinking. Entirely independent of special morphological adaptations, this factor becomes important because of the usually small, although varying, size of plankton organisms. Everyone knows that a large body sinks more rapidly than a smaller body of the same material, that powdered quartz or limestone, for example, makes water milky for a relatively long time before it settles to the bottom. *Stokes's law* for the free fall of small spheres in a fluid medium states that the rate of sinking varies directly with the square of the radius. If we apply this relationship to the most common sizes of plankton organisms, for example 0.1, 0.01, and 0.001 mm., we can see that under identical conditions of density and viscosity, a spherical cell of 0.01 mm. diameter (a flagellate, for instance) will sink 100 times more slowly, and a similar cell of 0.001 mm. (the size of bacteria) 10,000 times more slowly, than a cell 0.1 mm. in diameter. The specific surface area, however, also increases when the shape of the body departs from the spherical or when the surface is provided with horns, ridges, spines, setae, and the like. In this category belong all those well-known modifications interpreted as adaptations for flotation which occur in fascinating variety particularly among marine plankton, but also are often developed impressively in some of the freshwater forms.

The influence of the *orthogonal projection area* can be illustrated most simply by the observation that a stick sinks more slowly when in a horizontal than in a vertical position. A stick-like elongation (often to unusual length) is common among plankton algae, and in fresh water especially among diatoms (*Synedra*), desmids (*Closterium*), and Chlorococcaceae (*Ankistrodesmus*). There is always associated with the elongation a slight bending, which steadies the rod during sinking and maintains it in a horizontal position. Also common are disc-like forms (*Cyclotella*) and parachute-like forms produced by marginal developments or spines. Often such a flotation adaptation is accomplished by colony formation, as for example the little stars of the diatom *Asterionella* (Figure 31), which, like diminutive parachutes, dominate the winter plankton of many lakes. In this organism the elongated individual cells are remarkably arranged in a flattened spiral.

An example of the fact that different flotation adaptations can at times supplement each other and at other times replace each other is given in the photomicrographs (Figure 32) of a number of planktonic forms of the desmid genus *Staurastrum*, treated with India ink for differentiation. The floating ability of these forms is increased partly by the horny prolongations at the angles of the cells, which alter the form resistance, and partly by a gelatinous secretion, which in India ink preparations appears as a light corona on a dark background. It is apparent that as the length of the horns decreases, the thickness of the jelly increases, reaching considerable dimensions in the rounded, hornless *St. brevispinum*. It is likewise evident in this figure that the appendages of forms *d* and *e*, which are completely embedded in the jelly, can no longer function as flotation structures.

FIGURE 31. *Asterionalle formosa* var. *hypolimnica*. A diatom colony of the parachute type in the plankton, × 400.

The factor of *viscosity* has been shown on page 11 to be influenced markedly by temperature, much more so than is the density of the water. If we take the density, for example of a rotifer (according to Luntz, 1928), to be 1.025 g./ml., then the excess weight during a temperature increase from 0° to 25° increases from 0.025 to 0.028, or only 12 per cent (although for the sake of simplicity we have to assume that the volume remains constant). At the same time the viscosity decreases by half, so that the rate of sinking theoretically increases 100 per cent. In experiments, however, this theoretical value was never obtained; the increase in rate of sinking over the given temperature change was only 41 per cent for reasons as yet unknown. None the less, a given plankton species under identical conditions of form resistance and density has better flotation ability in winter than in summer.

For a long time the question has been argued in plankton investigations whether or not, and to what degree, plankton organisms possess the ability to compensate adaptively for these differences in flotation conditions resulting from temperature changes. When we observe, for example, that the species of *Ceratium* found in tropical oceans are often characterized by enormously elongated horns, we might interpret this phenomenon as an evolutionary adaptation to the unfavourable conditions of flotation in warm water (28 to 30°). However, the observations of the *Meteor* Expedition have shown that long-horned forms likewise occur in cold portions of the sea, and short-horned forms in warm portions. On the other hand, Luntz (1928) demonstrated experimentally that rotifers (*Brachionus*, *Euchlanis*)

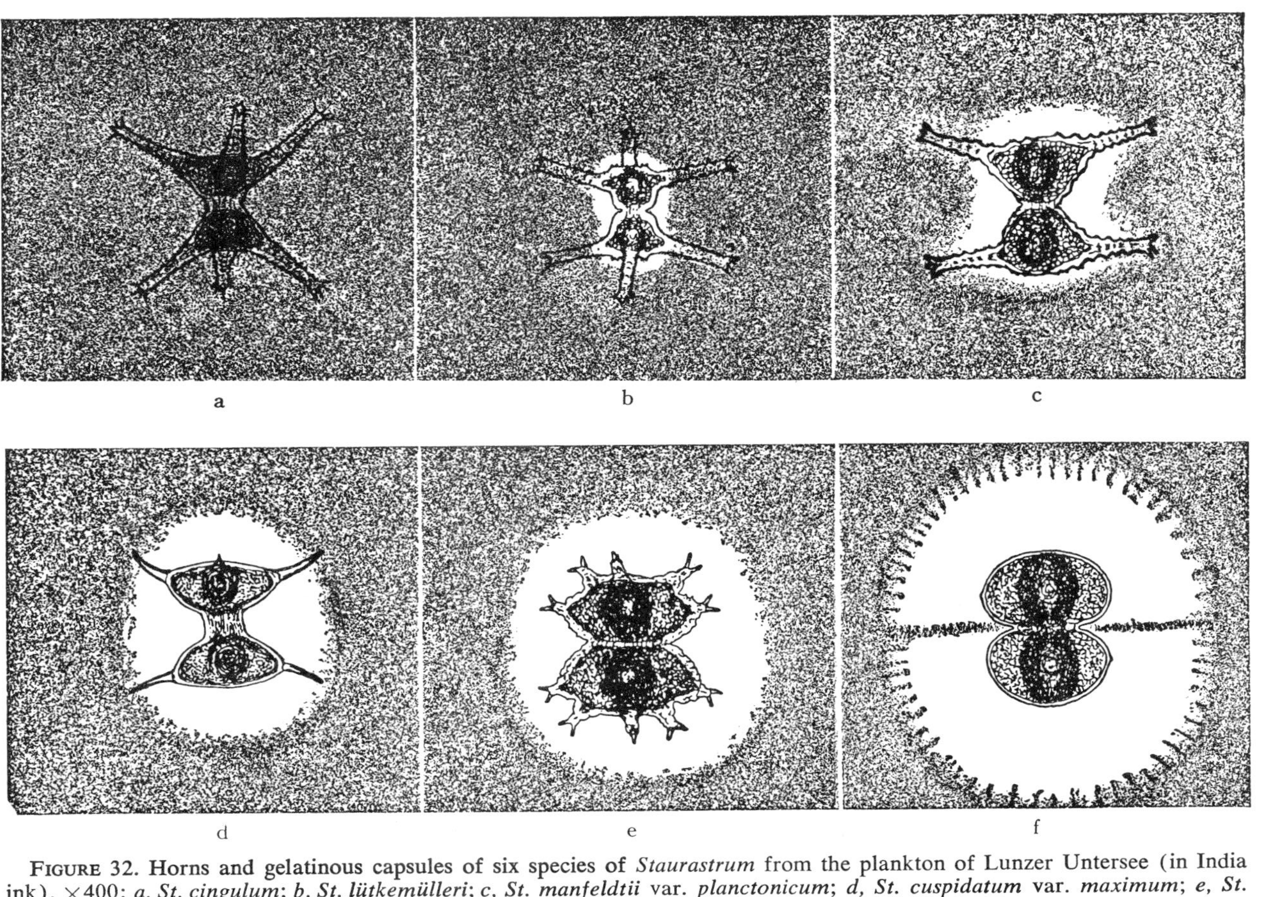

FIGURE 32. Horns and gelatinous capsules of six species of *Staurastrum* from the plankton of Lunzer Untersee (in India ink), ×400: *a, St. cingulum*; *b, St. lütkemülleri*; *c, St. manfeldtii* var. *planctonicum*; *d, St. cuspidatum* var. *maximum*; *e, St. furcigerum*; *f, St. brevispinum.*

show no adaptive change in flotation processes, but rather that animals raised in warm water have a lesser density than those cultured in cold water. But it is more difficult to determine if certain seasonal changes in the body shape of successive generations, designated as *cyclomorphosis*, are caused by fluctuations in water temperature. The observation that the horns of *Ceratium hirundinella* in a given lake are considerably longer in summer than in winter, that the length of the helmet of many races of *Daphnia* (Figure 33) varies in a strikingly parallel way, and that finally a similar variability occurs among certain rotifers, prompted Wesenberg-Lund (1900) to interpret all these phenomena as a means of compensating for the influence of temperature on flotation ability. Investigations by Lauterborn (1900) and later by Woltereck (1913) and his school have shown, on the other hand, that in tested instances a causal relationship between temperature

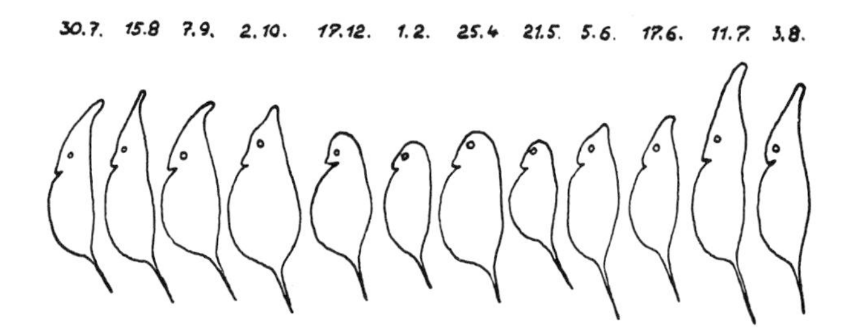

FIGURE 33. Seasonal changes (temporal variation) in the helmet length of *Daphnia cucullata* in Furesee (Denmark). From Wesenberg-Lund (1900).

and cyclomorphosis either is completely lacking (in certain rotifers) or plays only a secondary role (in *Daphnia*). The opinion, formerly widely held, which has found its way into the textbooks, that in the rotifer *Keratella quadrata* (*Anuraea aculeata*) the changes in form (variations in length of the spines of the lorica) are correlated entirely with the cycle of generations, that is, with the change between sexual and asexual reproduction, has been shown to be erroneous (Ruttner-Kolisko, 1949). This species encompasses rather a series of subspecies (*K. quadrata*, *hiemalis*, *testudo*) which on the basis of length of the spines and pattern of the lorica are genetically distinct and can co-exist. Of these forms only *K. quadrata* exhibits, although in limited extent, a change in body form that can be interpreted as a temporal variation. In *Daphnia* (according to Woltereck, 1909, 1928) the capacity for cyclomorphosis is determined genetically by means of the "Reaktionsnorm" of the particular race; that is, a race capable of cylomorphosis *can*, under favourable conditions of nutrition and temperature, produce forms with large helmets, although for the most part only in the middle generations of a sexual cycle. Under less favourable environmental conditions, however, even these *Daphnia* remain just as roundheaded in summer as those in which the "helmet potential" is lacking altogether (Figure 34). This interpretation by Woltereck, although very reasonable, has not remained unchallenged. In more recent times Brooks (1946, 1947) in particular has been concerned with cyclomorphosis in *Daphnia.* He concluded that temperature exerts an influence on the size of helmets only at the embryonic stage—a conclusion that was demonstrated experimentally by Coker and Addlestone (1938)—and that various other factors could be involved, especially the turbulence of the water. In animals that are capable of locomotion a much more fundamental control over maintenance of position must be considered: namely, the acceleration of movements with an increase in temperature, a reaction that is capable of largely compensating for the greater tendency to sink. The various appendages of the body, such as the helmets of *Daphnia* and the first antennae of *Bosmina*, play an important role as balancing organs during swimming, as Woltereck (1913) has convincingly pointed out.

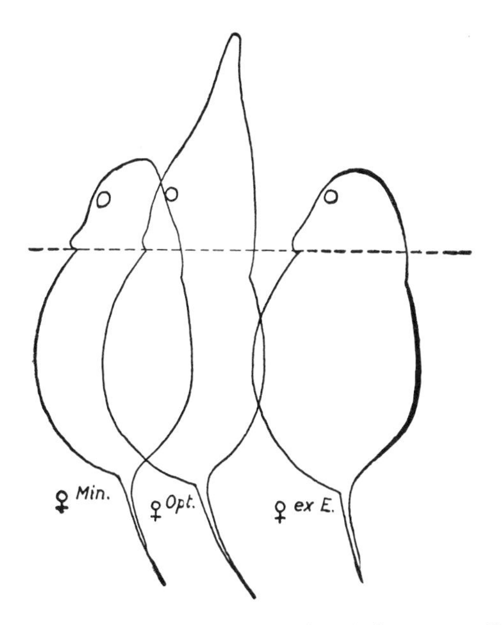

FIGURE 34. Limits of variation in helmet length in a pure line of *Daphnia cucullata.* At the left are two females of an intermediate generation (♀ Min., ♀ Opt.) under minimum and optimum culture conditions. At the right is a female (♀ ex E.) from the first generation after the ephippium (Dauerei or resting egg) under optimum conditions. From Woltereck (1913).

Even all these "flotation adaptations," however, with the exception of the reduction of the density of the organism to that of the surrounding water or even less, are not able *by themselves* to prevent a relatively rapid sinking, especially of the non-motile plankton. This fact is demonstrated by the common observation that the majority of living organisms in a plankton collection or water sample soon come to rest on the bottom of the container. That this settling does not occur in lakes, and that even fairly large and rounded forms remain suspended in the water means that the forces controlling flotation lie outside the organisms. This force is the *turbulence of the water* (the unorganized eddy diffusion currents described on p. 51), which maintains a state of continuous mixing action particularly in the epilimnion, which is where the non-motile phytoplankton chiefly occur (W. Schmidt, 1925; Utermöhl 1925). In the metalimnion and hypolimnion, vertical eddy diffusion is indeed very small, although always present to an extent sufficient to reduce further the rate of sinking already curtailed by low water temperature, or even to counterbalance it completely. We must, therefore, look upon the turbulence of the water as the most important factor influencing flotation of plankton, and may in consequence assign a purely supporting role to the so-called flotation adaptations.

These remarks are sufficient for the purpose of this presentation. Whoever wishes to inquire further into this many-sided problem of the flotation of plankton organisms can refer to the important study of Jacobs (1935).

2. *The Composition of the Plankton*

The ability to live "freely floating" determines the extent to which the individual families of the plant and animal kingdoms contribute to the composition of the plankton. The pelagial, which is the region of the open water, was certainly colonized originally from the shore (Naumann, 1927). Particularly in inland lakes close relatives of nearly all the species of plankton occur in the littoral. Morphological differentiation of these two communities in lakes has not proceeded so far as in the sea. Indeed, many of the modifications that are said to be "flotation adaptations," such as morphological changes that increase form resistance, gelatinous sheaths, and even the "gas vacuoles" of blue-green algae, are found in species that occur only in the littoral. The species that pushed forth into the free water from the endless variety of organisms living in the communities on the shore and on the bottom were essentially those that, because of their configuration and their physiological requirements, were able to cope with the environmental conditions prevailing in the open water. Only a portion of these organisms have developed further in the course of their phylogeny into manifestly well-adapted "floating forms."[35] This selection, controlled by the special

[35]This advancing of littoral forms into the plankton takes place continually, and to some extent before our eyes. In several large and deep lakes of the Sunda region species could be found abundantly in the plankton that in other lakes of the region occurred only in the littoral, for example, the diatoms *Cymbella turgida* (Ranu

environmental conditions of open water, has resulted in a much smaller number of species of plankton organisms than of littoral organisms in inland lakes. An example might make this clear. In an investigation of eleven lakes in the eastern Alps there were found 107 species (67 plants and 40 animals) which, because of their occurrence, must be regarded as genuine members of the plankton community of these lakes. Of course if one takes into consideration all the species present in the collections, even those occurring very rarely, the total number is markedly increased; but most of these latter species belong to the so-called *tychoplankton,* which consists of forms transported by currents from the littoral and affluents into the pelagial. These no more belong in this plankton community than does the dead organic and inorganic suspended matter, the *tripton,* which often is present in considerable quantities in the water.[36] Continuing with the example cited, the largest number of species present during the summer in a *single* lake was 73, the smallest 24. However 13 per cent of the species occurred in all eleven lakes and 45 per cent in more than half of them. Naturally the total number of true plankton organisms hitherto observed in inland lakes is considerably greater than in our example, but it is small compared with the abundance of organisms that inhabit the shore and the bottom of waters.

Pennak (1957) characterized the composition of the limnic zooplankton on the basis of his investigations in 27 Colorado lakes. In these lakes, likewise, the animal groups are poor in species—1–3 copepods, 2–4 Cladocera, 3–7 rotifers in the plankton of any given lake. Seldom does more than one species of a given genus occur at one time, but if so then one species is much more abundant than the other. The origin of this likely lies in the competition, especially with respect to food, among organisms with similar habits and requirements. Even within the three large systematic groups (copepods, Cladocera, rotifers) one species in each group is generally greatly dominant over the others.

The above-mentioned environmental selection not only restricts the number of species but even excludes many genera and entire families of plants and animals from the plankton. Considering the *phytoplankton* first, the significance of the increased floating ability of smaller volumes for the composition of the plankton is immediately apparent. The most predominant families by far are those in which the single-celled condition or the formation of

Pakis in east Java) and *Synedra ulna* (Ranau Lake in south Sumatra), as well as the crustaceans *Simocephalus serrulatus* and *Latonopsis australis,* which comprise the major portion of the plankton in Manindjau Lake (middle Sumatra) (Ruttner, 1952).

[36]The totality of living and non-living suspended matter (plankton + tychoplankton + tripton) is designated as *seston* (that which is capable of being removed by filtration).

small cell-aggregations (colonies) is the rule. Thus, the *flagellated species* of the Euglenophyceae, Chrysophyceae, Cryptophyceae, Xanthophyceae, Chlorophyceae (Volvocales), and Peridiniales are abundantly represented, as well as the *non-motile, single-celled species* of these groups (especially the Chlorococcales and their dainty colonial forms) and in addition the Bacillariophyceae, Desmidiaceae, blue-green algae, and finally the bacteria, the smallest representatives. Among these same groups of plants, however, the large many-celled species are *lacking* almost entirely from the true plankton, for instance the Zygnemataceae, the Ulotrichales and Oedogoniaceae of the Chlorophyceae; and also the Cladophoraceae and the Siphonales. The groups of the Charophyceae, Phaeophyta (brown algae), and Rhodophyta (red algae) on the whole are not represented, nor are the mosses, ferns, and flowering plants. The multitude of fungi is represented only by a very few primitive species, which are seldom encountered, a circumstance attributable less to the shape than to the mode of life of these plants.

Selection among the freshwater *zooplankton* is if anything even more severe. For the most part only three phyla of animals are concerned in its composition: the *Protozoa*, including the Sarcodina (*Difflugia* and the Heliozoa) as well as the colourless Flagellata and Ciliata; the *rotifers*; and from the phylum Arthropoda the *Crustacea*, especially the Cladocera and Copepoda. Other groups contribute representatives to the plankton only in isolated and usually infrequent instances. An exception is the larva of the midge genus *Chaoborus* (*Corethra*, *Sayomyia*), which is specially adapted for life in the open water through the possession of air bladders (p. 108); this organism, which is distributed over the entire world, is the only insect larva in the plankton. Of the few sporadic planktonic species from other groups of animals, especially noteworthy are the small medusae, representing the phylum Coelenterata that is so well developed in the seas. They were first observed in the lakes of tropical Africa (*Limnocnida tanganjicae*). Another species (*Craspedacusta sowerbii*), which occurs mainly in Asia and America but only seldom in Europe, has been found at Prague in a dammed-up portion of the Moldau. The sessile polyp generation lives in the littoral. Other organisms that might be mentioned are the following: among the flatworms the occasional rhabdocoele turbellarians, as well as the cercariae of a number of trematodes; among the molluscs the larva of the three-cornered mussel *Dreissensia polymorpha*, which has penetrated from brackish water; and among the arachnids scattered Hydracarina. Moreover, several species of fish might be included in the limnoplankton, although only with reservation, since with their active locomotion that is little affected by the weak currents in inland lakes, they scarcely fit the concept of "involuntarily drifting organisms," and belong to this community only to the extent that they feed upon zooplankton. The expression *nekton* (swimming organisms) has been coined for these special cases. Inhabiting the pelagic zone, and more or less completely dependent on plankton for food, are certain coregonids, which occur as countless local races in the alpine and northern lakes. This poverty of species sharply distinguishes the freshwater plankton from the corresponding pelagic community of the sea, in which nearly all classes of marine animals are abundantly represented, at least in their larval stages.

Most of the animal representatives of the limnoplankton likewise are characterized by their small size: the protozoans and rotifers seldom exceed microscopic dimensions; in fact the smallest known metozoan is a rotifer. The crustaceans can usually be seen with the naked eye, but even by far the largest of them, *Leptodora*, the cladoceran most highly modified for planktonic life, is scarcely more than 1 cm. long.

It would be a mistake to consider the composition of the plankton only on the basis of the floating ability of its members. Life in the open water makes additional large demands on the organisms, for example on their ability to carry on nutrition, reproduction, etc., so that we would not be correct in regarding shape as the only basis for selection in this community. Reduced volume not only results in a better floating ability, but also through an increase in the relative surface area markedly facilitates metabolism.

From the standpoint of *nutrition physiology*, the plankton is composed of *producers* and *consumers* of organic matter. The producers are simply the autotrophic plants, which take up the sun's energy in their green and yellow chromatophores and use it for the assimilation of carbon. First carbohydrate or fat is produced, and later the highly complex proteins and other portions of the plant cell are synthesized. Each animal in the plankton community, as everywhere in nature, is directly or indirectly dependent upon this production by plants. The phytoplankters, however, form a suspension of very small particles distributed through the entire water volume, and the animals that feed upon them require special modifications for obtaining them. It has been determined that the bulk of the phytoplankton is not, as a general rule, formed of the large obvious species that dominate the "net plankton," but rather of the diminutive nannoplankton species. Woltereck's (1908) observations have demonstrated that the *Daphnia*, which usually comprise the largest portion of the animal mass in the plankton of inland lakes, are directly dependent for their nourishment on these smallest members of the phytoplankton.[37] This is true also of the closely related

[37]Not to be underestimated in importance for the nourishment of many plankton animals is the very fine organic detritus suspended in the water, the "tripton" (cf. p. 115), which under certain conditions can in part replace the nannoplankton.

Pütter (1909) advanced the theory (unfortunately on the basis of inadequate calculations) that the quantity of phytoplankton is not sufficient to satisfy the nutritional needs of the plankton animals. He assumed that animals had the ability to take up the organic materials dissolved in the water, so that therefore in this respect they resembled plants in their nutrition. The possibility that there might be an uptake of dissolved substances by the metozoa, as there undoubtedly is by many Protozoa, is not argued here. Yet Pütter's theory is objected to on the basis that among the zooplankters there are no signs of any organization (for example, increase in surface area) that would indicate an adaptation for the uptake of materials from such extremely dilute solutions. On the contrary there are an unusually large number and variety of specialized and at times complicated modifications that serve exclusively for obtaining *particulate* food.

Bosmina and Sididae, of the most important family of copepods in the plankton, the Diaptomidae, as well as of most rotifers and ciliates. Varied as are the modifications for food getting among these animals, there is *one* feature common to all: the food particles are not caught individually through purposeful actions, but rather, by means of a current of water produced by the uninterrupted movement of certain organs, are brought *automatically* to the various types of feeding apparatus, where they are retained *involuntarily*. The movements that produce this current of water serve at the same time for swimming or, as in the Cladocera, for respiration; nutrition, therefore, is closely associated with these other activities. In order to obtain the necessary quantities of food, these animals must pass relatively large quantities of water through their food-catching devices. These devices are varied, and their function is only partially understood. Among the ciliates and rotifers we know that the coronas of cilia surrounding the mouth region whirl the food particles towards the mouth (Gossler, 1950). The complicated feeding apparatus of the daphnids and the Diaptomidae is best understood, primarily through the studies of O. Storch (1924, 1925).

In the well-known waterflea *Daphnia*, the five pairs of morphologically different thoracic limbs work together with the two shells like a valve pump. By means of the outward extension of these thin-walled limbs, which likewise serve as gills, the pump chamber is enlarged and in consequence water is drawn backwards at the upper anterior edge of the shell. By means of the subsequent flexion of the appendages, the water is forced out towards the caudal edge, and in so doing it must pass through a filter chamber formed by the delicate feathered hairs on the edges of the podobranchs. Here the particles contained in the water are strained out, pressed together by means of the automatic movements of specialized setae in the abdominal groove, and in the form of a sausage are pushed towards the mandibles and the mouth. By using a carmine suspension one can very easily study this fascinating procedure under the microscope. The frequency of the pumping movements is very great—200 to 300 beats per minute in *Daphnia*—so that in the course of a day a really considerable volume of water can be filtered. The feeding mechanism of *Diaptomus* is based upon a different principle, which can be compared with that of a suction pump. By means of the whirling movements of the anterior limbs (maxillae and maxillipeds) a stream of water is sucked forward on the ventral side of the animal, from which the nannoplankton is strained by means of a basket-like arrangement of setae.

Such *filter feeders* stand in contrast to *seizers*, which capture their particles of food individually. Among the plant-eaters in the plankton, seizers occur even among the protista, for example the flagellate *Bodo*, which feeds mainly on chrysomonads; and the colourless peridinian, *Gymnodinium helveticum*, which (presumably with its longitudinal furrow) can ingest even large cyclotellas. One of the most interesting examples is the rotifer *Chromogaster*, which is strictly "monophagic" on armoured peridinians (*Ceratium*), which it pierces with dagger-like jaws and then sucks out (Figure 35). In *Asplanchna*, on the other hand, the jaws are transformed into prehensile pincers, which by shooting forward grasp the algae and smaller rotifers whirled in

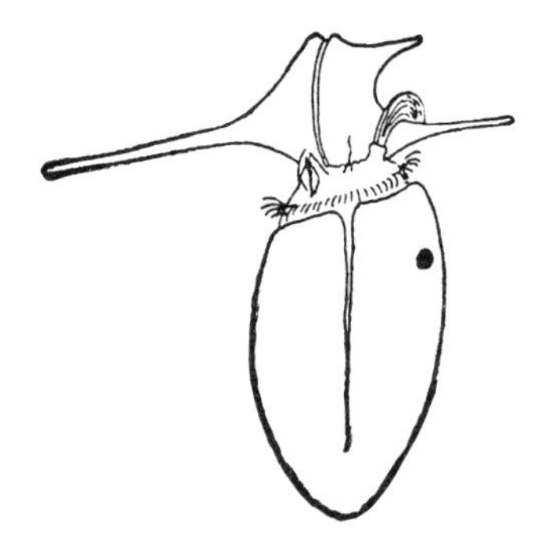

FIGURE 35. The rotifer *Chromogaster testudo* with its prey *Ceratium hirundinella*. From Ruttner-Kolisko (1938).

by ciliary action. Just as *Asplanchna* and other rotifers do to some extent, so the seizers among the entomostraca as well as the phantom midge larva *Chaoborus* feed entirely upon other animals and thus are decided predators. In some species the limbs are partly modified into spine-covered prehensile organs, which capture the prey suspended in the water in much the same way as the legs of a dragonfly capture a flying insect. This modification is particularly true of the cladocerans *Polyphemus*, *Bythotrephes*, and *Leptodora*. Among the copepods the genus *Cyclops* is predacious and hunts especially the equally large plant-eating copepod, *Diaptomus*. It is worthy of note that the nauplii (larvae) of *Cyclops* are still phytophagous and possess a primitive mechanism for catching nannoplankton.

3. *Spatial Distribution*

Each physiological activity, such as assimilation of food, respiration, movement and reproduction, is affected by conditions of the environment—always by temperature, and frequently also by light, oxygen content, and other physico-chemical properties of the water. This dependence is shown by the fact that a particular physiological activity begins at a certain *minimum* level of the external factor. At a higher level, the *optimum*, it reaches its peak value. With a further increase of the factor it again gradually declines until at the *maximum* it reaches zero. Respiration is an exception in that with increasing temperature it also continues to increase until death from heat occurs. The organism as a whole also responds to a gradient of an external factor in the same way as each of its individual functions. Its *activity curve* results from the combined effect of the curves of its individual life functions. But since the demands that the various species of plants and animals make upon their environment often deviate markedly from one another, their activity curves likewise are very different in appearance. This has been illustrated schematically in Figure 36. On the vertical axis is represented the inten-

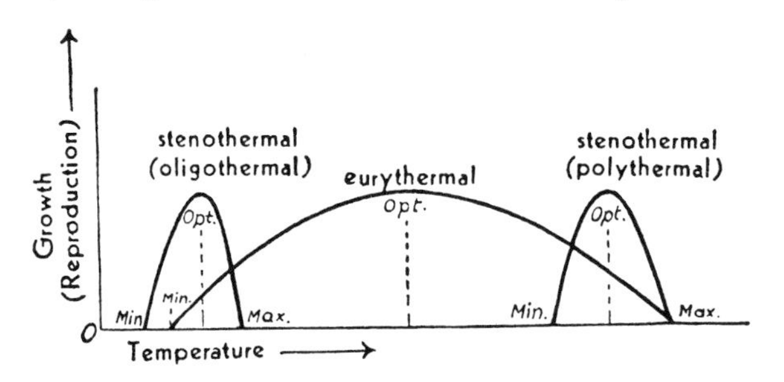

FIGURE 36. Schematic representation of the activity curves of stenothermal and eurythermal organisms.

sity of growth (or reproduction) as an expression of physiological activity, and on the horizontal axis the external factor, in this case temperature. An organism whose reaction is like that of the middle curve is able to thrive over quite a range of values, as of temperature. An organism with such a broad activity curve is said to be *eurytopic* with reference to the particular environmental factor, or specifically in relation to temperature, *eurythermal*. The curves to the right and left both show a much more limited region of activity; organisms that are restricted to small ranges of the environmental factor are *stenotopic*, or in relation to temperature, *stenothermal*. Here minimum, optimum, and maximum lie close together; and a small shifting of environmental conditions, which in eurytopic forms might produce almost no change in physiological activity, can make the existence of a stenotopic species questionable in a particular niche. Between these two extremes of activity curves there are all conceivable intermediates.

In nature, however, plants and animals are continually under the influence of a *great number of environmental conditions*, and the success of a species in any particular biotope is determined by their combined effects.

Through various combinations of environmental conditions certain modifications of the activity curves can be brought about. For example, it has been observed that many animals at a higher salt content are able to get along with a smaller quantity of oxygen, and that a plant's optimum activity can occur at different temperatures under different light conditions.

As the algebraic sum of the external conditions in a biotope approaches the optimum of the activity curve for a species, an increase in the rate of growth or reproduction results; on the other hand, a departure from the optimum brings about a decrease and finally a cessation of reproduction. We are not able in the majority of cases to measure the rate of reproduction itself; instead we have to be satisfied with determining changes in the numbers of individuals. Special caution is necessary in drawing general conclusions from such observations, because the number of individuals of a species present at a particular time is determined not only by the *rate of reproduction*, but also by another factor with just the opposite effect, the *rate of depletion*. The latter is governed by various conditions, such as natural death, consumption by other species, or mechanical removal from the biotope (in the case of plankton, for example, through sinking or being carried away), and consequently its magnitude is subject to very significant fluctuations. It is possible that, in spite of a high rate of reproduction under otherwise optimal conditions, a decline in the number of individuals can occur because the rate of depletion is preponderant (in all these problems cf., e.g., Edmondson, 1946, 1960).

Before we attempt to relate these general concepts to the existence of plankton in an inland lake, we must realize that a change in the environmental conditions in a lake basin capable of influencing this community arises in a twofold manner: first, with *time*, as a result of the changes in weather conditions during the course of a year; and second, with *depth*, as a result of the physical and chemical stratification described in detail in the first part of this book. On the other hand a change in external conditions has no appreciable effect in a horizontal direction within a particular depth stratum of the pelagic zone proper. Since many phenomena in the annual cycle can be understood only through a knowledge of the *vertical distribution of the organisms*, we shall begin our considerations with this.

But first of all a few words must be said about the *methods* used in such investigations. In order to ascertain the distribution of organisms in space or time, a statistical method must be used, which requires the determination of the number of individuals of each species of organism present in a unit volume of water. Hence a procedure to accomplish this might be recommended here. Quantitative methods that enable evaluation of the total plankton without reference to the component species (e.g., determination of dry weight, nitrogen content, chlorophyll content) and that permit consideration of total production by itself will be discussed later (p. 161). We shall skip over the procedure that attempts to get the desired information from counts of net plankton because of the considerable errors involved, some of which have already been mentioned, and shall confine ourselves instead to a presentation of the method used at the present time. This method is based on the sedimentation or filtration of samples that have been treated with a fixative in suitable containers. For enumeration of the *nannoplankters*, which occur in large numbers, the chamber method introduced by Kolkwitz (1911) has proved best. A water sample is treated with iodine–potassium iodide as a fixing agent until a wine-yellow colour is obtained and is then poured into a plankton chamber

FIGURE 37. Plankton counting chambers: *a*, a shallow 1-ml. chamber, the original type of Kolkwitz (1911); *b*, a 2-ml. chamber with a cover glass bottom, from Utermöhl (1935).

of 0.5 to 10 ml. capacity (occasionally even larger), which if possible should be provided with a cover glass bottom (Figure 37). Precipitation of the plankton occurs within several hours. A number of strips of known length and width are then enumerated with the "inverted" microscope developed by Utermöhl (1931), or, if such is not available, with the use of a water immersion objective (after removal of the upper cover glass). If a lower power objective is used, the entire bottom of the chamber is examined.[38] An appropriate calculation gives the number of individuals in the chamber. Even with careful preservation, many of the delicate nannoplankters are changed beyond recognition. In order to facilitate the identification of the fixed material, it is therefore necessary, before making the counts, to determine from the living material which species are present. This identification is accomplished by centrifuging fresh-water samples in tapered tubes. Early in nannoplankton investigations the centrifuge was also used for quantitative studies following Lohmann's (1908) example. More recently this method has been replaced in quantitative studies by the easier and more accurate chamber procedure, although centrifuging remains indispensable for qualitative investigation.

The larger forms, which occur in smaller numbers, can likewise be precipitated by fixing with iodine–potassium iodide solution in flasks of 0.5 to 1

[38]Species that are lighter than water rise in the chamber and collect beneath the cover. If such species are present they must be counted separately.

litre capacity. After the supernatant liquid has been decanted, the entire sediment, or an aliquot portion of it, is transferred to a chamber of sufficient size where the counting can be undertaken. In addition, for a more accurate determination of crustacea and similar forms, one can strain say 5 litres of water from the depth being investigated through a bolting silk net, since with these large forms no losses through the net are to be feared. This material can be preserved as usual with formaldehyde.

Sedimentation takes place more and more slowly as the plankton organisms decrease in size, as explained on page 109. Also, sedimentation is very much delayed by convection currents set up by changes in temperature. These currents are of greater magnitude in a large than a small water volume, where they are curtailed by the relatively large surface area of the chamber walls. Hence, in a study of small nannoplankton a chamber of small volume and especially of low height (1 cm. or less) is recommended.

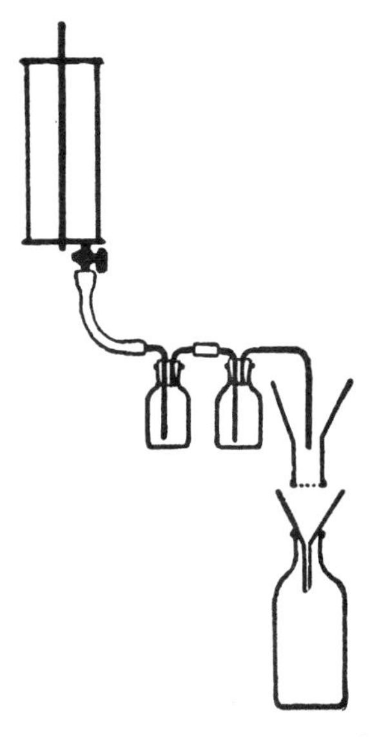

FIGURE 38. Sketch of the arrangement for drawing off samples for chemical and biological analysis from a single water sample.

For certain investigations (for example, the dependence of plankton distribution on the content of O_2, Fe, or H_2S) it is desirable to determine the plankton content of, and carry out certain chemical analyses on, the *same* sample of water. This can be done very simply in the manner shown in Figure 38. First a small flask (approx. 20 ml.) is filled from the sampling bottle, and this water is treated in chambers with iodine–potassium iodide for the examination or counting of the nannoplankton as described above. The water remaining in the sampler is washed through two Winkler bottles in tandem by means of a suitable combination of rubber and glass tubing; the overflow water is directed over a bolting silk filter, which retains the larger plankton, and finally is collected in a bottle. The overflow water, amounting to about 1 litre, can be used for the determination of electrical conductivity, alkalinity, and pH, as well as for other chemical analyses (for example, Fe, Mn, P, N), whereas the water in the Winkler bottles affords reliable samples for gas determinations (O_2, CO_2, or H_2S). By following this procedure one can be certain of measuring the environmental conditions under which the plankton was actually living at the moment the sample was collected; this relationship is not obtained with certainty in the case of samples collected *successively*.

The application of statistics to the distribution of plankton in a lake assumes that the individual species are distributed *horizontally* over the entire basin in approximately the same population density, and that therefore the observations obtained for a particular place are valid for the lake as a whole. In a basin that is not too extensive, not divided into separate basins, and not too shallow, this assumption can be made with certain qualifications. Even so, there is no uniformity in a mathematical sense, and the counts obtained do not conform to the yardstick of physical or chemical exactness. Variations of 10 per cent or 20 per cent have no significance in plankton statistics.

In the graphical representation of plankton distribution, the practice has become widely adopted of plotting, not the individual counts, as such, in a co-ordinate system, but rather their cube roots (the "spherical curve" of Lohmann, 1908). By so doing there is represented not the *number* of individuals in a given volume but rather the *density* of individuals along the radius of a sphere of this volume or along an edge of a cube of this volume ("linear concentration," according to Kohlrausch) (cf. the explanation of Figures 27 and 40). Since we are concerned with a spatial distribution, this method of representing the data is certainly logical, and at the same time it offers the advantage of minimizing non-significant fluctuations and not permitting small values to be completely overshadowed by large values in the same graph. The diagrams used in this book are of this type.

Just as in the ascent from sea level to high altitudes in the mountains the plant and animal kingdoms exhibit a step-like stratification (well known to every traveller), so likewise the *vertical distribution* of plankton is by no means uniform. There are several important differences, however, from the altitudinal stratification found on mountains. First of all, the gradient of environmental conditions in water is incomparably steeper, and as a result the stratification of organisms is compressed into a much shorter distance than in the mountains. In addition, and this is of particular importance, land plants, and also a large proportion of the animals associated with them, are closely restricted to their place of occurrence, whereas the plankton of lakes comprise a free floating community, which can offer little or no resistance to mechanical transport and whose distribution therefore is effected in large degree by the dynamic processes in the water.

Hence, when we consider the agents controlling this vertical distribution, we must take into account not only the *biotic factors* which influence life processes and which in other habitats are of exclusive importance, but also *mechanical factors*. The latter in their action are not concerned with living organisms alone; they distribute dead suspended material in exactly the same way as they distribute living plankton, provided that the size of the particles and the rate of sinking are identical.

Before we concern ourselves more closely with these relationships, however, we must attempt to provide a *concise survey* of the variety of phenomena exhibited in the vertical distribution of plankton. The habitat of the plankton is bounded at the top by the water-air interface. Many organisms have established themselves precisely at this abrupt transition between the two media. The floating plants of our pools, which send their roots into the free water and elevate their leaves and blooms into the air —for example, the frog-bit and the duck weed, or the water hyacinth (*Eichhornia crassipes*) and *Pistia*, which often crowd the shores of tropical waters in prodigious masses—certainly belong more properly to the littoral. However, in the ocean there are true pelagic animals, such as *Physalia* and *Velella* in the order Siphonophora, which are driven over the surface with the help of large gas bladders projecting out of the water. The expression *pleuston* has been coined for this life habit. Closely related in their manner of living, but restricted to miscroscopic dimensions, are the *neuston*—that surface film community consisting mainly of protista, which use the surface tension for stabilizing their position.

As Geitler (1942) has shown, two types must be distinguished among the organisms of the neuston: those that live *upon* the upper surface of the water—the *epineuston*—and those that live *beneath* the surface film —the *hyponeuston*. The former are essentially *aerial organisms*, which by means of a flotation disc (as in the heterokont *Botrydiopsis arhiza* and the chlorophycean *Nautococcus*) or by means of a stalk (as in *Chromatophyton rosanoffii*[39]) rest upon the surface film (Figure 39). The hyponeuston, on the other hand, are submerged forms which are scarcely different from other plankton organisms in their metabolism, and consequently are not strictly limited to the water surface film. The majority of the true neuston forms appear to be epineuston. On most

[39]According to Bourrelly (1954) this is now *Ochromonas vischeri*.

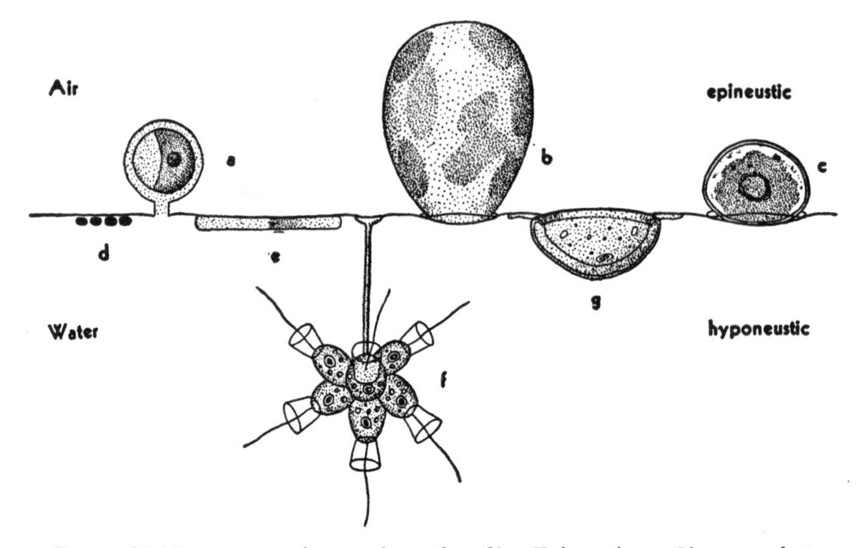

FIGURE 39. Neuston organisms at the surface film. Epineustic: *a, Chromatophyton rosanofii* (Chrysophyceae); *b, Botrydiopsis arhiza* (Heterokontae); *c, Nautococcus emersus* (Chlorococcaceae). Hyponeustic: *d, Lampropedia hyalina* (Coccaceae); *e, Navicula* sp. (Bacillariophyceae); *f, Codonosiga botrytis* (Craspedomonadaceae); *g, Arcella (Pyxidicola)* sp. (Rhizopoda).

small, quiet bodies of water they form a more or less cohesive covering resembling a film of mould, often of conspicuous colour ("water bloom"), which reveals its epineustic nature by its dry appearance.

The attentive observer surely knows the golden lustre that shines forth at times from the surface of shaded woodland pools when viewed from a certain direction. The small flagellated chrysomonads, *Chromatophyton rosanoffii*, which produce this phenomenon, penetrate through the water film and surround themselves with a spherical or pyriform gelatinous envelope, which rests upon the upper surface of the water by means of a little stalk. Within this envelope the flagellar movement is preserved, so that the flagellates, at the time of wetting and dissolving of the envelope, are immediately capable of swimming and can penetrate the surface film at another place. The cells resting on the upper surface of the water orient their chromatophores, which are shaped like concave mirrors, at right angles to the incident light. The light rays are concentrated on the chromatophore by the cell as by a lens and are completely reflected by the outside wall next to the air. If one changes his angle of observation the lustre fades, and a grey dust covers the water (Molisch, 1901). Especially striking are the blood-red, mould-like films, which are formed by *Euglena sanguinea*, particularly on the surface of alpine pools ("blood lakes"), and the radiant green coverings which may be composed of various kinds of algae. The neuston (and also the detritus floating on the water surface) form the food of a number of specialized plankton organisms. The cladoceran *Scapholeberis* and the ostracod *Notodromas* glide along on the underside of the surface film by means of the ventral edges of the carapace which are shaped like runners, and from it remove everything adhering in order to eat it.[40] The larvae of the malaria mosquito *Anopheles* also feed on the neuston.

The *lower boundary* of the vertical distribution of *autotrophic phytoplankton* is determined by their ability, which is dependent on light penetration, to accomplish an excess assimilation. At a depth where the balance of CO_2 assimilation over a 24-hour period is negative, that is, where less organic matter is produced than is used in the dissimilation process (respiration), the permanent survival of this plant community is no longer possible. Because animals, in the final analysis, are dependent on plants for their nutrition, the bulk of the zooplankton likewise does not extend markedly below this depth. Even the lowermost portions of very deep lakes, however, exhibit a generally sparse population, which feeds on the organic remains that dribble down from the trophogenic layer; yet these organisms consist entirely of scattered individuals of species that also occur farther up (if we disregard the oligoaerobic bacteria and related forms). *A counterpart of the specialized pelagic deep-sea fauna that occurs in the ocean is lacking in fresh water.*

Even in the clearest alpine lakes the layer of water inhabited somewhat densely by plankton is scarcely over 100 metres thick. In the majority of cases, if no other conditions such as lack of oxygen are limiting the distribution, we find that at depths of 30 to 50 metres the numbers of individuals already have declined to very small values. In this respect also there is a marked difference as compared with the ocean where the habitat of the "surface-plankton," including the phytoplankton, is usually several hundred metres thick, a condition that can be explained not only by the greater transparency of sea water but also more significantly by the far greater thickness of the turbulent layer affected by vertical eddy diffusion currents.

Thus the habitat of the limnoplankton is relatively small in vertical extent, even where the depth of the lake would permit otherwise. That very marked differences in the composition of the community occur even within this limited depth, however, can be demonstrated to one's satisfaction by a simple experiment in any lake of moderate depth. If one first draws a plankton net only through the surface layers, and afterwards sends the net to the bottom and draws it up vertically, one

[40]The recent investigations of Fryer (1956) have demonstrated that two similar entomostracans in the tropics—*Dadaya macrops* (Cladocera) and *Oncocypris mülleri* (Ostracoda)—behave in the same way as our two native neustic species.

will usually find a very different composition in these two samples: in the vertical tow, species will be present that are lacking in the surface sample, from which it is clear that only a vertical sample can furnish a reliable picture of the plankton composition of a lake.

As examples Figure 40 shows the commonly occurring types of distribution of a variety of phytoplankters and zooplankters during the summer in Lunzer Untersee, which is roughly 30 metres deep. Some

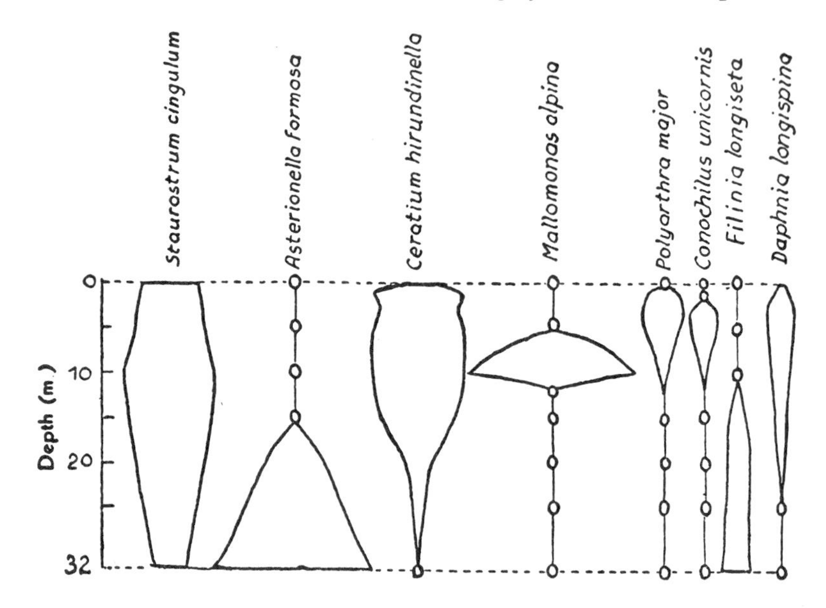

FIGURE 40. Examples of the distribution of planktonic species in Lunzer Untersee. The diameter of the "spherical curves" corresponds to the cube root of the number of individuals per litre, hence to the number of individuals occurring along the diameter of a cylinder of water (or a sphere) or along an edge of a cube containing 1 litre.

species are distributed through the entire water column (*Staurastrum*), although in varying concentrations; others prefer the upper and middle layers of water and are lacking in the depths (*Ceratium*, *Polyarthra*, *Conochilus*, *Daphnia*); still others are restricted to the deep layers (*Asterionella*, *Filinia*); and finally there is a species that is sharply confined between depths of 5 and 10 metres (*Mallomonas alpina*).

The *one* type of distribution not included in these examples is that in which the greatest population density regularly occurs at the surface. This type, which is not rare and is especially prevalent among the blue-green algae provided with gas vacuoles (cf. p. 108), is nevertheless much less common than one would expect on the basis of the light and temperature conditions prevailing at the surface. Indeed, the range of a great many species extends to the water surface, but in most instances the greatest concentrations of these organisms lie deeper.

Some figures, which were obtained in the investigation of eleven lakes in the eastern Alps, will serve to illustrate. Forty-four per cent of the species[41] had their *upper limit* of distribution at the surface, 37 per cent between 0 and 3 m., 13 per cent between 3 and 10 m., and 6 per cent between 10 and 20 m. On the other hand, the mean positions of *maximum population densities* were the following:

Depth	*Phytoplankton* (%)	*Zooplankton* (%)	*Total* (%)
0–1 m.	6	9	7
1–5 m.	40	40	40
5–10 m.	27	37	32
10–20 m.	12	9	10
20–30 m.	9	6	7
30–50 m.	6		3
Average depth of greatest population density	9.7 m.	7.0 m.	7.8 m.

It is apparent from these figures that the stratum between 1 and 10 m. is especially preferred: 72 per cent of all species of the plankton reached their greatest population density here. It is worthy of note that the mean depth of maximum population density for the plants is somewhat greater than for the animals, and that below 30 m. no representative of the zooplankton exhibited its greatest abundance, although 6 per cent of the species of phytoplankton did.

What are the conditions that bring about these varied types of depth distribution of the limnoplankton? When we examine one of the characteristic distribution diagrams, we immediately see its similarity to the previously described "activity curve." The idea is inescapable that the upper and lower boundaries of the zone of occurrence of a species probably coincide with one of the cardinal points, maximum or minimum, of the conditions for physiological activity, and that at the depth of greatest population density the optimal conditions (the resultant of all factors) might be expected. But mechanical factors, as previously explained, also play an important part in the distribution of plankton. Without an exact knowledge of all conditions one can no more determine *a priori* in a particular instance of stratification how much of this distribution has resulted from the physiological activities of living

[41]A total of 68 species was taken into consideration, of which 33 were plants and 35 animals.

organisms and how much from the dynamics of the water, than one can recognize which of the biotic factors (in the former case) determine the boundaries of a zone of distribution or bring about the formation of a population maximum. The causal analysis of these phenomena is consequently very difficult, and we still cannot give a complete explanation of them.

With respect to their mode of action the *mechanical factors* are easier to comprehend and less ambiguous than the biotic factors, and they will therefore be considered first. As already explained, these factors work on living and non-living suspended matter in the same way, and their effect is dependent solely on the physical properties of the water and the particles suspended in it. *Density* by itself, for instance, can create very significant types of stratification, especially when the density of the plankton organisms is less than that of the water, so that they become buoyant. In calm weather the blue-green algae of the plankton rise to the surface in consequence of their "gas vacuoles" and there aggregate into thick masses, often forming a "water bloom" recognizable from afar by its blue-green coloration. *Anabaena flos aquae* and *Aphanizomenon flos aquae* reveal by their names that they belong to the group of algae that can produce this striking phenomenon. Figure 41*a* illustrates such a surface maximum of the species *Gloeotrichia* (*Rivularia*) *echinulata* occurring in the Baltic lakes. This species consists of small spherical colonies a millimetre in diameter, formed of radially arranged whip-like threads (cf. Figure 30*c*). The same phenomenon is produced by colonies of the green alga *Botryococcus*; in this case the density is reduced by a fatty oil, which imparts a yellowish-red colour to the water bloom. But even in those species that are heavier than water, delimited aggregations can be brought about by purely mechanical means *at such times*, for instance, when the organisms are so nearly in balance with respect to the water that they sink through the warm epilimnion, yet remain suspended within the colder metalimnion. This effect is reinforced by the greater *viscosity* of the cold water. Yet, the commonly occurring metalimnetic population maxima can seldom be explained on a purely mechanical basis, because biotic factors can also be important. Likewise the concentration of entomostraca often occurring immediately above the mud surface cannot be attributed for certain to curtailment of rate of sinking, since in all probability factors concerned with nutritional physiology are of greater importance.

Without a doubt the most important agent affecting the distribution of plankton is the movement of the water, especially the thorough mixing action of the *turbulent eddy diffusion currents*. We have already seen that these currents constitute the primary means of keeping the non-actively moving plankton suspended. Accordingly it is clear without further explanation that in strata with a high degree of turbulence, stratification of these organisms cannot occur. Under these conditions an optimum rate of reproduction prevailing at a given depth does not lead to a population maximum at that level, but rather, as a result of mixing,

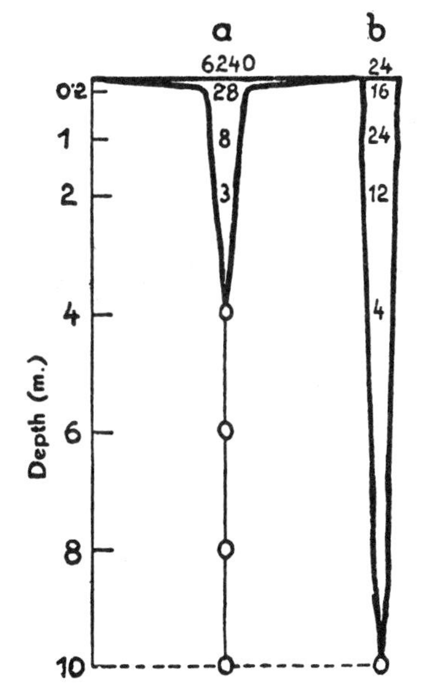

FIGURE 41. The distribution of the blue-green alga *Gloeotrichia echinulata* (containing gas vacuoles) in Gr. Plöner See during calm weather (*a*) and windy weather (*b*). According to Strodtmann, from Ruttner (1914). The figures are the number of individuals per litre.

to a general increase in the number of individuals in the entire turbulent layer. Only those species that possess active movements of their own, especially the entomostraca and rotifers, can be independent of the eddy diffusion currents and can select the depths most suitable for themselves. The consequence of all this is that the phytoplankton seldom exhibits persisting stratifications within a well-developed epilimnion.[42]

[42]But since water movements vary with time and with the vertical distribution of current velocities (cf. Figure 16), there occur even within the epilimnion both times and layers of stronger and weaker turbulence alternating with one another. As Einsele and Grim (1938) have pointed out, this alternation can lead to temporary differences in population density, as well as to a certain amount of sorting according to specific gravity, in that forms with a greater excess weight (for example many diatoms) sink out of weakly turbulent layers into more strongly turbulent layers having a greater supporting ability, and there accumulate.

Important differences in the composition even of the phytoplankton community occur first in the region of the temperature-controlled density gradient (the metalimnion), where the eddy diffusion currents are greatly curtailed. The fact that even the non-motile phytoplankters usually show a decided decrease in numbers just below the surface in no way contradicts this interpretation, because the surface of the water also acts as a brake on the eddy diffusion currents, thereby bringing about a depletion of the uppermost layer through sinking. The effect of turbulent eddy diffusion upon the distribution of plankton in the epilimnion is illustrated in Figure 41, which shows the distribution of the previously mentioned blue-green alga *Gloeotrichia* on a calm and on a windy day. In calm weather the alga, because of its low specific gravity, had accumulated in large numbers at the surface and was no longer encountered at a depth of 4 metres. In windy weather, on the other hand, as a result of turbulence the distribution down to 2 metres was almost uniform, and the alga extended downward almost to 10 metres.

With the beginning of autumnal circulation, eddy diffusion becomes more pronounced and involves increasingly deeper strata, and its influence on the distribution of plankton becomes continually stronger. At this time even those animals with active movements of their own cannot completely escape its effect, as is readily apparent from a comparison of the diagrams in Figure 42*a* and *b*. (Notice in particular the good swimmers *Polyarthra*, *Asplanchna*, and *Daphnia*.) During complete circulation (*b*) the vertical distribution of plankton is virtually uniform, but under the winter ice cover stratification phenomena again occur (*c*).

Water renewal, which is associated with the amount of inflow-outflow in a lake, affects the vertical distribution of plankton to as great depths as turbulence does. But while the latter merely equalizes vertical distribution, water renewal is one of the most radical depletion factors, in that it carries away a part of the plankton from the lake and causes an impoverishment of the strata concerned. As was described on page 56, the depth to which water replacement extends as a result of inflow-outflow depends on the relative temperatures of the affluent stream and of the lake itself. Since the metalimnion is generally the lower boundary of throughflow, high water periods in mountain lakes through which a large amount of water flows are associated with an impoverishment of the epilimnial plankton. This impoverishment is especially evident during the long period of snow melting in spring. That growth conditions at this time are otherwise entirely favourable is indicated by the fact that after the ice thaws on the lake but before the snow begins to melt in the mountains, a mass development of phytoplankton, particularly diatoms, frequently occurs. Later, under the influence of high water, these forms disappear completely, but only from the epilimnion. Such a radical reduction within a short period of time could not possibly result merely from a decline in, or cessation of, the rate of reproduction. Moreover, the water renewal that has taken place is indicated by an abrupt change in the chemical stratification (cf. p. 71). To be sure, the entire plankton is not affected by the throughflow. The good swimmers,

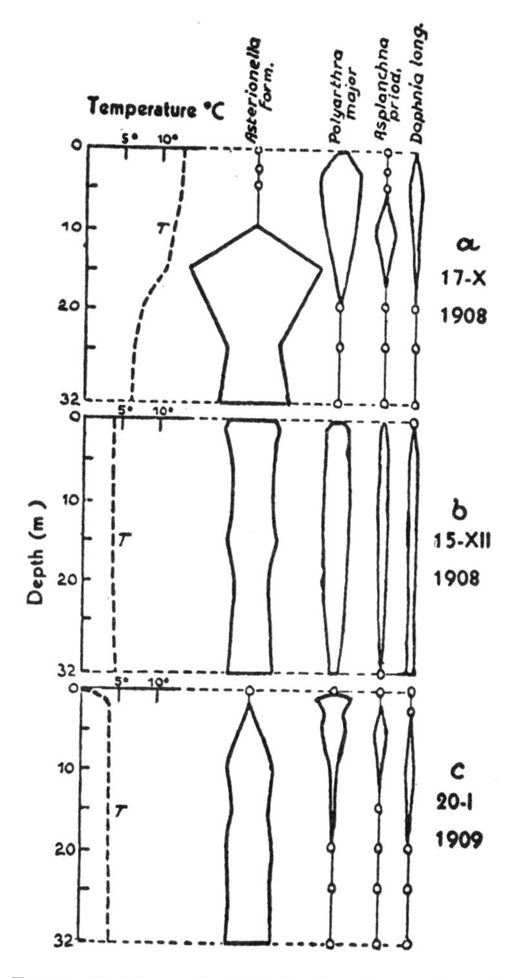

FIGURE 42. The vertical distribution of several species in Lunzer Untersee: *a*, at the end of summer stagnation; *b*, during complete circulation; *c*, during winter stagnation. From Ruttner (1914). "Spherical curves" are used as in Figure 40.

especially the entomostraca, are frequently able to avoid the outlet and so to escape being carried out of the lake (p. 150).[43]

Internal seiches, as Demoll (1922) has pointed out, can also exert an influence on vertical distribution. By means of these standing waves, which at times can attain amplitudes of several metres, the water of the metalimnion along with the plankton it contains is raised and lowered, and the population densities occurring at these depths can come to lie at greatly different levels in the course of a few hours.

Since all the phenomena and processes considered thus far that influence the stratification of plankton in a purely mechanical manner (specific gravity, viscosity, turbulent eddy diffusion, inflow-outflow, and internal seiches) are dependent on temperature to a considerable extent, we may consider the *thermal stratification* of a lake as indirectly a very effective mechanical factor in itself.

The action of the *biotic factors* is essentially more complicated and much more difficult to analyse, not only because many of them are always present, closely interrelated and often scarcely separable, but also because each of them can influence both the *reproduction* and the *active movement* of organisms, those two physiological processes that are most important in determining vertical distribution. Hence, only under especially favourable circumstances are we in a position to ascribe a particular phenomenon to the influence of a *single* factor with any degree of certainty. The chief environmental conditions that we have to consider in our inquiry are temperature, light, and finally that group of factors which are included under the designation "chemical nature of the water."

It is natural to regard *temperature* as one of the most effective regulators of vertical distribution, and there is scarcely a doubt that in reality it plays a controlling part in many instances. If the temperature gradient that occurs in temperate lakes in summer were the *only* variable and all other factors remained constant, then the population maximum of each species would necessarily coincide with the depth of its optimum temperature. In reality, however, the reaction curve for temperature is overlaid by those of the other factors, especially light, and as a result of this overlapping the depth of the population maximum can differ considerably from that of the optimum temperature. This is the reason why in the comparison of *isolated observations* one often encounters the

[43]The two peaks in the annual production curve of phytoplankton described on page 152 can also be observed in lakes in which vernal high water is not an appreciable factor. In these instances there is a temporary exhaustion of nutrients by the mass development of diatoms in spring.

greatest numbers of a species at considerably different temperatures. Hence, it is all the more surprising to discover that in a large number of observations from various lakes the *average* temperatures at which the population maxima occur are, within certain limits, almost constant for the individual species.

A comparison of investigations during several years in Lunzer Untersee with observations made 20 years later in 10 additional alpine lakes yielded mean values which in 60 per cent of the 25 species studied differed from one another by less than 1°, and in 90 per cent by less than 2°. This can be explained by the fact that the totality of the environmental conditions in lakes of the same type is quite similar under average summer conditions, and for this reason the magnitude of the mean temperatures at which the population maxima occur is likewise restricted. Although these temperatures do not necessarily correspond to the absolute temperature optima of the particular species, still we can consider them as *relative optima* for ecologically equivalent species and races, which at an average spectrum of other factors guarantee the best success of the species.

In answer to the question of how the population maxima of the species are distributed with respect to temperature, we shall again draw upon the investigations of the several alpine lakes already referred to a number of times. The reaction of 68 abundant species is evident in the following table:

Mean temperatures of maximum abundance	*Percentage of the species*		
	Phytoplankton	*Zooplankton*	*Total*
5–7°	15	12	13
7–9°	6	3	4
9–11°	15	23	19
11–13°	30	31	31
13–15°	33	31	32

In this summary it is apparent first of all that the number of species occurring in the individual two-degree classes at first decreases with decreasing temperature, then after a distinct minimum between 7° and 9° increases again suddenly and very considerably. The former sharply differentiated group includes all those plants and animals that are confined to deep water in summer. That it is not light or other factors that brings about this condition follows indisputably from the fact that all these species populate the surface waters in winter and multiply abundantly there. Hence, in this instance, we can confidently designate *temperature* as the factor that controls the vertical distribution. To this group of *cold stenothermal* or *oligothermal* organisms (cf. p. 120) belong several of the most abundant diatoms of the plankton, such as *Asterionella formosa* var. *hypolimnica* (Figure 31) already referred to several times previously, the delicate needles of *Synedra acus* var. *delicatissima*, the little discs of *Stephanodiscus astraea* and *St. alpinus* armed with small spines; and in addition the rod-shaped desmid *Closterium polystictum*, the blue-green alga *Oscillatoria rubescens*, and the chrysomonad *Mallomonas akrokomos*. Among the animals that prefer such low temperatures are the ciliate *Stokesia vernalis*, several rotifers (for example, *Filinia longiseta*), and even isolated entomostraca, such as the copepod *Limnocalanus*, which is abundant in the northern lakes.

The much more numerous species whose population maxima occur at higher temperatures are much less distinct and interesting as to behaviour. They are distinguished from the oligothermal organisms mentioned above not only on the basis of temperature preference; the almost transitionless separation of these two ecological groups is determined rather by the fact that the oligotherms are strictly limited to a narrow zone of low temperatures and cannot inhabit warmer strata, whereas the others are far less selective and can occur abundantly at temperatures or depths markedly different from the optimum. Consequently, the latter group, which includes most of the plankton species of our latitude, must be considered *eurythermal*, although it should certainly be remembered that with respect to temperature each individual species possesses a certain optimum of physiological activity and a more or less extended range of activity.

Temperate lakes in general are little suited to accommodate strictly *polythermal* species (those confined to higher temperatures), because temperature reversals occur continually, even during the warm part of the year. Yet we recognize a considerable number of species that show a clear preference for high temperatures through their occurrence only in the summer months and their inhabiting the warm layers of the epilimnion. Included here are several Myxophyceae and Chlorococcales among the phytoplankton, and many rotifers (for example, species of *Brachionus* and *Trichocerca* as well as *Chromogaster*) and entomostraca (*Diaphanosoma, Moina*) among the animals. The centre of distribution of the decidedly polythermal species is naturally in tropical waters. But even in temperate regions, shallow lakes, pools, and residual pools ("Altwässer") along rivers, especially those in a warm continental climate, in particular can temporarily attain "tropical" temperatures and afford favourable conditions for polythermal species with a short developmental cycle. In fact the phytoplankton and rotifer fauna in waters of this type strikingly resembles that in tropical lakes.

The reaction to temperature of different populations of a species is not always the same, although only in isolated species do very striking differences occur. Thus *Asterionella formosa* is strictly oligothermal in many alpine lakes, but eurythermal in those of the alpine foothills and the plains. We are dealing here with races or varieties that cannot be distinguished systematically on the basis of their morphology, although they are very different in their physiological characteristics. These so-called *ecotypes* naturally can differ among themselves by *other* features than just their temperature requirements.

Temperature, as already explained, affects not only the success of plankton but also their *intrinsic movements*, the latter in a two-fold manner: first, through an influence (acceleration or retardation) on the activity of the organs of locomotion (cf. p. 113); second, through the orientation of the direction of movement within a thermal gradient, the so-called *thermotaxis*. The latter phenomenon has been little studied in plankton up to the present, and it is not yet known if it contributes significantly to the stratification of lake plankton.

The *light* penetrating into the water is at least as effective as temperature in this complex of phenomena. It was emphasized at the beginning of this chapter that light, or at least the lower boundary of photosynthetic assimilation, is the factor that limits the habitat of the plankton to an upper, not very thick, layer of water. But even within this space the penetrating radiation affects the distribution of organisms in a variety of ways. Although we could definitely establish that certain phenomena, such as the restriction of particular species to the deeper water strata, are caused by temperature, there are indeed many other facts that are explainable only through the influence of light, for example the characteristic stratification patterns of rotifers and entomostraca in the epilimnion.

Like temperature, light also influences both the reproduction and the movements of plankton, but there are essential differences in the mode of action of these two factors. Whereas temperature plays a part in *all metabolic processes*, light is the source of energy and the controlling factor only in the photosynthesis of plants. *Multiplication* of the zooplankton, therefore, is only indirectly controlled by light through the nutritional dependence of these organisms on plants. With respect to *mobility*, on the other hand, the orienting influence of light (*phototaxis*) is substantially more important than its effect upon the intensity of movement.

The fundamental process of life, the *assimilation of carbon*, is dependent not only on the quantity of carbon dioxide present and on the energy of the incident radiation, but also to a large extent on temperature and on the quality or wave-length of the light. The close connection between the influences of light and temperature is of particularly great significance for the living conditions of the plankton. At a constant low

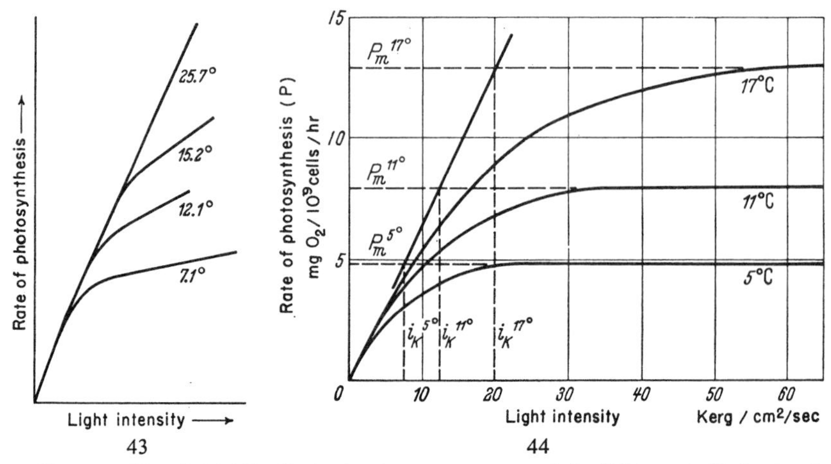

FIGURES 43 and 44. The limitation by temperature of the light-controlled rate of photosynthesis: 43, in *Elodea canadensis*; 44, in *Asterionella* (from Talling, 1957).

temperature, as the light intensity is increased the rate of assimilation responds at first by a straight-line increase. The investigations with *Elodea* demonstrate that above a certain light intensity the rate of photosynthesis increases more slowly. The curve (Figure 43, below) shows a more or less sharp inflection, which means that a further increase in light intensity has relatively little effect. Only with an increase in temperature does the rate of photosynthesis again increase. That phytoplankton behave in essentially the same way as *Elodea* is shown by Talling's (1957) investigations on *Asterionella* (Figure 44).

When we apply these experimental results to conditions in lakes, we must expect *a priori* that this *limiting effect of temperature* within the epilimnion will become important in most instances at the high light intensities prevailing there. This means that especially at low epilimnetic temperatures (as in spring) the light intensity in the upper layers, which rapidly increases towards the surface, has no influence on the rate of assimilation, providing the latter is limited by temperature rather than by light intensity, which is usually the case during the bright daylight hours. With increasing temperatures (in summer) the limiting effect naturally takes place at a higher level of assimilation, but it is almost always demonstrable. Light becomes a limiting factor only in greater depths, and here its decreasing intensity brings about a rapid reduction in assimilation. In the upper layers, therefore, the utilization of the existing *light surplus* is determined by *temperature*. When the latter is uniform, the light gradient has no effect on the intensity of reproduction or consequently on vertical distribution, except negatively through the regularly observed reduction in the amount of assimilation at very high light intensities as a result of an inactivation of chlorophyll. Investigations in which flasks containing cultures of plankton algae were suspended at different depths in a lake (cf. the summary by Printz, 1939) have likewise demonstrated that the CO_2 assimilation in the upper layers of water suffers no decrease (in spite of a progressive weakening of the light intensity with increasing depth), but in most instances even increases at *first*; only at a greater depth is there a decline controlled by light.

As an example we shall consider the instructive experiments with pure cultures of green algae (*Coccomyxa simplex* and *Chlorella pyrenoides*) that Schomer and Juday (1935) have conducted at various depths in several lakes of very different transparency. The most important results are summarized in the following table:

Lake	*Trans-parency* (m.)	*Colour of water* (mg. Pt/l.)	*Maximum photo-synthesis* (at depth in m.)	*Compensation point* *Depth* (m.)	*Radiation* (cal./cm.² in 3 hours)	*Per cent of surface radiation*
Crystal	13.5	0	5–6	17	1.2	0.9
Trout	5.9	14	1–2	10	1.6	1.0
Mud	1.9	33	0.5	4.5	0.8	0.4
Helmet	1.5	168	0.25	1	2.9	1.5

From these data it is apparent that in the clear water of Crystal Lake the decline in the rate of assimilation began first at the considerable depth of 5 to 6 m., where 12 to 15 per cent of the incident surface radiation was present. In the three other lakes this depth decreases steadily with decreasing transparency. These differences are especially evident in the depth of the compensation point, which is the lower limit of a positive assimilation balance. The thickness of the assimilation layer rapidly decreases with increasing dissolved colour (hence decreasing transparency), and in the brown-coloured Helmet Lake amounts to no more than 1 m. Worthy of note are the relatively small differences in the radiation energy present at the compensation level in the various lakes.

That the vertical distribution of photosynthesis accomplished by the phytoplankton occurring naturally in lakes agrees with the results of these investigations (particularly concerning the noteworthy condition that maximum photosynthesis does not occur at the surface in full daylight, but at some depth below the surface which increases with the transparency of the water) has been substantiated by numerous observations, as for example by the investigations of Edmondson (1956) in lakes of the State of Washington, and by measurements of primary production by the C^{14} method (Figures 50 and 51), which will be discussed later (pp. 167ff.).

In addition to this physiologically determined property of the upper epilimnetic layers, however, there is a *mechanical* factor that necessitates considering the *epilimnion* as *homogeneous* with respect to the utilization of light. The turbulent motion causes a phytoplankton cell to float in lesser depths at one time and in greater depths at another. Over a sufficient period of time, therefore, *all the phytoplankters of this layer will enjoy light equally*. Hence, a well-formed epilimnion can be designated as the *lighted region* (photic zone) of the lake. By comparison, in the metalimnion the non-motile organisms are retained within a given layer as a consequence of the stable stratification of the water, and in addition the light intensity in this layer is already greatly reduced, and its effect on assimilation is strongly influenced by the temperature gradient. Here begins the *twilight zone* (dysphotic zone), the lower boundary of which can be defined metabolically as the cessation of a positive balance of CO_2 assimilation by the phytoplankton community. Below the twilight region is the *region of light deficiency* (aphotic zone), for at the light intensities prevailing here photosynthesis remains below the compensation level in the alternation of day and night. The thicker the epilimnion is the lesser is the average amount of light available to a suspended phytoplankton cell. Especially in arctic seas or in very deep inland lakes with a thick zone of circulation it is conceivable that in such non-stratified water masses of great thickness the assimilation balance can be negative as a result of the long time spent by the algal cells at depths having low light intensities (Steemann Nielsen, 1939).

The *qualitative* changes light undergoes as it penetrates into a lake are of great significance. As is apparent from the detailed description on p. 18 the long wave-length "red" radiation is absorbed in the upper layers of water, and at a depth of 15 to 20 metres in a clear lake "green" light predominates. We know that the chromatophores of plants absorb wave-lengths of light that are largely complementary to their own colours and make it usable in CO_2 assimilation; hence "red" light is absorbed by the green chromatophores, and "green" light by the red (cf. also p. 189). Accordingly, those phytoplankters that live in deep water possess, with few exceptions, chromatophores coloured brown (the diatoms) or red (for example, the flagellate *Rhodomonas*), or have a chromatoplasm coloured red with phycoerythrin, as in the cyanophycean *Oscillatoria rubescens*.

However, it is difficult to relate details of phytoplankton distribution to the influence of light *alone* with any degree of certainty. The dependence of the radiation effect on temperature is too close for that, and the physiological requirements of the individual species, which undoubtedly are often very different, have been much too little investigated experimentally. But there can be no doubt that light not only limits the downward distribution of phytoplankton but also considerably influences the distribution of species within the region they inhabit.

In the behaviour of the plankters towards light, just as in their response to temperature, there can be distinguished species with narrow and species with broad ranges of activity (*stenophotic* and *euryphotic*), and, within the first category, *weak-light forms* and *strong-light forms*. But since in temperate lakes during summer both the light gradient and the temperature gradient usually decrease from the surface downwards, one can scarcely decide in many instances which of the two factors, light or temperature, is to be considered responsible for an observed stratification of plankton. Nevertheless, Findenegg (1947), through careful analysis of an extensive series of observations in Carinthian lakes, has succeeded in separating the influences of light and of temperature and in demonstrating the interplay between them (cf. p. 155).

This interplay will be made clear by an example. Figure 45 shows the vertical distribution of the blue-green alga *Oscillatoria rubescens* in Wörthersee from spring until autumn. In spring, as a result of high turbulence, the alga is distributed almost uniformly throughout the entire water column. As soon as a stable thermal stratification has been formed, the alga disappears from the epilimnion and at the same time increases in abundance in the upper hypolimnion. Here it seems as though a temperature of about 6° corres-

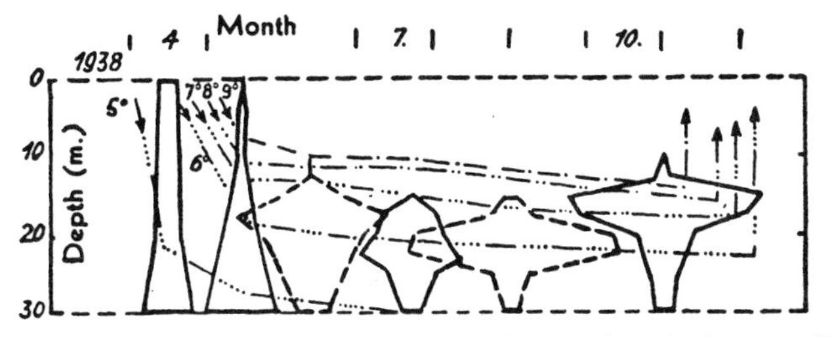

FIGURE 45. The distribution of *Oscillatoria rubescens* in Wörthersee (Carinthia) in relation to temperature and the enjoyment of light. From Findenegg (1947).

ponds to the optimum throughout the summer, according to the depth of the population maximum. In autumn, when light conditions become less favourable as a result of shorter days and a lower sun, the maximum density of population shifts from 20 m. and a temperature of about 6°, to 15 m. and 8°. The alga evidently prefers that depth within its temperature-controlled range of activity (between 3° and 10°) in which it finds conditions most favourable for its assimilation. A similar upward migration of the maximum in autumn has been shown by earlier observations in Lunzersee for the oligothermal *Asterionella formosa* var. *hypolimnica*.

The situation with respect to the influence of light on the *locomotion of plankton* is substantially clearer. Here nature herself, through large-scale experiments, offers us a glimpse at the causal relationships. When a layer of snow of sufficient thickness overlies the winter ice cover of a lake, the incident radiation is thereby reduced to physiological ineffectiveness, without the other factors, especially temperature, experiencing any change. We find under these conditions that the zooplankton is distributed more or less uniformly from the uppermost water layers down to great depths. Under ice without snow cover, on the other hand, most rotifers and entomostraca withdraw from the surface, bringing about a pattern of distribution very similar to the type we are accustomed to seeing in summer (Figure 42*c*). One can also produce this effect experimentally by creating a "window" of several square metres through removal of the snow cover. Under the window there promptly develops the type of stratification of zooplankton just described, whereas the distribution under the intact snow cover extends unaltered up to the ice. These observations demonstrate beyond a doubt that *light* keeps these zooplankters away from the surface layers of water (Ruttner, 1914).

Another natural experiment is the influence of the alternation of day and night. If, under a bright midday sun, a white disc of the type used in determining transparency is slowly lowered from a boat, not until a depth of from one to several metres do the shadows of the plankton crustacea first appear as dark spots against the brightly illuminated surface. But if this experiment is repeated in the evening, as soon as the disc is lowered just beneath the surface planktonic crustaceans scurrying about in large numbers can be seen by the last rays of the setting sun. These animals have ascended to the surface at the beginning of evening twilight. This striking phenomenon of the *diurnal vertical migration of plankton* has engaged hydrobiologists since early times, and is still being debated even now.[44]

On closer examination this migration proves to be exceedingly varied and complicated. First of all, by no means do all the plankton participate. Those members of the *phytoplankton* possessing their own means of locomotion (the flagellated forms) only rarely exhibit regular vertical movements of any extent, for example the colonies of *Volvox*. In this case, however, the movement follows an entirely contrary course: the colonies wander downwards at evening and rise towards the surface at morning. Another member of the Volvocales, *Eudorina elegans*, exhibits similar behaviour in Ranau Lake in South Sumatra. But even among the *animals* only a portion ascend to the surface at night. The phenomenon is relatively less pronounced among the rotifers, where it has been definitely demonstrated for only a few

[44]In this connection consult the paper by Cushing (1951), which summarizes in detail the existing observations in the ocean and in inland waters.

species. It is quite common among the entomostraca, and yet even here there are species that do not migrate, for example the surface forms *Scapholeberis* and *Polyphemus*. Many species appear at the surface at nightfall, but sink down again towards midnight. Moreover, it has been determined that members of a particular species can behave differently according to sex and stage of development. For example, the nauplii of *Diaptomus gracilis* and of *Cyclops strenuus*, in contrast to the adults, do not migrate; and the males of the first-named copepod congregate in large numbers immediately beneath the surface at the beginning of evening twilight, only to descend again at nightfall, whereas the females do not appear until later and populate the upper water layers uniformly. A similar "twilight migration" repeated at dawn has been observed in *Hyalodaphnia kahlbergensis* in Grosse Plöner See. A peculiar movement was determined in a North American lake for the copepod *Limnocalanus*; during the day it remains at the considerable depth of 50 m., but at night ascends only to the metalimnion (15 m.). Several oligothermal cladocerans of the Carinthian lakes behave similarly. The only insect larva of the plankton, *Chaoborus* (*Corethra*) *plumicornis*, likewise clearly exhibits a nocturnal ascent. On the other hand some rotifers, for example *Polyarthra*, and sometimes also the nauplii of copepods exhibit a *downward* displacement of the maximum at night. (Further details are given by Ruttner, 1914, 1929, 1952.)

The phenomenon becomes even more complicated when the behaviour of a single species in different lakes is compared. Then it is learned that many species usually migratory, for example *Daphnia*, are found during the day in the surface layers as well, or that the extent of migration in one lake amounts to from 20 to 30 m., whereas in another it is confined to a layer barely a metre thick.

At the present time it is not possible to relate all the individual instances of so complicated a phenomenon to a simple scheme. This variability is probably determined in large degree by the great modifiability of the animals' response to light through external and internal factors. Yet there can be no doubt that light is the controlling factor. This is demonstrated by the fact that migration ceases when the ice is covered with snow, but is immediately resumed beneath a "window" shovelled out of the snow covering.

When we attempt to analyse the main features of this phenomenon (cf. also Ullyott, 1939), we must distinguish between two phases: (1) the downward movement at dawn, and (2) the upward movement at evening. The *downward movement* in most instances depends upon *negative* phototaxis, or, to use more general terms, upon a stimulation produced by light towards the increasing light intensity.[45] It occurs

[45]That there are other possibilities besides simple phototaxis has been demonstrated by the investigations of Munro Fox (1925), showing that ciliates and sea urchin larvae swim downwards on illumination and upwards on darkening, no matter from which direction the light rays come. Other investigators have demon-

even in species that do not ascend at night, as investigations under the ice have demonstrated, and it leads to an accumulation of the animals at a certain depth where the orienting effect of the light rays ceases. The *upward movement* at evening, on the other hand, certainly is not to be explained on the basis of only *one* causal agent. The repopulation of the space from which the animals are displaced by light during daytime occurs automatically during the night after the cessation of the light stimulus that influences movements, provided the mobility of the organisms is great enough. This *equalizing migration* is facilitated when the mechanism of locomotion favours upward movement, as for example among the *Daphnia*, which frequently comprise the bulk of the zooplankton. Vertical convection currents in the water aid in this upward movement, especially among species of low mobility. In many species, however, the ascent at evening is doubtless initiated by a *positive phototaxis* of the bright-adapted animals towards the decreasing light intensity. This is the case in the males of *Diaptomus gracilis* and *Hyalodaphnia*. To what exent phototactic reactions, which are easily modified by previous adaptation and by environmental influences (for example, the CO_2 content), play a role in other forms is not yet known.

Recently two papers have appeared that are concerned in detail with the problem of the vertical migration of zooplankton. Schröder (1959) investigated the relationships of several plankton crustaceans in the lakes of the Schwarzwald, and established through a comparison with the results in other lakes that the depth at which the population maximum occurs depends on the transparency of the water. He supplemented the observations in the field with ingenious experiments in the laboratory,[46] and particularly on the basis of the latter concluded that both phototactic as well as geotactic reactions are involved in the phenomenon of plankton migration.

Siebeck (1960) working on the Lunzer lakes placed chief reliance on careful observations in the field. At the time that each plankton sample was obtained by means of a 4-litre water sampler, the light intensity prevailing at the depth of sampling was measured by a photocell attached to the water sampler. The fundamental result of Siebeck's observations is that migration is directed not by the stopping of an animal at a certain preferred (spectral) level of illumination, but by the reaction to *changes* in light intensity, whether they occur at high or

strated experimentally the influence of light upon the geotropism of crustaceans and the larvae of *Chaoborus*.

[46]Harris and Wolfe (1955) were the first to succeed in following the course of migration phenomena in *laboratory investigations*.

low light levels. For this reason the previous history of the illumination is of decisive importance.

As a last group of factors that influence the vertical distribution of plankton we shall consider briefly the *chemical stratification* in lakes and the *distribution of food materials*. In the space inhabited by autotrophic phytoplankton, wherein the major portion of the zooplankton necessarily lives also, vertical differences in the chemical properties of the water occur rather uncommonly and hence would scarcely be expected to have an influence on plankton distribution. However, in lakes of high transparency where conditions favourable for assimilation prevail even in the upper hypolimnion, a downward shifting of the production maxima of certain algae into considerable depths can occur during the summer, along with a progressive depletion of essential nutrient materials from the surface downwards, as Findenegg (1947) has shown in the lakes of Carinthia. Through the decline of population density in the upper layers (because of the scarcity of nutrient materials) the permeability to radiation becomes greater, making possible an active assimilation even at a depth of 20 metres. Sufficient nutrients come up to this depth from below to permit a rich development of phytoplankton. This example also shows the close interaction of two environmental conditions, light and nutrition, in their influence on plankton stratification.

At the boundary between the trophogenic and tropholytic zones, particularly in stratified eutrophic lakes, changes occur that vastly influence and modify the composition of the plankton community. At this boundary it is the *decrease in oxygen content* that has the most striking effect. It is worthy of note that under normal conditions the oxygen content does not exert a distinct limiting action until relatively low values. Even among animals, which generally are more sensitive in this respect than plants, the crustaceans in Lunzer Obersee, for example, occur below the 0.5 mg. O_2/l. boundary, and the rotifers even below the 0.25 mg./l. boundary. To understand this condition one must bear in mind that the respiration of organisms is dependent on temperature according to *van t'Hoff's law*, as already explained on page 76. Thus at a temperature of 5°, which commonly prevails in the deeper portions of lakes in the temperate zone, under otherwise uniform conditions only half as much oxygen is used as at 15°; the *respiratory value* of the same quantity of oxygen is approximately twice as great at a temperature 10° lower. The fact, then, that the majority of plankters in the metalimnion or hypolimnion can still respire at such low concentrations of oxygen is explainable by the low temperatures prevailing there. But as soon as the oxygen content decreases to several tenths of a milligram, the aerobic plankton, that is, those plankters utilizing dissolved oxygen, is suddenly terminated (Figure 46). In tropical lakes as a result of their higher hypolimnetic temperatures this lower limit for aerobic plankton occurs at higher O_2 levels: for the zooplankton of the Sunda lakes (Ruttner, 1952) it occurs very sharply at about 1 mg./l. However, one should remember that individual species react differently. Thus it deserves to be emphasized that the larva of *Chaoborus*, already referred to several times, is able to exist at especially low oxygen tensions, and for that reason of all the metazoa it descends farthest into the hypolimnion of eutrophic lakes.

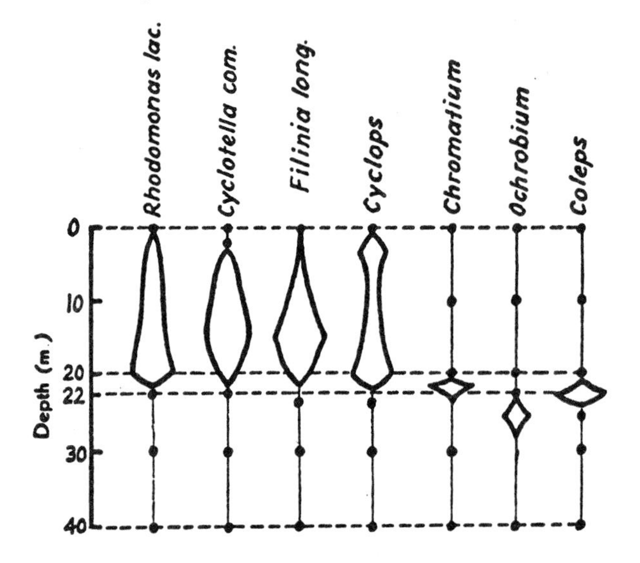

FIGURE 46. The influence of oxygen stratification on the distribution of plankton in Krottensee (Salzburg). Oxygen disappeared between 20 and 22 m. (cf. Figure 27).

It would be false to assume, however, that all life ceases to exist below this oxygen boundary. Quite the contrary, for below it we often find mass developments of certain organisms, of a magnitude that occurs only occasionally in the epilimnion. Individual counts of 100,000 or more per cubic centimetre are not uncommon. To be sure, this community has a very special character and is completely distinct from the epilimnetic community. It consists mainly of organisms with a specialized metabolism which do not obtain their life's energy, as do other plants and animals, by the oxidation of organic carbon linkages, but rather from inorganic materials. In temperate lakes it is represented primarily by the physiological groups of the *sulphur* and *iron organisms.* Many of these are oligoaerobic, that is, they require a certain low oxygen tension and at the same time small quantities of oxidizable inorganic substances, such as hydrogen sulphide or ferrous linkages. Since these substances and larger quantities of oxygen are mutually exclusive, it is understandable that organisms requiring both can thrive only within a boundary zone of often very limited thickness, in which hydrogen sulphide or iron from below and oxygen from above are furnished to them. Thus there arises at these depths an extremely

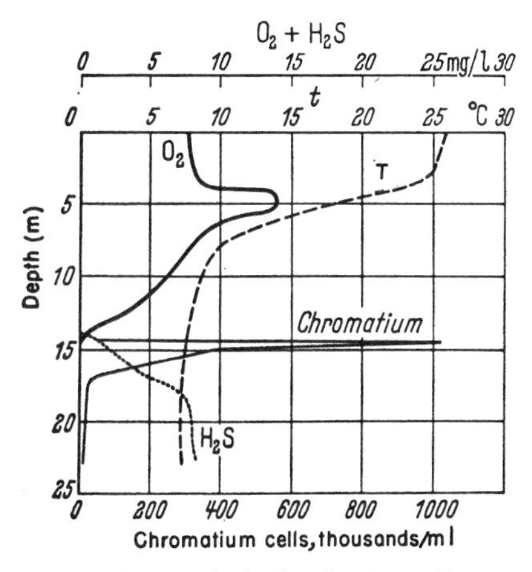

FIGURE 47. The vertical distribution of temperature, oxygen, hydrogen sulphide, and of *Chromatium* in Belovod Lake on July 30, 1938. After Kuznetsov (1959).

sharply stratified "bacterial plate," whose population density and composition, according to the different requirements of the individual species, can often change from decimetre to decimetre. Examples of this change are given in Figure 46, which depicts the distribution of plankton in Krottensee, a meromictic lake in Salzkammergut. The concomitant oxygen distribution can be seen in Figure 27 (cf. also p. 85 and Figure 24). An especially beautiful example is the stratification in the meromictic Belovod Lake (Figure 47), described by Kuznetsov (1959). Here the development of the *Chromatium* "plate," which is only 20 cm. thick, occurs precisely at that depth where the O_2 and H_2S curves come into contact with one another.

That this does not apply in all instances is demonstrated by the observations on Lunzer Mittersee (Ruttner, 1955), a spring lake which is only 3 m. deep, normally completely unstratified and extremely oligotrophic. However, in an autumn with low precipitation there can occur a density stratification propagated from the mud surface upward, which produces temporary conditions that completely resemble those in meromictic lakes. In the deeper layers stabilized by the concentration gradient there develops a rich plankton, which otherwise is completely lacking in this lake, and in which sulphur organisms (*Chromatium*, *Lamprocystis*, *Thiospira*) occur in concentrations of roughly 10,000 cells per ml., even though the mean oxygen content of this layer amounts to 3–4 mg./l. and only immediately above the bottom declines to about 0.5 mg./l. Hence, the sulphur organisms here are by no means restricted to low O_2 content, and one can designate them merely as facultative oligoaerobes. Indispensable for their economy, however, is a continuous supply of H_2S. Although this gas does not occur in the free water of Mittersee, any more than in many other lakes that often contain a rich development of sulphur organisms, yet there is no doubt concerning the continuous supply of H_2S from the sediments. This means that as in the case of any true minimum substance, the amount of H_2S made available is immediately incorporated into the metabolism of the sulphur organisms, even before its oxidation can occur by purely chemical means.

Among the sulphur bacteria, which first oxidize H_2S to elemental sulphur, stored as black droplets inside the cell, and then this material to H_2SO_4, there are two groups: the *colourless* sulphur bacteria, and those *coloured red* by bacteriopurpurin, the so-called purple bacteria or Thiorhodaceae. It is these latter, and among them especially the very active members of the genus *Chromatium*, that in many lakes can impart a very noticeable colour to the stratum of water they inhabit. For example, in Lunzer Obersee one can at times obtain water samples of a peach-red colour from a particular depth, while only half a metre above or below this level the water is clear and colourless. Besides *Chromatium* the non-motile hollow spherical colonies of *Lamprocystis* and the lamelliform colonies of *Thiopedia* are abundant, but usually they are less sharply stratified. According to Van Niel (1936) the red sulphur bacteria are capable of photosynthesis, whereas the colourless ones, of which the genera *Macromonas* (*Achromatium*) and *Thiospira* occur commonly in the plankton, assimilate carbohydrate by means of chemosynthesis (p. 201).

Among the iron bacteria those inhabiting the oxygen-poor layers are chiefly single-celled forms such as *Ochrobium* and *Siderocapsa* as well as the diminutive colonies of *Leptothrix echinata* resembling morning-stars. They often occur in unusually large numbers. Their vertical distribution, however, is not generally so restricted as that of *Chromatium*. The iron bacteria are claimed to have the capacity to utilize the energy set free in the oxidation of ferrous to ferric compounds, but this type of chemosynthesis has not been completely substantiated.

Besides these two groups there are certainly a large number of other specialists that find suitable living conditions within the boundary layer and in the oxygen-free hypolimnion. Here there are often observed enormous numbers of various bacteria, which are difficult to characterize morphologically and physiologically. The denitrifying species, which cause the disappearance of nitrate in these layers as described on page 89, evidently belong to this group, and likewise the sulphate reducers, which in contrast to the metabolism of the sulphur bacteria reduce sulphate back to H_2S. It is noteworthy that, especially towards the upper boundary of this oligoaerobic community, there occur considerable numbers of organisms with assimilation pigments, especially Euglenophyceae and cryptomonads, which are lacking in the overlying epilimnial layers. Evidently these are mixotrophic forms, but this designation by no means explains their occurrence at the oxygen boundary. Only in the case of the iron-storing genus *Trachelomonas*, which is frequently found here, is the presence of iron organisms in this community understandable. In these aggregations of bacteria there often live, in addition, large numbers of ciliates which are obligate or facultative anaerobes (those not requiring free oxygen), such as *Coleps* (Figure 46), *Uronema*, and *Caenomorpha*, which here find a rich supply of food.

Although the distribution of these ciliates is probably determined in fact by the sharply stratified occurrence of their food plants, and although immediately above the aggregations of bacteria there are often present population maxima of higher zooplankters, such as *Daphnia*, copepods, and especially rotifers, which presumably must also be there for the sake of food, nevertheless an *influence of nutrition* on the vertical distribution of zooplankton cannot be demonstrated with certainty. Aside from the fact that the vertical range of distribution of the zooplankton corresponds in general with that of the phytoplankton, closer relationships are not evident. In this connection one should also consider that a large number of the zooplankters perform diurnal vertical migrations, and "graze" in different layers at night than during the day. Conversely, there are many indications that the distribution of phytoplankton is affected by that of the zooplankton, in that the layers thickly populated by animals are gradually "eaten free" of algae, even when the latter are actively multiplying (Edmondson, 1957).

There is much less to be said about the *horizontal distribution* of plankton than about the vertical distribution at the present level of our knowledge. If we consider it established that the qualitative and quantitative composition of the floating community in each place is dependent upon the physical and chemical conditions and processes in its environment, then, in view of the fact that in a lake of regular shape and moderate depth these environmental conditions within the same stratum are quite uniform over the entire area of the lake, we should not expect any greater variation in distribution of the plankton. However, no very exact yard-stick can be applied to this uniformity in the horizontal distribution of plankton. Even differences in environmental conditions that are slight and at times not even measurable, such as those that develop during periods of calm weather especially, can be very effective biologically and lead to irregularities in distribution. Such irregularities were demonstrated quite precisely first by Tonolli (1949) in Lago Maggiore, and more recently by Schröder (1959) in lakes of the Schwarzwald. Schröder (1961) even more recently, through the interesting application of echo sounder and television, has obtained in these lakes as well as in Bodensee very impressive pictures of the irregular structure of the horizontal distribution of planktonic crustaceans over both very short and longer distances. It is assumed that the rotifers likewise exhibit such irregularities in horizontal distribution, but that the phytoplankton, by virtue of having no or only little locomotion in the open water, does not. At any rate these observations demonstrate that especially in studies of production biology a single series of observations is inadequate and that mean values must be obtained (cf. here the observations as well as the discussion of their statistical evaluation in the works of Ricker [1937] and of G. W. and J. J. Comita [1957]). Yet as a rule these phenomena, the cause of which it is difficult to ascertain, are temporary conditions, which are quickly altered or destroyed by resumption of horizontal turbulent currents. It is evident that these environmental conditions can attain significance in shallow (e.g. in Lake Balaton [Sebestyén, 1960]) and irregular basins and can produce lasting differences in plankton composition in isolated portions of lakes.

The situation in the *vicinity of shore* is markedly different. In many instances, especially in the alpine lakes, a striking change in the composition of the plankton occurs in this region, resulting from the fact that entomostraca in particular avoid the shallow water. Thus, along the margin of the littoral shelf (water depth 1.5 metres) in Lunzer Untersee, virtually no *Daphnia*, *Bosmina*, or adult *Diaptomus* are observed. *Daphnia* begin to occur first where the lake reaches a depth of 7 metres, and attain their full population density at a distance of 50 metres from shore where the water is approximately 20 metres deep. Lindstrøm (1957) has recently related a similar behaviour for *Daphnia* in shallow lakes of Sweden. On the other hand, *Polyphemus*, for example in Lunzer Obersee and in the Swedish lakes mentioned, exhibits a distinct tendency to concentrate in the littoral zone. The phytoplankton and some rotifers occur in unreduced numbers even in the shallowest water, and indeed *Synchaeta* exhibits a noticeable increase here at times. This *avoidance of the shore by plankton* has not yet been explained. It is natural to assume that the cause is a stimulus emanating from the shore or from the shallow bottom, and yet this can scarcely be optical, because the phenomenon also occurs under snow-covered ice at the end of a long winter night (Figure 48). Just as vertical migration is complicated by a number

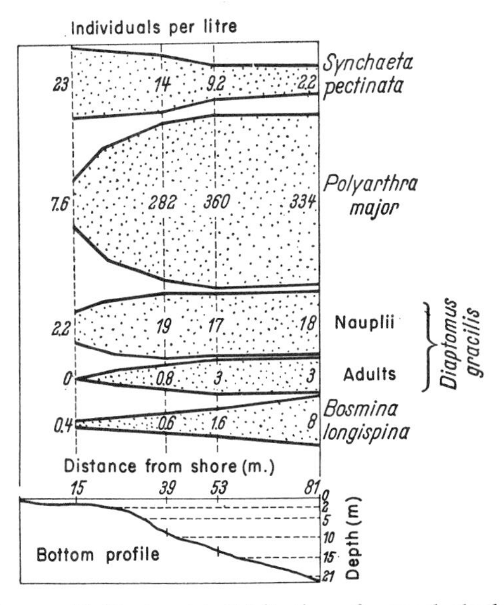

FIGURE 48. Horizontal distribution of several plankton animals in Lunzer Untersee at 1 m. depth in the vicinity of shore at 6 a.m. on February 19, 1913, under the ice. From Ruttner (1914).

of factors, so in the present consideration the same species that exhibit avoidance of shore in large and deep lakes (for example *Daphnia*) often can be obtained in the immediate vicinity of the shore in small and shallow bodies of water.[47] Also ascribable to the avoidance of shore is the fact that the plankton crustacea stay away from the outflow of a lake, although here a negative rheotropism might also be operating.

4. *Temporal Distribution*

The same factors (although not exclusively) that determine the spatial distribution of the plankton are also effective in their *seasonal occurrence*, their so-called *periodicity*.

[47]Observations that Siebeck (1960) obtained on the horizontal distribution of plankton crustaceans in Lunzer Obersee deserve to be mentioned at least briefly. He found that ephippial females of *Daphnia longispina* and egg-bearing females of *Diaptomus denticornis* were much more numerous in the shore zone than in the middle of the lake. It is tempting, although still premature, to associate this phenomenon with egg deposition.

When a lake is investigated continuously a striking change in the composition of the plankton is found, which is repeated each year in approximately the same manner. This change comes about in two different ways: first, *perennial* species (those present throughout the year) occur in different densities at different times; and second, numerous species appear only for a varying period of time in the plankton and at other times are completely lacking. The prerequisite for this *intermittent* occurrence is the ability to form resistant stages, which sink to the bottom of the lake or are washed up on shore, whence they repopulate the open water upon the recurrence of favourable environmental conditions. In Lunzer Untersee 39 per cent of the phytoplankters and 50 per cent of the zooplankters (although only one-fourth of the rotifers, compared with three-fourths of the crustaceans) are perennial. The periodicity curves of the individual species show, as a rule, more or less well-marked maxima, which in Lunzer Untersee, for example, occur 52 per cent of the time during the high summer period of the lake (July to September). But even in the apparently unfavourable winter months from January to March, 10 per cent of the species attain their highest numbers of individuals. For the majority of species the peak of development in different years occurs at the same time of the year. In our example, 67 per cent of the species had their maxima in the same quarter of the year, and 23 per cent (for example, *Ceratium*) repeatedly in the same month. There are other species that do not show such regularity, and in a few extreme cases, for example several rotifers (*Polyarthra*) and *Bosmina* in Lunzersee, development appears to be completely independent of the time of year. Moreover, many species either regularly or else in most years show bimodal curves, the peaks of which are sometimes separated by an interval in which the species is completely lacking. In these instances the cause is usually the succession of two cycles of reproduction (dicyclism). Concerning the course of development of the biomass of the total plankton, especially the phytoplankton, in most lakes there are two maxima (in spring and autumn) and two minima (in winter and midsummer). This phenomenon is controlled by the fact that the diatoms, which include the most important dominants in the limnoplankton, have their major development in spring and autumn (McCombie, 1953).

Fluctuations in population density over a period of time are determined through the interaction between the *rates of multiplication and depletion.* A beautiful example of a *biotic* depletion factor is provided by the investigations of Canter and Lund (1951), which demonstrate how the population of *Asterionella formosa* in Lake Windermere can be reduced by an attack of the fungus *Rhizophidium planktonicum.* Similarly, each year in Lunzer Untersee the rotifer *Conochilus unicornis* is almost completely wiped out by an infection with *Plistophora* (Sporozoa). In addition to biotic factors, *mechanical* factors are highly effective. Foremost among these mechanical factors is the loss, through sinking into the hypolimnion and ultimate sedimentation upon the lake bottom, of heavy types of phytoplankton. The magnitude of this loss, which is related inversely to the floating ability of the organisms and to the turbulence of the layers in which they live, has been determined quantitatively for several phytoplankters by Grim (1950). In mountain lakes through which a considerable amount of water flows and elsewhere in river lakes, inflow-outflow frequently plays a decisive role (cf. p. 55). The phytoplankton inhabiting the uppermost layers of water and those zooplankters that do not exhibit a pronounced avoidance of shore are trapped by the outflowing water, and are removed from the lake in proportion to the quantity of water flowing through at the time. Inflow-outflow is the factor that brings about the spring or early summer population minima of almost all plankton species in the alpine lakes, in spite of the fact that their rate of reproduction is high at these seasons because of favourable environmental conditions. In addition to the mechanical factors that bring about depletion, *consumption by animals* must also be considered, although it is difficult to comprehend its influence clearly (Edmondson, 1957).

The *magnitude of the rate of multiplication*, which is the productive basis of these periodicity phenomena, is influenced by the seasonal variations in environmental factors. The mode of action of these factors must correspond essentially with that which has been discussed in detail in the description of vertical distribution. Whereas in vertical distribution there is a *juxtaposition* of environmental differences at the same site, in seasonal distribution a *temporal succession* of environmental differences occurs, whereby the influence of restricting conditions in the latter situation is much easier to visualize than in the former (Fry, 1947, and McCombie, 1953).

Of the various factors important for life, *temperature* exhibits the greatest variations during the course of a year in lakes of the temperate latitude. Its influence on the periodicity of plankton is extensively substantiated by knowledge gained through the study of vertical distribution. If, for instance, we investigate the temperatures at which the individual species usually attain their maximum of development, we find a close agreement with the values for the stratification maxima. Thus, the species which during summer stagnation are found in cold deep water at temperatures of from 4° to 6°, and which have been designated as *oligothermal*, immediately command attention because they attain their maximum of development at the same temperatures within the surface layers during *winter* or *early spring*. On the other hand most of the plankton organisms both in summer and in winter are more or less eurythermal, although each undeniably exhibits a consistent preference for a certain temperature range. The fact that in Lunzer Untersee and in many other lakes, particularly in the northern Alps, *only* oligothermal and eurythermal species occur, without any decidedly polythermal species, can be easily explained. With the variable summer climate of this region, the periods of warming in summer, which often can lead to rather high surface temperatures of more than 20°, are always interrupted by cold reversals. Cold stenothermal organisms are able to avoid warmed surface layers by retreating into deeper water, but warm stenothermal species cannot withdraw from a cold reversal, and as a consequence they do not find the proper conditions for existence in lakes with inconstant temperature conditions. Only very large deep lakes with little inflow-outflow in relation to their area can guarantee, even in the climate of the northern Alps, the constancy of temperature conditions necessary for polythermal species. Thus, for example, many of the water-bloom-forming blue-green algae, which constitute the characteristic midsummer flora of the eutrophic lowland lakes, also occur in the large oligotrophic basins of the northern Alps and in the lakes of the Carinthian basin, which is characterized by its constantly high summer temperatures. The proper habitats of polythermal species, however, are the continually warm lakes of the tropics, which have high temperatures at all levels. Here the facies of the plankton is dominated by many species that occur only during midsummer in the temperate lowland lakes, along with cosmopolitan eurythermal species and relatively few exclusively tropical species. Understandably, the oligothermal species, which comprise a considerable part of the plankton in lakes of the temperate region, cannot occur at any time of the year or at any depth in the tropics. Little is known concerning the annual periodicity in tropical waters, although such a phenomenon could not be determined by temperature, because the annual fluctuations in this factor are insignificant.

The seasonal variations of physiologically effective *radiation* concern less the intensity than the duration of illumination. Yet the influence of the latter factor is by no means great enough to produce a continual minimum of plankton in winter. In many instances, it is true, we can establish a clear retrogression at this time, but not infrequently mass developments of oligothermal species occur in the winter half-year (cf. p. 152). These relationships take on a different aspect in frozen lakes but only when the ice is covered with opaque snow (cf. the figures for light transmission on p. 23). Such winters are characterized by a relative poverty of plankton, especially phytoplankton.

Overlapping of the reaction curves for light and temperature is clearly encountered in the annual course of plankton development. Because both factors increase from a winter minimum to a summer maximum and then again decline, one might expect that the same conditions in spring and autumn would provide the basis for the occurrence of communities having a similar composition: in short, that the spring forms would recur in autumn. This, however, is true only in exceptional instances. Usually the spring and fall plankton are very different, and the explanation of this difference is to be sought in the fact that the seasonal temperature curve in waters lags considerably behind the curve for the quantity of light. Thawing of lakes in spring takes place when the sun is already fairly high, and at even lower temperatures there is already considerable light. On the other hand, in autumn, when the sun is already low and the days are short, lakes usually still exhibit relatively high temperatures. As a consequence of these relationships, the forms occurring in the upper strata are selected on the basis of their individual light requirements, according to the following scheme proposed by Findenegg (1947):

	Weak-light forms	*Strong-light forms*
Cold-water forms	Winter plankton	Spring plankton
Warm-water forms	Autumn plankton	Summer plankton

The dependence of plankton periodicity on the *chemical properties* of the water, especially on the content of the nutrients essential for life (minimum substances), is not easy to demonstrate for reasons already specified on page 101. No direct relationship exists between the quantity of total plankton or of individual species present at any one moment and the content of nutrients measurable at the same time, for the reason that any quantities of these production-limiting materials are completely utilized as they are introduced. Only when it becomes possible to measure the magnitude of the supply from all sources of any account—inflow, atmospheric precipitation, and the turnover in the lake itself—will the existing relationships be capable of exact investigation. In addition since the nutritional physiology of plankton organisms is still very incompletely understood, we shall have to be content to perceive only the gross outlines of the results of these interactions.

Circulation in winter and early spring undoubtedly supplies considerable quantities of nutrients to the free water through the moving upward of the deep water carrying with it the decomposition products that accumulated at the mud-water interface. The blooms of diatoms, which are often immense at these seasons, as we frequently have had the opportunity to observe, might with great probability be related to this increased supply of nutrients. After summer stratification has been developed, however, the stock of nutrients can in most instances be supplemented only insufficiently. The generally cold inflowing water usually intercalates itself in the metalimnion, hence below the trophogenic zone; and as a result of the greatly curtailed eddy diffusion, the addition of nutrients to the epilimnion is primarily by precipitation and by the decomposition of organic substances in the littoral zone and in the open water. An extensive impoverishment can result. Nutrients, for example nitrate, which were present earlier in more than minimal quantities, can almost completely disappear in midsummer (cf. p. 89), causing a significant decline in reproduction. Especially warm dry summers, in which a sharply marked stratification almost completely eliminates eddy diffusion from below, are for this reason generally not distinguished by a high production of plankton organisms, in spite of favourable temperature relationships. In lakes of high transparency this impoverishment of the epilimnion in nutrients can be accompanied by a downward displacement of many plankton forms into deeper layers richer in nutrients, as already described on page 145 for the Carinthian lakes. This is true also of warm seas. Only in late summer when vertical convection gradually extends deeper, and the nutrient-rich water of the depths is transported upwards, does there begin a population expansion, often very suddenly. Even in tropical seas the regions of upwelling deep water are characterized by a mass production of phytoplankton visible even to the naked eye (for example, in the Gulf of Aden).

It is to be expected at the outset that the succession of the individual populations in the course of a year is determined not only by environmental conditions external to the lake, but also by changes that the populations themselves bring about in their environment. We have seen how a bloom of phytoplankton can greatly modify the chemical composition of the water, for example how after a diatom maximum the silicon content is much reduced, etc. These are especially extreme and noticeable changes in the quantity of materials; the smaller changes that escape observation can scarcely be of lesser physiological significance. We know at any rate that under the influence of life the nutrient content of lakes undergoes a continuous change in composition. In the course of this process the original combination of nutrients, more or less optimal for the form dominant at that time, gradually deteriorates, and in consequence the rate of reproduction and the number of individuals decline. At the same time, however, the conditions become favourable for other species, as their optima for physiological activity are approached. A new population appears, for whose development the requisite conditions were provided primarily by its predecessor. Thus the studies of Pearsall (1932) and of Hutchinson (1944) along these lines have shown that populations of blue-green algae very likely do not reach bloom dimensions until there has been an extensive depletion of mineral nutrients and at the same time a copious increase in organic substances (a condition that normally occurs in late summer and autumn). But investigations in this field are only in their infancy. Unequivocal results can be expected only from a close association of field observations with exact laboratory experiments in nutritional physiology, a procedure that has been successfully undertaken recently (cf. especially Rodhe, 1948).

5. *The Bacteria of the Plankton*

At this point a few words might be added concerning the occurrence of bacteria in the plankton of inland lakes. That our knowledge of this subject is still extremely fragmentary is adequately demonstrated by several observations that can in no way correctly indicate the importance of this group of organisms. The book by Kuznetsov (1959), which fortunately has also appeared in German, summarizes our present understanding of this subject.

The content of schizomycetes in a sample of water is generally determined by means of the "plating procedure," especially in hygienic and technical water testing. Nutrient agar is inoculated with, for example, 1 ml. of the water being investigated, and poured into a sterile petri dish; after incubation at room temperature the colonies that have appeared are counted. It stands to reason that by this procedure only those bacteria can be obtained that are able to grow on the nutrient agar employed. These, however, are not the true water bacteria but for the most part putrefaction bacteria, generally widespread, which have reached the water secondarily through pollution. For the quantitative determination of the true water bacteria, which are frequently highly specialized in their metabolism, this plating procedure is just as unsuitable as would be the attempt to determine the content of phytoplankton by inoculating agar plates. Likewise, in the latter instance, only colonies of readily growing and generally distributed green algae and diatoms would appear, and the plates would in no way present a picture of the composition of the phytoplankton.

In recognition of this source of error investigators have repeatedly avoided determining the number of bacteria, and have confined themselves to demonstrating the presence of certain organisms with specialized metabolism (nitrifiers, denitrifiers, sulphate reducers, etc.) by inoculation of a suitable nutrient broth with the water being investigated and subsequent determination of changes arising in the medium. By inoculating a series of media with graduated amounts of water ("dilution method") data can also be obtained concerning the number of the particular organism in a unit volume (Klein and Steiner, 1929).

In more recent times successful attempts have been made to determine the number of bacteria not by the culture method but rather by a direct microscopical examination of stained material. The first procedure of this kind—filtration through a membrane—was described by Cholodny (1929) and in the time since then has been improved by Russian investigators until now it is the method most widely used. Another method elaborated by Kuznetsov and Karzinkin (1930) and adopted by Bere (1933) in Wisconsin is based upon the vacuum evaporation of water samples to about one-tenth their original volume, from which aliquot volumes are dried on microscope slides, stained, and the bacteria counted. By this direct microscopic procedure Bere (1933) showed that the number of bacteria in Wisconsin lakes is 20 to 335 times greater than that obtained by the plating technique. The investigations of Razumov (1932) gave even greater differences. A few figures taken from Kuznetsov's book are given below:

	Total number of microorganisms per ml.		
Kind of water	*a, by direct count*	*b, by the plating procedure*	*Ratio a:b*
Mausly Lake (unpolluted)	256,000	13	18,200
Bolshoye Bagodak Lake (slightly polluted)	2,240,000	125	17,900
Effluent from a canal	307,000,000	3,400,000	93

It is apparent that in non-polluted lakes the ratio *a*:*b* is much greater than in sewage effluent, which indicates that true water bacteria do not thrive on a peptone substrate, whereas the putrefying bacteria dominant in effluent water do. Similarly, the investigations of Beling and Jannasch (1955) in various sections of the polluted Fulda River showed relatively small ratios between numbers obtained by direct count and by the culture method.

The following table abbreviated from Kuznetsov gives information concerning the number of bacteria determined by *direct* count in lakes of different tropic levels:

	Bacteria in 10^3/ml.
Lake Baikal	50–200
Lake Beloye	2,230

(These results pertain to the epilimnion at the time of summer stratification.)

The seasonal fluctuations in the eutrophic Lake Chornoye are given in Figure 49 (after Kuznetsov).

Because of the slight morphological differentiation of bacteria the microscopical method permits the determination of species only in special instances (iron bacteria, sulphur bacteria, etc.). Quantitatively,

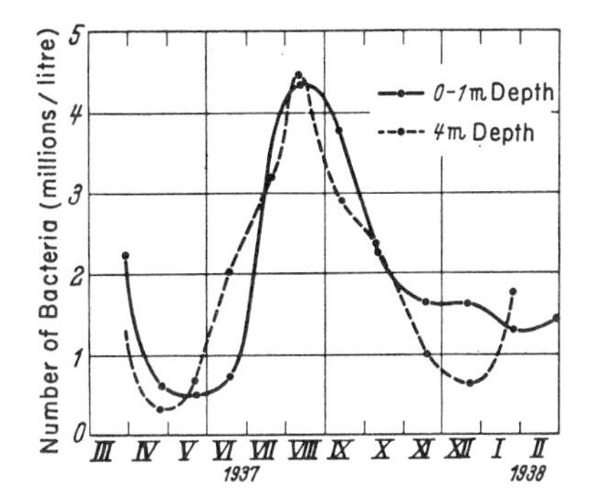

FIGURE 49. Seasonal fluctuations in the number of bacteria in Lake Chornoye at Kossino in 1937. From Kuznetsov (1959).

bacteria are usually relatively insignificant: in the Wisconsin lakes they amounted to only about 1 per cent of the total organic seston. However, in lakes where sulphur and iron bacteria occur abundantly in the hypolimnion their contribution to the biomass is considerably greater.

6. *The Problem of Production*

For the meaning of the term "production" in limnology one can refer to the presentation by Thienemann (1931) as well as to the further discussion of this question by Grote (1934) and Münster-Strøm (1932). In agreement with the definitions given in these references we shall here consider the "production" of a community to consist of the total amount of organic matter (representing the balance between assimilation and dissimilation) that is formed within a certain period of time from the raw materials supplied to the community. Transformations taking place in the organic realm (for example, nourishment of animals by plants) are not included. Nor is the reconstruction of organic matter by the immediate re-utilization of mineralization products arising in the trophogenic layer through the decomposition of plankton that have died there; this reconstruction must be considered solely as a transformation within the biocoenose—in the same way, applying this idea to the individual organism, the re-assimilation by a plant of the CO_2 arising from its own respiration is not a gain in assimilation. Of the more recent papers concerning this and similar general questions, which have been much discussed recently, there might be mentioned the study by Lindeman (1942), which is well worth reading (cf. also Ohle, 1955, and Elster, 1958).

Whereas we are able to determine the annual yield of, say, a field from the material harvested at the end of the growing season, we cannot by the same means arrive at the *annual* quantity of plankton produced beneath a unit area of lake surface. Figures obtainable refer solely to the population present at the moment of sampling, and we are unable to determine how much of the previously existing production has been removed from the free water habitat by decomposition, by sinking to the bottom, and by inflow-outflow. If we are satisfied to determine the *instantaneous* living population ("standing crop"), it is evident that the statistically derived individual counts of plankton species, which differ so greatly in size, are of no value.

Earlier attempts to determine the instantaneous living population on the basis of the "settling volume" of net catches did not prove fruitful because of the many sources of error. It therefore became necessary to employ more exact procedures, which permitted the determination of the total biomass in a unit volume of water or under a unit area of lake surface.

In this matter, too, the Wisconsin Lake Survey has carried out a representative study. Their procedure was based upon the centrifuging of a very large volume of water (up to more than 1 $m.^3$)[48] that previously had been filtered through a plankton net in order to extract the zooplankton. The filter residue and centrifugate were combined, dried, weighed, and analysed chemically. One must realize, however, that *by this procedure one obtains not only the living plankton but also the organic and inorganic tripton.* The efficiency of the method is approximately 95 per cent of all suspended materials larger than bacteria, and the centrifuge even collects from 25 to 50 per cent of the bacteria floating in the water.

The dry weight of the suspended organic matter in Wisconsin lakes, according to this method, varied from 0.23 to 12 mg. per litre. For Lake Mendota, which with respect to vertical distribution has been studied in greatest detail, a mean value of not quite 2 mg. per litre was established on the basis of numerous determinations. The "standing crop" beneath 1 hectare (without taking into account the shoreward region) varied

[48]A high-speed continuous centrifuge having annular rotating discs, similar to the type used in separating milk, was employed, with a capacity up to 10 litres per minute.

between 258 and 522 kg. dry weight organic matter; in this crop the weight of the centrifuged "nannoplankton" was five times greater than that of the "net plankton." The contribution of the nannoplankton to the assimilation accomplished can be even greater. In Lake Erken in spring this amounted to 95–98 per cent, according to Rodhe, Vollenweider, and Nauwerk (1956), whereas in autumn in the same lake the net plankton accounted for more than 50 per cent of the production.

In conjunction with the discussion of the centrifuge method widely employed by Birge and Juday (1922, 1934) there might be mentioned two other procedures used extensively in recent times, which by indirect means seek to ascertain the biomass or at least a portion of it suspended in the water. One is the calculation of organic matter from its *nitrogen content*. The total nitrogen content of a sample of water is determined before and after filtration through a good quality filter. The difference between the two values is the nitrogen content of the suspended matter. Multiplication of this figure by 6.25 yields the amount of raw protein corresponding to the nitrogen content. The dry weight of the total organic matter in the filter residue is approximated through multiplying the nitrogen value by 20. This method, like the preceding, determines the total seston, hence the tripton in addition to the plankton.

The second method applies solely to the phytoplankton and is based on the determination of the *chlorophyll content* of filter residues. It is hoped thereby to obtain a measure of the quantity of phytoplankton present in a unit volume of water or of its capacity for photosynthetic assimilation. The investigations of Rodhe (1948, 1958) have demonstrated, however, that in cultures of algae the chlorophyll content, quantity of substance, and cell multiplication are to a certain extent independent of one another, and that consequently knowledge of the chlorophyll content *alone* cannot provide a sufficient basis for investigations in production biology. Recent investigations by Wright (1960) likewise have shown that the assimilation capacity of lakes is by no means proportional to their chlorophyll content, but rather that at a high density of phytoplankton, and hence at a correspondingly high chlorophyll content, a curtailment of the rate of photosynthesis can be demonstrated, possibly through the limiting action of the CO_2 content.

Another factor to be considered is that the same species of alga produces different amounts of chlorophyll depending on whether it has been growing in strong or weak light. This can be readily demonstrated by a comparison of algae that have grown at the surface with those that have grown at some depth. Nevertheless, the determination of chlorophyll content, which has been practised for more than 20 years especially in marine investigations, is well suited for advancing our knowledge of production in waters when these data are carefully related to those obtained by other procedures. It would lead us too far afield to enter into a closer discussion of the already quite voluminous literature on this subject. One can refer instead to the detailed presentation by Gessner (II, 1959, pp. 491 and 618 ff.).

A procedure with an entirely different basis is the one recommended by Lohmann (1908): the volumes of the individual species are computed and then these figures are multiplied by the observed number of individuals (cf. p. 122) to yield the total volume of plankton per litre. Since the specific gravity of plankton is not much greater than 1.0, the total volume so obtained can also be considered the "wet weight." The results of this method are not directly comparable with those of the Wisconsin Lake Survey, even when the "wet weight" is calculated from the dry weight by use of the generally applicable assumption that plankton contains approximately 90 per cent water. The reason for this is that the "calculated volume" comprises only the *plankton*—the biomass present in a living condition at the moment of sampling—whereas by means of the centrifuge, as already explained, all suspended materials are sedimented out. Since the quantity of "tripton" can be several times greater than that of the living plankton, as we know from the investigations of Rylov (1931), an agreement of the results obtained by these two methods is not to be expected. As a matter of fact, investigations carried out in several alpine lakes have yielded calculated volumes that are disproportionately small, even considering the fact that Lake Mendota is eutrophic and the alpine lakes oligotrophic; the dry weight of the calculated volume averaged only about one-eighth of that determined in Wisconsin by the centrifuge procedure. Moreover, in the Wisconsin lakes the quantity of the centrifuge sediment (nannoplankton) was four to five times as great as that of the net plankton, whereas in the alpine lakes the quantity of nannoplankton averaged only half as much as that of the net plankton (zooplankton). The latter condition in particular indicates that these differences are the result of including the tripton in the centrifuge method.

Both methods prove useful when applied with their limitations in mind, and indeed they accomplish different tasks. The *centrifuge method* introduced by Birge and Juday yields the total content of particulate matter in the water and makes possible its chemical analysis. The *tripton* obtained along with the plankton is a factor that cannot be disregarded in considering production biology. These organic and inorganic particles, arising in part from decomposing plankton and from the littoral region, but mainly from the watershed, play an important role in metabolism. In so far as they are organic they can serve as nourishment for animals and bring about an additional production, which in part does not derive from the photosynthetic primary production of the plant life in the lake, but rather is allochthonous, based on organic matter imported from the watershed and from the atmosphere. In many waters this can be quite considerable and under certain circumstances can even exceed primary production in magnitude (cf. Steinböck, 1958). In addition the organic matter of the tripton can appreciably influence the metabolism especially of small lakes by its decomposition. In contrast, the *calculated volume method* yields only the mass of living plankton and the extent to which the individual species contribute to its composition.

In the lakes of the northerly limestone Alps the wet weight (volume) of total plankton beneath 1 hectare amounts to from 120 to 600 kg., averaging 300 kg. If one now investigates which of the numerous species that may at times be present make up the major portion of this biomass, one finds as a rule both among the plants and among the animals that only one or a few species clearly predominate. Indeed the major portion of the plants is almost always formed by a few members of the nannoplankton (mostly by the minute diatom *Cyclotella comensis* and the flagellate *Rhodomonas lacustris* in the alpine lakes), whereas the noticeable, large species that dominate the composition of the net plankton make up much less of the weight. Among the animals, on the other hand, the large species themselves are most important, notably the entomostraca and especially *Daphnia*. The ratio of the total volume of plants to animals is subject to great fluctuations. On the average (in alpine lakes in summer) the biomass of animals has been found to be twice as great as that of the plants, although in individual lakes the phytoplankton predominated (2 to 1). On the other hand instances were observed in which the mass of the animals was 15 times greater than that of the plants. This disparity is explained by the fact that the dependence of the zooplankton on the phytoplankton assumes the character of a succession. In these instances the animals had probably "eaten the lakes empty" and were living on the fat stored in their bodies. In interpreting the significance to production biology of the phytoplankton:zooplankton ratio it is important to know if the individual members of the production chain are directly related to one another. Thus, we know that precisely the dominant forms of the zooplankton—the rotifers (with few exceptions), *Daphnia*, *Bosmina*, and the diaptomids—are completely dependent on the nannoplankton and cannot ingest the larger forms.

When the volumes of plankton beneath a *unit surface area* of diverse lakes are compared with one another, the conclusion is reached, which initially is surprising, that the quantities present do not always bear the expected relationship to the trophic level of the lakes as deduced from other properties. Thus Findenegg (1942) was able to demonstrate that in the markedly oligotrophic lakes of Carinthia the mean total quantity of biomass present during the course of a year was scarcely less than that in lakes of a more eutrophic character, and indeed was even greater in certain instances. A comparison of the biomasses in the tropical lakes of the Sunda Islands led to the same conclusion, and likewise the investigations of Riley (1940) and Deevey (1940) on lakes in Connecticut.

On the other hand if we compare the biomasses contained in a *unit volume* of the trophogenic zone rather than those present beneath a unit surface area, the picture is changed: the content per litre and the trophic level are in good agreement. This apparent contradiction is immediately clarified by a comparison of the thickness of the productive zones. In eutrophic lakes the zone of production is not as thick as in oligotrophic lakes, for the reason that in the former the large population density in the upper layers shades out much of the penetrating radiation, and as a result the lower boundary of effective assimilation, the compensation level, lies at a lesser depth than in oligotrophic lakes.

From these conditions it can be seen that two different things must be distinguished in considering production biology: (1) the momentarily existing plankton production of the *lake*, characterized by the biomass present beneath a *unit surface area*, and (2) the fertility of the *water*, which is indicated by the average biomass of a *unit volume* of water within the zone of production. In the quotient V/O, which is a constant for any particular lake, Rodhe (1958) has related the two quantities assimilation capacity beneath 1 $m.^2$ surface area (O) and optimum assimilation in 1 $m.^3$ of water (V) to one another. The fertility of a body of water is to a certain extent a potential quantity, controlled by the content of nutrient substances limiting production (or by the extent of their addition); their conversion into organic matter occurs in proportion to the illumination and temperature prevailing at the individual depths. The energy reaching the surface, which in the final analysis is necessary for any production, can in general be considered approximately the same for all lakes in the temperate latitude. Differences in the energy factor, therefore, are to be sought entirely in the different transparencies of the lakes, and these, as we have seen, depend first on the particular absorption (coloration) of the water and second on the scattering of the radiation by suspensoids. Considering first of all the absorption loss by itself, waters coloured with humic materials have a very thin assimilation layer under conditions of stable stratification because the compensation level occurs at a slight depth (cf. Helmet Lake in the table on p. 139). Within this layer, to the extent that the water is fertile, very considerable biomasses per litre can be produced, which on the basis of production related to surface area, however, are correspondingly low. (This is one of the explanations of the low production of dystrophic lakes; cf. p. 226.)

But even in a lake in which the water *by itself* shows a high transparency, a compression of the zone of production is brought about on the basis of the quantity of nutrient materials present, as mentioned above. This is effected not through the absorption characteristics of the water but rather through the shading effect of the plankton population or through the scattering of radiation by the organisms suspended in the water. If the content of nutrients decreases, then the population density also becomes smaller, permitting light to penetrate deeper and thereby increasing in thickness the zone of production. Hence, even though the content per litre decreases, the biomass present beneath a unit area of surface can remain the same; or, in other words, the *decrease in population density is compensated for by a corresponding increase in thickness of the zone of production.*

We see, therefore, that the *population density* of the assimiliation layer depends on the fertility, that is on the *trophic level* of the *water.* The *volume of plankton* beneath a unit of surface area, and hence the biomass of the entire pelagial region, is a function of the *transparency. This holds true, however, only within certain limits*: if the available nutrients decline still further and the population density becomes so small that its effect on the transparency of the water is of less importance than that of the absorption by the water itself, then a decrease in the plankton content can scarcely increase further the thickness of the zone of assimilation. In such a case not only the population density but also the biomass beneath a unit of surface area is limited by the nutrients available, a condition realized in strongly oligotrophic lakes.

In the discussion up to this point we have not considered the influence of *turbulent water movements* on the photosynthetic capacity of a lake. In a mass of water mixed by turbulence, all the phytoplankters enjoy the same average conditions of light, as explained on page 140. As soon as the layers lying below the compensation level are involved in the circulation, each individual phytoplankter, having been carried into the depths by eddy diffusion currents, necessarily exhibits a negative assimilation balance so long as it remains under conditions of insufficient light. With further advancement of the mixing into deep water, these periods of negative balance become progressively longer, and this must lead to a deterioration of the total assimilation balance and to a decline of the plankton population density (cf. p. 140, Steemann Nielsen,

1939). The frequently observed decline of the phytoplankton at the time of complete circulation might be referable to this condition.

The biomasses (standing crops) in a lake and deductions concerning their activities do not furnish a reliable basis for judging the *annual production* of a lake. This is a problem of over-all complexity, dependent upon a great many factors difficult to measure and varying from one case to another. It is scarcely possible to obtain such pertinent data as the life span of the various species, the rate of reproduction, the rapidity of decomposition, the loss through sinking out of the trophogenic layer and through outflow, etc., all of which are necessary for calculating the total annual production on the basis of the biomasses alone, even when the latter are measured at short intervals of time during the course of the year. Hence, at the present state of our knowledge, although all such investigations based on biomass alone will certainly enable comparison of different bodies of water, quantitatively they will continue to be only approximations.[49]

Recently attention has been hopefully centred on another procedure, which selects as its point of departure in considerations of production biology not the biomass present but rather the assimilation accomplished by the phytoplankton within a unit of time, hence the *primary production.* In plant physiology the amount of photosynthesis accomplished by autotrophic plants can often be determined without difficulty from the quantity of carbon dioxide taken up or of oxygen given off. For aquatic plants the application of this procedure is limited by the sensitivity of the analytical methods—in the case of *oxygen* given off in photosynthesis, by the Winkler method for the determination of dissolved oxygen. If one wishes to use this procedure for measuring the photosynthesis accomplished by the phytoplankton contained in a given

[49]A very noteworthy attempt to determine *directly* the annual production of at least several components of the phytoplankton is the procedure employed by Grim (1950) in Schleinsee to recover in submerged vessels the shells of diatoms sinking out of the zone of production. The total production in the diatoms studied was from 8 to 10 times greater than the highest concentration of cells of the particular species beneath a unit surface area at the time of the developmental maximum. In this manner Grim could determine even the daily loss through sinking out of the zone of production, and, by continuous observation of the population densities present, he was able to arrive at reliable conclusions concerning the intensity of reproduction of the individual species. This procedure, however, exhibits considerable sources of error in lakes with a large inflow-outflow, as shown by the investigations on Lunzer Untersee (p. 196).

Through detailed consideration not only of photosynthesis but also of respiration, mortality, and predation among the individual members of the entire biocoenose, Teal (1957) recently has attempted to comprehend the annual production of a small spring-fed pool.

volume of water under natural conditions, one can employ an experimental arrangement similar to that already described on page 69, with the difference that rather than placing submersed plants or suspensions of algae in the bottles, water samples with their natural plankton populations obtained from various depths are used as the experimental objects. In this procedure each water sample, after being raised to the surface, is distributed to two bottles of 100 to 200 ml. capacity, which are then returned to the same depth from which the water sample was obtained. One of the bottles, which is enclosed by a light-tight covering, serves for the determination of the O_2 utilized by the respiration of the phyto- and zoo-plankton and by the bacterial breakdown of organic substances. The clear glass bottle is used for ascertaining the assimilation surplus. This method has yielded significant results particularly in eutrophic waters, hence in those rich in phytoplankton (e.g. Nygaard, 1955). But even in the plankton-poor lakes of the Alps the method has proved usable with certain limitations, as the detailed investigations of Vollenweider (1956) in Lago Maggiore have demonstrated. Nevertheless, Vollenweider does not believe this procedure ought to be recommended for small amounts of photosynthesis, because the Winkler method of O_2 determination is not sensitive enough in such instances.

More recently, however, an isotope method has been introduced, which enables us to determine the assimilation accomplished with much greater precision. This is based on "marking" the photosynthate with *radioactive carbon* C^{14}, a method Steemann Nielsen (1952) devised and applied during extensive investigations in the ocean (*Galathea* Expedition) and in a number of freshwater lakes.

The exposing of water samples at the same depths from which they were obtained is accomplished in the manner described above. Previous to this a known small amount (in relation to the total C content of the water sample) of C^{14} in the form of bicarbonate is added to each bottle, whose content of free CO_2 and bicarbonate C has been determined. It is known that the bicarbonate of C^{14} is assimilated at almost the same rate as that of C^{12}. Hence, if the amount of C^{14} assimilated is determined, one can calculate the total assimilation from the ratio $C^{12}:C^{14}$ in the water. After a suitable period of exposure (see p. 170 for a discussion concerning length of exposure) the samples are raised and filtered through a membrane filter with a pore size of 0.5 μ, which retains the total plankton. The assimilated carbon, both C^{14} and C^{12}, is contained in the cells of the filtered phytoplankton. After the sample has been dried the amount of C^{14} that has been fixed can be measured with a suitable Geiger counter, and then the total amount of C assimilated can be calculated. Since the half-life of C^{14} amounts to about 5000 years, one does not need to hurry in processing the dried samples.

The basic assumptions enabling the calculation of the assimilation surplus or *net production* by the C^{14} procedure differs somewhat from those of the oxygen method. The O_2 excess at the end of an experiment compared with the O_2 content at the beginning gives an immediate measure of the assimilation gain. Only the net O_2 increase is evidenced, since the total amount of O_2 set free by the assimilation of the plankton algae contained in the sample bottle suspended at depth has already been reduced by concurrent oxygen utilization through respiration of the algae and the zooplankton as well as through the bacterial breakdown of dissolved organic matter.[50] If one wishes to determine the *gross production* for the period of exposure he must add to the oxygen excess in the light bottle the amount of oxygen consumed in the dark bottle.

The C^{14} method does not measure the oxygen given off in assimilation or used up in respiration, but rather the *carbon* incorporated in the photosynthate. Hence, it is important to know whether the respiration of the plant cell is based on the carbon presently being assimilated, part of which is labelled with C^{14}, or whether older carbohydrate deposits are being drawn upon for this purpose.[51] In the former instance the quantities of assimilated labelled C will be reduced, whereas in the latter instance they will remain unchanged. In other words, C^{14} determination would give net production in the first instance (related to plant cells) and gross production in the second.

Steemann Nielsen (1955), relying in part on the older investigations of van Norman and Brown (1952), has undertaken an experimental proof of this, with the results that "the direct measurements lie in between gross and net production" (Steemann Nielsen, 1959). If one assumes with this author that "the respiration intensity amounts to 10% of the optimal intensity of photosynthesis," then one must employ a correction of plus or minus 5 per cent, according to whether one wishes to measure the gross or net production in relation to cell metabolism during the period of exposure. This correction is relatively minor, and in many instances can be neglected (Steemann Nielsen, 1960). Not included in this correction are the losses of bound organic C arising

[50]For the calculation of the CO_2 assimilated from the O_2 given up, according to Ryther (1956) one should not use the theoretical value of 1, but rather the mean value of 1.25 for the ratio O_2:CO_2 (O_2 given off to CO_2 taken up).

[51]It should be mentioned that under certain circumstances the exchange of C^{14} and C^{12} between the cell and the surrounding medium, which proceeds even in the dark (the so-called "night assimilation"), can play a significant role (Steemann Nielsen, 1955).

through the respiration of animals and through bacterial decomposition, the inclusion of which gives the true net production of the investigated water samples as biotopes ("community production" according to Steemann Nielsen, 1960). These losses can attain a high value especially in waters with a considerable content of organic matter. In these instances it is recommended that consumption rates be determined by the oxygen method. Moreover, for the determination of the magnitude of respiration during the night one can scarcely rely on approximations, since among other reasons it is known that respiration is most intense immediately after sunset during the first hours of night and later falls off considerably.

Although the C^{14} method still exhibits a variety of difficulties and although we have not yet reached the state where we can determine by its use alone the annual production of a body of water, nevertheless it still offers an unusual advantage, and in the short time of its application to numerous inland waters it has already led to significant advances in our understanding, as is apparent from the works of Steemann Nielsen, Rodhe, Vollenweider, Goldman, and others. For our discussion it will suffice to present several typical examples of the magnitude and vertical distribution of primary production in a variety of lakes.[52]

Figure 50 shows first of all the vertical distribution of the intensity of production in Ossiacher See (Carinthia) on a sunny summer day. It is apparent that at the strong light intensity at this time the assimilation maximum, in agreement with the *Chlorella* experiments of Schomer (p. 139), is not at the surface but rather at a depth of 2 m. Below this pronounced maximum, production decreases in rough proportion to the decreasing light intensity, and the compensation point is reached at approximately 10 m. In the autumn curve for Wörthersee (Carinthia) there is a very weak maximum in the epilimnion, but there is in addition a very pronounced maximum at a depth of 10 m., where at this time in the metalimnion there is a very sharply stratified maximum of *Oscillatoria rubescens* (Figure 45, p. 141). This assimilation maximum occurring at such a considerable depth is very remarkable; subsequent investigations on the physiological relationships of this red-coloured blue-green alga will have to decide whether or not this is a case of *chromatic adaptation* (p. 190). The meromictic Krottensee harbours in its O_2-free depths an oligoaerobic biocoenose of bacteria capable of chemosynthesis; the increase in the assimilation curve below

[52]Herr Professor Findenegg has graciously made available from his rich store of as yet unpublished observations the curves for Lunzer Untersee, Ossiacher See, Wörthersee, and Krottensee, for which I wish to thank him heartily at this time.

the photosynthetic compensation point might well be caused by these organisms. Finally, Lunzer Untersee shows a summer distribution of assimilation that is typical for the oligotrophic lakes of the Alps—a very weak epilimnetic maximum and a production extending deep into the lake, gradually declining with depth. In spring the assimilation accomplished in Lunzer Untersee is significantly greater than that in the summer condition shown in Figure 50 because of a maximum of diatoms.

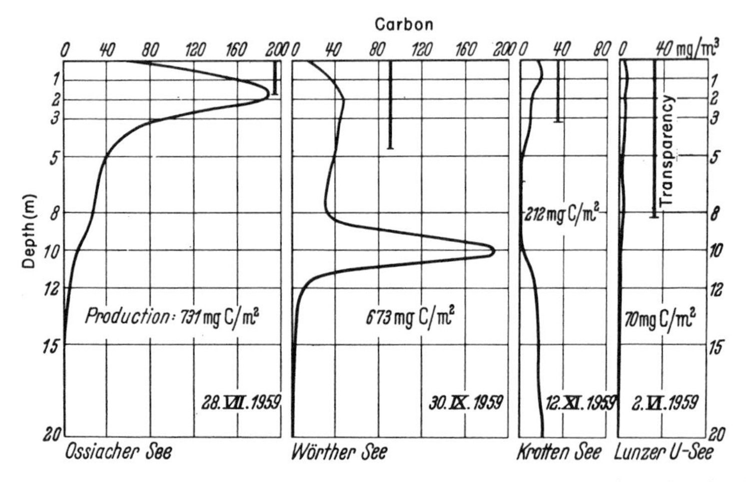

FIGURE 50. Primary production in several alpine lakes determined by the C^{14} method. From Findenegg.

Figure 51 illustrates the measurements of one day from the extensive investigations of Vollenweider on Lake Erken, according to the publication of Rodhe (1958). In addition to the curve for total radiation the figure shows the results of short observations approximately 4 hours long, which followed one another directly, and a comparison of their summation curve with a single 19-hour exposure covering the same period of time. From the short observations it is clearly apparent that the decline of the curves towards the surface is controlled by light, because during the morning and evening series the maximum lies at the surface. At the same time with decreasing light intensity the compensation point clearly moves upward. It is noteworthy that the summation of the five short observations gives a considerably larger value than the concurrent long observation lasting 19 hours. The origin of this phenomenon is still unexplained. One can presume that at a longer exposure time a sedimentation of the phytoplankton (because of a lesser

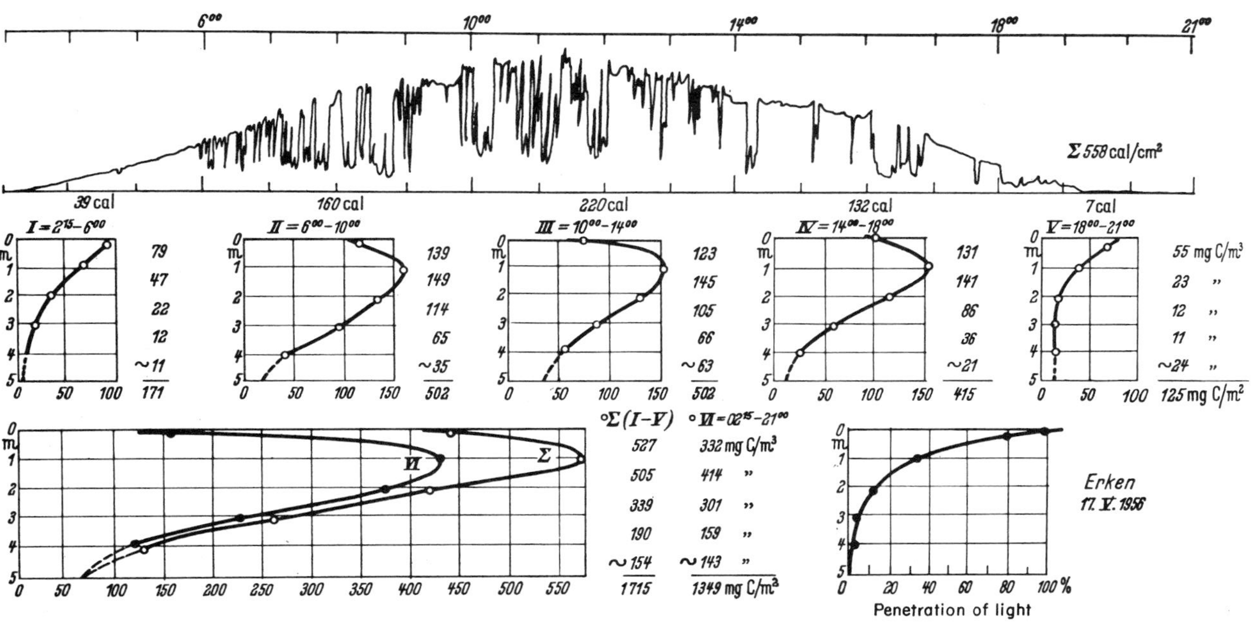

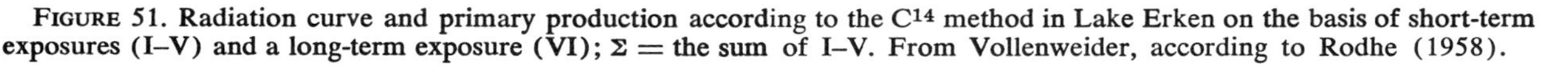

FIGURE 51. Radiation curve and primary production according to the C^{14} method in Lake Erken on the basis of short-term exposures (I–V) and a long-term exposure (VI); Σ = the sum of I–V. From Vollenweider, according to Rodhe (1958).

turbulence within the water sample bottle) can adversely affect the assimilation. At any rate freely floating algae exposed to turbulence live under more favourable conditions than those in bottles, and hence in order to minimize these sources of error short exposure times are preferred.

In spite of these very promising results the C^{14} method still needs a many facetted testing on numerous waters under the most varied conditions along with simultaneous comparison with the O_2 method in order to reach an understanding of the many aspects of the method that are not yet explained, and in order to develop the procedure into a completely reliable tool for the quantitative measurement of primary production.

Other attempts have been made to obtain usable approximations of production by indirect means, at least for a certain portion of the year (from the beginning to the end of summer stagnation). These are based on measurable *changes produced in the dissolved materials in the water as a result of organismic production*, which have already been discussed in detail in Part A-II. Changes in the trophogenic layer (for example, biogenic decalcification) are less suitable for this purpose (because of the impossibility of excluding atmospheric influences) than those in the tropholytic layer. Changes in the latter include the *decrease of oxygen* and the *increase of carbon dioxide* or (with certain limitations) of ammonia.[53] The quantity of organic matter formed during a certain period of time within the zone of production is computed from the accumulation of its decomposition products in the hypolimnion or from the utilization of oxygen in its oxidation. In order to arrive at the total production within the period of time under consideration there must be included with the above value obtained indirectly by calculation, the standing crop present on the sampling days and the undecomposed sediment remaining on the lake bottom. This latter quantity is difficult to measure and can only be approximated.

Of these procedures the most usable, according to Einsele (1941), is the one based on the *CO_2 increase* (including both the free CO_2 as well as that held in the bicarbonate form), which is applicable even in those instances where oxygen is completely lacking in the hypolimnion. It yields the largest and presumably the most accurate values when the fact is taken into consideration that under conditions of anaerobic de-

[53]If the nitrogen cycle is to be used in such calculations the possibility of denitrification to elemental nitrogen under certain conditions must be reckoned with. The accumulation of phosphate in the tropholytic layer does not lend itself to calculations of biological production because of the involvement of phosphate in the iron cycle (p. 90) and because of its storage in microorganisms.

composition beginning with the complete lack of dissolved oxygen, only half of the carbon is oxidized to CO_2. The detailed application of this procedure has been established by Ohle (1952) in a comprehensive work.

The hypolimnetic *decrease in oxygen*, which indeed has already led to the formulation of the concepts of oligotrophy and eutrophy so important for limnology, has been used, since Thienemann's (1928) fundamental investigations, as a measure of the total production during a period of stagnation. The theoretical assumptions of this method, the factors that must be taken into consideration and the regularities that the changing proportions of planktonic (that is, occurring during sinking) and post-sedimentary decomposition exhibit in different lakes, depending on the morphology of their basins, have been the subject of numerous investigations, of which particularly those of Thienemann (1928), Grote (1934), Alsterberg (1935), and Hutchinson (1938) might be referred to. If one wishes to calculate the quantity of organic matter oxidized from the amount of oxygen used, he can assume as Einsele (1941) has done that with an average composition of the plankton (60 per cent carbohydrate, 35 per cent protein, and 5 per cent fat) 1.5 g. of oxygen will be required for the complete decomposition of 1 g. dry weight organic matter.

All these procedures contain sources of error and are applicable only under especially favourable conditions. Otherwise they can yield only approximate values, generally speaking. Above all, the effect of the material exchange between the trophogenic and tropholytic layers, which is dependent upon the position of the thermocline and the stability of stratification, can scarcely be determined. In addition there is the loss of organic matter through the outflow of the lake, and, on the other hand, the condition where a portion of the material decomposed in the tropholytic layer has not resulted from production within the lake, but rather has been brought in from outside as allochthonous matter. The difficulty in determining the amount of material remaining undecomposed on the lake bottom has already been pointed out.

On the basis of the CO_2 accumulation in the hypolimnion Einsele (1941) calculated a total production of 5,600 kg. dry weight organic matter (= 380 kg. per hectare) in Schleinsee during the period April to September 1937. Schleinsee is especially suited for such studies because of its very small inflow-outflow. The suspended biomass, on the basis of numerous observations during this five-month period, averaged 1,350 kg., hence approximately one-fourth of the total production. The ratio of these two figures calculated for *one* month yields the "turnover coefficient" $5600/(1350 \times 5) = 0.83$, a value of undoubted importance in production biology, which indicates by how much the total production has increased each month. It is obvious that this coefficient is subject to great variations with respect to time, and from one lake to another.

These indications will probably suffice to point out the difficulties in determining accurately the total production of plankton in a lake and to explain why our knowledge in this field is only just beginning.

Of great interest both theoretically and practically is the question, much discussed recently (Einsele, 1941; Hasler and Einsele, 1948), of how a lake reacts to changes in its nutrient spectrum, how its production can be affected by the addition of nutrients (*fertilization*) essential in the synthesis of living matter. It has already been reported (p. 101) that in the fertilization experiments in the eutrophic, *nitrate-free* Schleinsee the phosphate that was added disappeared from the water through accumulation in the phytoplankton, but it did not lead immediately to an increase in production because in this case the nitrogen content was the limiting factor. Increased production began after a certain *latent period* only when the nitrogen deficiency had been eliminated as a consequence of the phosphate stimulating the fixation of free nitrogen (by bacteria and blue-green algae). A year after fertilization, conditions in the lake were scarcely distinguishable from those before fertilization; all the phosphorus (because of its unfortunate involvement in the iron cycle, p. 91) had been precipitated from the lake and incorporated into the sediments. Einsele's fertilization experiments in the oligotrophic, *nitrate-rich* Nussensee (Salzkammergut in Austria) yielded different results. In this instance the dose of phosphate *immediately* set off a mass development of algal vegetations following one another in rapid succession, and at the same time exhausted the supply of nitrate present. Yet even here in the further course of the experiment a nitrogen fixation could be demonstrated resulting from the phosphorus fertilization. Associated with these reactions there occurred in the lake, as expected, a significant decrease in the hypolimnetic oxygen content, and there could then be demonstrated in the deep water a previously unobserved occurrence of iron and manganese. In Bare Lake, Alaska, Nelson and Edmondson (1955) obtained an increase in photosynthesis up to 7 times the initial value within 10 days after fertilizing with nitrogen and phosphorus together. The zooplankton likewise showed an increased rate of multiplication, and the fish population exhibited a greater growth in the years following fertilization.

Hutchinson and Bowen (1950) used radioactive phosphorus P^{32} to follow in detail the fate of quantities of phosphorus added to a small lake (Linsley Pond), its sedimentation by means of seston as well as its return to the water from the sediment, and its distribution.

Investigations of this sort give hope for deep insight into the material mechanism of our lakes, entirely apart from their significance for practical fisheries. They provide a basis for anticipating the consequences that might be associated with the fertilization of lakes, and at the same time they point out the dangers that can result from interference with the metabolism of a body of water when such procedure is not sufficiently understood in theory (for example, restriction of the inhabitable layer through oxygen depletion, enrichment of iron in the hypolimnion and its detrimental consequence for the phosphorus cycle).

II. SURVEY OF THE OTHER COMMUNITIES OF LAKES

Within the limits of this presentation the plankton was reviewed in detail as an especially suitable example of aquatic life limited by its environment. In the following portrayal of the other biocoenoses of a lake we shall limit our consideration to the most important features.

Whereas the environment of the plankton exhibits a *unity* and is characterized by an extreme orderliness in the change of physical and chemical properties scarcely known elsewhere in nature, all other communities of the water exist at the interface between *two* basically different media and depend more or less both on the influence of the open water and on the influences, difficult to control, of the often nonhomogeneous lake bottom or of its varied types of covering. The causal investigation of the dependent relationships in these biocoenoses is therefore incomparably more difficult than in the plankton.

1. *Origin of Lakes and Arrangement of Depth*

Since the substrate now enters as a new factor in the compass of our considerations, it might be opportune to insert a few words about the *origin* and *morphology of lakes.*

The water-filled depressions in the earth's surface that we designate as lakes have arisen in very different ways. First of all we can distinguish the *tectonic lakes*, which are formed primarily by the mountain-producing forces of the earth. To this uncommon type belong several of the largest, deepest, and oldest lakes on earth, for example Tanganyika (area 35,000 square km., depth 1435 m.) in the Rift valley of Africa, and Baikal (area 33,000 square km., depth 1522 m.) in Siberia. Much more abundant are the basins formed secondarily by *deposition* of loose materials or by *excavation*. To the former class belong the *kettle lakes* (*Wallseen*), for example, the crater lakes of extinct volcanoes; the *ground moraine lakes* of northern Germany; and the *dam lakes*, especially abundant in the Alps, which have arisen as a result of the damming of a side valley by the moraine of a Pleistocene glacier flowing in the main valley, or by a mountain slide. Excavation basins may have originated through the caving in of caverns in the earth's crust; or they have been produced by the individual explosions of volcanic gas chambers, forming the small, usually circular and often very deep *Maare lakes*; or they owe their ultimate origin to erosional forces working from above (*removal basins*). Of the removal basins the most common type among the alpine valley lakes and gorge lakes is that produced by glacial erosion, which can readily form these excavated basins, especially where there are differences in the hardness of the contiguous rocks. Yet there are persons who maintain that glacial erosion by itself is not sufficient to explain the formation of the deep alpine lake basins, and that tectonic movements must have assisted.

As already explained in the introduction, inland lakes cover only a relatively small portion of the earth's surface—only about 1.8 per cent or 2.5 million $km.^2$ according to present estimates. The number of lakes, however, in several countries (e.g. Sweden and Finland) is very large, and hence the dimensions of most lakes are necessarily relatively small. Thus in Europe, for example, lakes with areas of more than 1000 or even 100 $km.^2$ are exceptional. In Figure 52 a number of the large inland lakes of the world have been assembled together at a uniform scale, in which the largest—the Black Sea—should be interpreted rather as a part of the oceans. If one disregards the Caspian Sea—a salt lake with a surface area of 438,000 $km.^2$—then the St. Lawrence Great Lakes of North America with an area of 242,000 $km.^2$ and a volume of almost 35,000 $km.^3$ constitute the greatest continuous mass of fresh water on earth. By comparison Lake Geneva (= Genfer See, 582 $km.^2$) and Lake Constance (= Bodensee, 538 $km.^2$), which are the largest lakes of Central Europe, are almost insignificant. Many of the lakes that have been investigated intensively limnologically are relatively small: Lake Mendota, 39 $km.^2$; Grosser Plöner See, 30 $km.^2$; Wörther See, 19 $km.^2$; and "tiny" Lunzer Untersee, 0.68 $km.^2$ The greatest depths of the inland lakes do not begin to approach those of the oceans, and yet in the examples given above for Baikal and Tanganyika they are still very considerable.

The basins originating in the ways described above do not remain unaltered after being filled with water. Erosion by waves sets in and

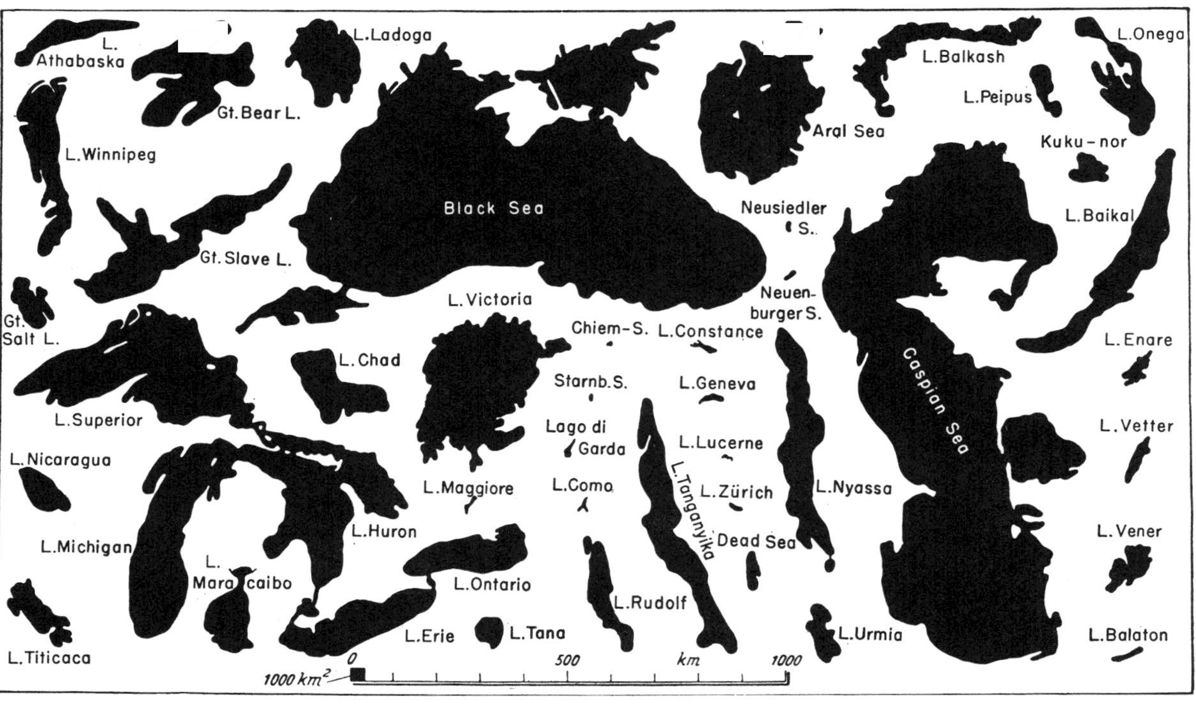

FIGURE 52. The surface areas of a number of the large inland waters of the world, all drawn to the same scale. From the journal *Die Umschau*, vol. 60, p. 23 (1960).

modifies the original shore profile. In addition the rock formations making up the basin become covered over usually to a great thickness with lake sediments generally designated as "ooze" (*Schlamm*), and thereby the depth of water progressively decreases. The picture that the shore profile modified in this manner exhibits in many places, particularly in the alpine lakes, is shown somewhat diagrammatically in Figure 53. On the outermost shore the wave action has created a steep "cliff" and an *erosional terrace*, which is usually quite narrow. To this is joined

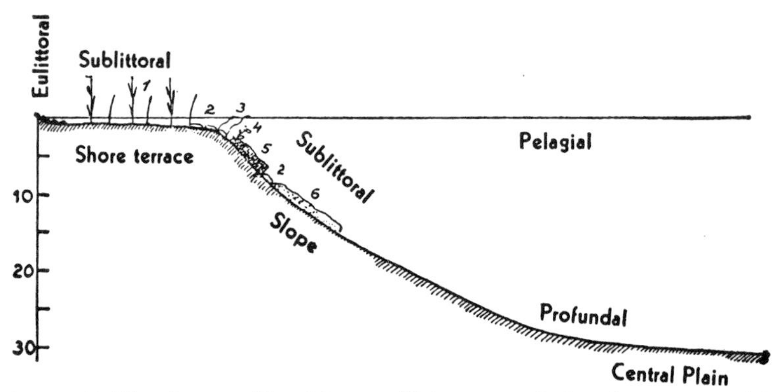

FIGURE 53. The shore profile of Lunzer Untersee; vertical scale exaggerated 2×. 1, quaking beds of *Phragmites* and *Scirpus*; 2, *Chara*; 3, *Potamogeton natans*; 4, *Potamogeton praelongus*; 5, *Elodea*; 6, *Fontinalis.*

the *depositional terrace* (known as *Wysse* at Lake Constance) consisting of lake sediments the level surface of which is controlled by wave movements. This surface extends lakeward often for 100 metres or more, whereupon it is suddenly interrupted by a steeply descending *slope* (*Halde*). The slope then gradually changes into the *central plain* (*Schweb*), the region of level deep-water sediments occupying the middle of the lake.

The subdivision of the habitat is based on this morphometrical organization: the open water portion, the *pelagial*, which is bounded by the slope, central plain, and lake surface, is the habitat of the *plankton*; the shore terrace, slope, and central plain themselves are inhabited by the biocoenoses of the *benthal.* The separation of the biotopes within the latter region is subject to different interpretations. In accordance with physiological principle we shall regard the *littoral* as that portion of the shoreward profile inhabited by autotrophic plants. This corresponds to the trophogenic layer of the pelagial; hence, its lower boundary is likewise the limit of a positive assimilation balance (over a considerable period of time).

Within the littoral the water level relationships provide a basis for further classification: the region of fluctuating water level (hence, changing moisture conditions) between the high and low water marks, in which at the same time the beating of the waves is effective, is designated the *eulittoral*, and the remaining, much more extensive portion, the *sub-littoral*. At a greater depth there follows the region of the *profundal*, poor in light and inhabited entirely by heterotrophic organisms (if we disregard the chemo-autotrophic bacteria that may be present). It corresponds to the tropholytic layer of the open water.

2. *The Shore Flora*

Before we proceed to characterize briefly the plant and animal life of the benthal region, it seems appropriate to consider first a description of the higher aquatic vegetation—the *marsh and water plants* of common usage—belonging mainly to the phanerogams and in lesser degree to the pteridophytes and mosses. These plants occupy a unique position in a number of respects. First of all, because of their size and the extent of their beds, they have become part of the habitat of the aquatic microflora and microfauna, and thereby have become the bearers of a biocoenotic sub-organization of the littoral. But above all, they stand completely apart from other aquatic plant life on the basis of their phylogeny and consequently of their morphology, and frequently of their mode of life. Whereas the representatives of the microflora and also the larger algae are true children of the water, that is, they have never left this medium in the course of their phylogeny, the higher aquatic plants are fugitives from the land. In the course of evolution their ancestors came out of the water and were transformed into aerial organisms, and only subsequent to this adaptation, with its associated far-reaching modifications, have individual plastic members of this line of descent again returned to the water. With *re-adaptation*, however, the characteristics of aerial life have been largely preserved. These forms, therefore, are *outposts of the land flora in the aquatic habitat.*

But precisely for these reasons the study of the higher aquatic vegetation turns out to be especially instructive, since here the control of morphological characteristics by aquatic conditions is more clearly apparent than usual. Indeed this is especially true because many of the forms concerned are amphibious, that is species in which a *single* individual is able to form both aerial and aquatic processes. Adaptations to life under water encountered in the morphology and anatomy of these plants can be summarized in the following principal features. The cuticle and all other modifications for curtailing transpiration are reduced, since they are superfluous in a medium of saturated humidity. The floating organs carried upward by the current are not affected by bending but only by pulling; for this reason the vascular bundles and supporting elements are not arranged peripherally as in the stems of land plants, but rather centrally as in their roots (cf. Figure 54). The gases serving in respiration and assimilation are absorbed in dissolved form through the epidermal cells of the leaves; consequently stomata are lacking on the submersed portions, and in the floating leaves are shifted to the upper surface. Aerial leaves tend to have a reduction in their outer surface to prevent an injurious loss of water (in so far as this is possible without impairing the uptake of light), and in compensation the inner surface that functions in gaseous exchange (respiration chambers, spongy parenchyma) is markedly increased. On the other hand, the thin aquatic leaves consisting of from only two to several layers of cells are often very finely dissected in order to facilitate the exchange of materials (compare, for example, the aerial and aquatic leaves of the water crowfoot *Ranunculus*

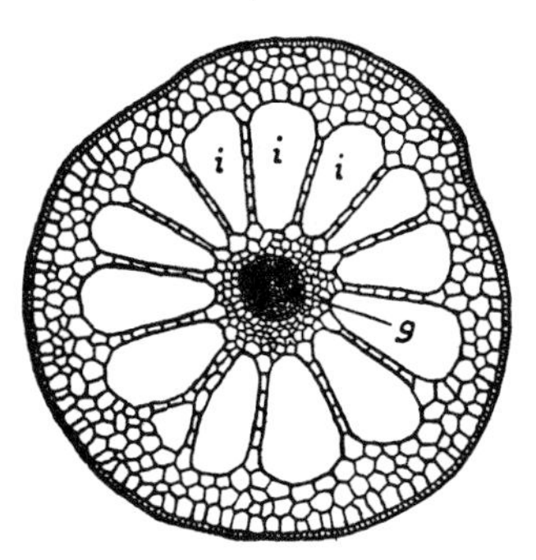

FIGURE 54. Cross section through a stem of *Myriophyllum spicatum*: *g*, core of centrally arranged vascular bundles; *i*, air canals. From Vöchting.

aquatilis). The obtaining of oxygen is much more difficult in water than in air, first because of the smaller content of oxygen in the former medium and second because of the lesser mobility of water. Aquatic vegetation obviates this disadvantage through the development of a very large, gas-filled, intercellular system, which dominates the entire anatomy. These air canals, which also accumulate the oxygen of assimilation, extend through the entire plant and conduct gas for respiration particularly to the roots, which extend into the oxygen-free ooze. Specialized organs for the performance of ventilation occur in many plants, such as the often extensively developed aerenchyma in our willow weed (*Lythrum salicaria*) and the pneumatophores of tropical aquatic plants, for example, the genus *Jussieua* and many arborescent species of mangrove. Through the aeration system more than anything else aquatic plants disclose their origin; by means of it they provide to a certain extent an internal aerial habitat.[54]

[54]One can convince onself of the high efficiency of this aeration system by an interesting experiment. If in a flat dish containing 2 to 3 cm. of water an isolated

Proceeding lakeward from the shore, one encounters all degrees and types of the adaptations described. Many representatives of the meadow and forest plant communities push their way close to the lake shore and into the region of well-moistened ground without belonging to the aquatic vegetation in their morphological features. Not until the substrate is continually flooded, or at least continually saturated, do the *marsh plants* (helophytes) occur. These plants are distinguished from land plants in their life habits only by the fact that their roots, growing in ooze free of oxygen, require an increased aeration. As a consequence the intercellular system is well developed in them, but the supporting elements are still arranged peripherally (for example, in the stems of *Scirpus lacustris*) because of the requirements for bending. The horizontal composition of this helophytic community is determined by the increasing depth of the water. The forms that extend farthest into our lakes are the largest ones—the reed (*Scirpus*) and the rush (*Phragmites lacustris*). Seidel (1955, in Thienemann, *Die Binnengewässer*, vol. 21) has recently considered in great detail the biology and cultural significance of these plants. Nevertheless, marsh plants can seldom advance beyond a water depth of 1 metre. They are replaced by a type having flexible firmly attached stems with leaves floating on the surface of the water, the *floating-leaf plants*, which in our region are represented by the white and the yellow water lilies and by the floating pond weeds. But on the basis of their metabolism they are still aerial plants, as indicated by the shifting of the stomata to the upper surface of the leaves. At a water depth of about 3 metres the critical point is reached. In order to develop in deeper water, a plant has only two possibilities, either to withdraw its roots from the bottom and become freely floating

---

water lily leaf is floated without wetting the upper surface, and the dish then is slowly warmed with a flame, at first scattered bubbles arise from the cut surface of the petiole, but soon a continuous effervescing stream of air (100 cc. or more per minute) flows out by the hour. The stream of air immediately comes to an end when the dish is loosely covered with a plate of glass, a small board, or something similar, and begins again when the dish is uncovered. This indicates that this bewildering phenomenon, which has been known for a long time but is still by no means completely explained, is associated with transpiration from the upper surface of the leaf. The outer air evidently flows into the leaf through the stomata faster than the inner air escapes, so that there develops within the leaf an excess pressure, which can exceed that of a column of water 8 to 10 cm. high. Similarly constructed leaves of other genera, such as *Limnanthemum*, exhibit the same phenomenon. Hence, particularly on warm days, air under considerable pressure is forced from the upper surface of the leaves into the intercellular system of the plants. The attempts up to now to explain the significance of this phenomenon are summarized by Gessner (II, 1959). Berger has carried out an experimental investigation (as yet unpublished) of the phenomenon, which gives a plausible explanation.

in the water, or to submerge its organs of assimilation and thereby assume a completely aquatic life habit. Both methods have evolved. The *floating plants* (for example the frog-bit, *Hydrocharis*, and the duck weed, *Lemna*) have detached themselves from the bottom and float about on the surface of the water as littoral "pleuston." The *submersed plants*, among which are *Elodea* (*Anacharis*), *Ceratophyllum*, *Myriophyllum*, and most species of *Potamogeton*, cover the slope down to a depth of from 6 to 7 metres, where decreasing light intensity evidently becomes a limiting factor.

That the lower boundary of aquatic plants is not solely dependent on the transparency of the water is demonstrated convincingly by the fact that in clear lakes colonization by submersed phanerogams already has ceased in depths that still exhibit a marked positive assimilation balance, as has been demonstrated by the suspension at various depths of bottles containing sprigs of these plants. On the other hand, in lakes with a significantly lesser transparency the lower limit of phanerogams can occur not at all or only a little higher than in lakes with very transparent water. This discrepancy has been clarified by the experimental studies of Gessner (1952) and his pupil Ferling (1957). Detailed studies on the effect of *hydrostatic pressure* on submersed phanerogams have demonstrated that already at an excess pressure of about 1 atmosphere these plants cannot thrive because of a curtailment of various functions. This condition, which is evidently associated with the ventilation system, explains why phanerogams, even in lakes of great transparency, do not reach a depth of 10 m., whereas plants without a gas-filled intercellular system, such as *Fontinalis*, *Chara*, *Nitella*, and many microphytes, can extend considerably deeper. Especially noteworthy is the occurrence in many alpine lakes at depths of 10 to 15 metres of closed stands of *Fontinalis antipyretica*, for which the presence of free CO_2 below the thermocline provides favourable living conditions (p. 69).

The aeration system, which distinguishes the phanerogams that have reverted to an aquatic existence from all other plants in the water, affords to a number of *animals* the opportunity for maintaining their original aerial life habit beneath the water. These are always the larvae of insects, hence of air-breathing animals, which, in so far as their developmental stages live in the water, are comparably in phylogeny to the submersed phanerogams. Living as "respiratory parasites" on the water plants are, for example, the larvae and pupae of the familiar reed beetle (*Donacia*), whose brilliant metallic imagoes can be found everywhere on reeds or on the floating leaves of water lilies and pond weeds.

Their maggot-like larvae bore into the aerenchyma of their food plants by means of two chitinous projections on the abdomen which bear stigmata at their tips, and thus provide themselves with the necessary respiratory gas. The pupae enclosed in water-tight, air-filled cocoons usually occur at the base of the reed stems and respire in the same manner. The larvae and pupae of the fly genus *Hydrellia* and the mosquito *Mansonia* behave similarly. Certain midge larvae (*Cricotopus*), which eat galleries in the mesophyll of floating leaves, for example, of *Potamogeton natans*, can be regarded as specialists of a similar type. Some of these "miners" let the epidermis remain as a protective covering and hence breathe exclusively the intercellular air of the leaves. Others eat down from above, that is they eat away the epidermis, after they have taken care, by means of a small opening directed downwards, that the mined gallery fills with water by capillarity and does not dry out.

The *littoral zone*, with its growths of macrophytes, and the *profundal zone*, without plants, form by means of the varied nature of their substrate and environmental conditions the manifold biotopes of the benthic communities, among which we distinguish two main groups: *the biocoenoses of the Aufwuchs*[55] and *those of the ooze.*

3. *The Communities of the Aufwuchs*[55]

By the term *Aufwuchs* we mean all those organisms that are firmly attached to a substrate but do *not* penetrate into it (in contrast to plants rooted in the bottom or certain parasites). In the terrestrial environment, development of the plant *Aufwuchs* is dependent on the adequacy of a water supply. For this reason a rich and varied development, chiefly in the form of epiphytes, occurs only in moist climates, especially in the tropical rain forests. In the aquatic environment there is no limitation on the basis of moisture content. In water itself the developmental possibilities of the *Aufwuchs* are unlimited, and hence in waters of the temperate zone there is scarcely a stone, scarcely a dead or living part

[55]Translator's note: The German term *Aufwuchs* has a much broader connotation than the closest English equivalent "periphyton." *Aufwuchs* comprises all attached organisms (except the macrophytes), including such forms as sponges and Bryozoa, which are usually considered as benthos by American authors; also included are the various forms living free within the mat of sessile forms. In American usage the term "periphyton" is frequently restricted to the sessile communities on plant stems (although used in a broader sense by Young, 1945); the adjective "epilithic" has been applied to those on inorganic substrates. However, even in its broadest sense the term "periphyton" cannot very well be applied to corresponding aerial organisms. The more general concept of *Aufwuchs* is so useful that it has been taken over in its entirety.

of a plant that is not populated more or less thickly with sessile organisms.

The environmental conditions of this biotope resemble to a certain degree those of the plankton in that the organisms of the *Aufwuchs* still live essentially under the conditions of the open water. Yet the physical characteristics of the substrate to which these organisms are attached are of the greatest significance in the formation of the *Aufwuchs* and furnish a suitable basis for the sub-division of this biocoenose, as will be described later.

Just as among plankton organisms it is the "adaptations for floating" that are especially characteristic, so among the species of the *Aufwuchs* it is the modifications for attachment. There are a great variety of these, of which several types will be described presently. It is noteworthy that the same principles can recur among both the plants and the animals. The attachment ability is a protection against being washed away by the action of currents and waves. Where these forces are less demanding, as in *quiet* water, relatively weak adaptations are sufficient (Figure 55). Among the single-celled organisms, particularly the diatoms, there occur gelatinous stalks of the most

FIGURE 55. *Aufwuchs* on a *Myriophyllum* leaf from Lunzer Untersee, showing various means of attachment, ×50 approx. With their broad surfaces applied directly to the leaf, the diatoms *Cocconeis* (*l*) and *Epithemia* (*h*); with gelatinous buttons or short stalks, the diatoms *Synedra* (*i*), *Tabellaria* (*f*), and *Achnanthes* (*k*); in gelatinous tubes, the diatom *Encyonema* (*g*); with long stalks sometimes branched, the diatoms *Cymbella* (*d*) and *Gomphonema* (*e*), and the peritrich *Vorticella* (*a*); with algal threads firmly attached by hold-fasts, *Oedogonium* (*b*) and *Bulbochaete* (*c*).

diverse structure, from the small pin-like type in *Achnanthes* to the slender, branched structures in *Cymbella* and *Gomphonema*, and to the long gelatinous tubes enclosing the cells in *Encyonema* (*Cymbella prostrata*). In the animal kingdom the rigid or contractile stalks of the bell animalcules (*Vorticella*, *Epistylis*, *Opercularia*) correspond to this method of attachment. Very abundant in the plant and animal kingdoms are rigid or gelatinous cases, which are attached to the substrate by little stalks or by a broader base. An especially interesting example of a method of attachment occurs in the tiny flagellate *Chrysopyxis*, whose case, shaped like a broad flask, is attached by a gelatinous thread around an algal filament. We should mention here too the stalked or unstalked loricas of many infusoria, and among rotifers the gelatinous cups of the Collothecidae and the often elegant tubes of the Flosculariidae constructed of regularly arranged fecal pellets. Among the filamentous algae the basal cell, which must bear the entire pull, is firmly fastened by means of a lobated attachment disc closely applied to the substrate, for example in the Oedogoniaceae. In *agitated* water, more resistant types of attachment are found. Among these are the following: a flattened development of a thallus broadly applied to the substrate, seen particularly well, for example, in *Coleochaete*; hemispherical gelatinous cushions often reinforced with lime, for example in the Rivulariaceae, in *Chaetophora*, or (without lime) in the infusorian *Ophrydium*; finally, thick and shortened gelatinous stalks, for example in many species of *Gomphonema*.

In addition to attached species there belong to the *Aufwuchs* biocoenose a large number of free living forms, which crawl upon the substrate, swim about in the dense confusion of the sessile species, or even undertake farther temporary excursions into the open water. These are members of the most diverse branches of the plant and animal kingdoms occurring in fresh water.

The *influence of the substrate* shows up most clearly when the *Aufwuchs* on a stone is compared with that on a portion of a living plant. A substantial difference is apparent both with reference to the *over-all impression* offered even to the unaided eye, and also to the *species* participating in its composition. On stones, crust-like growths often of considerable thickness and firmness predominate, whereas on living leaves and stems the coverings are light and flocculent. Since under certain conditions this flocculent *Aufwuchs*, often consisting of filamentous algae, can encroach upon the solid substrate as well, the distinction is not always entirely sharp, yet in many biotopes it is of great significance. The origin of this striking difference is primarily two-fold. Living portions of plants serving as a substrate for the *Aufwuchs* are usually *transitory*, very short-lived formations. Since they survive scarcely longer than a summer and then are destroyed through decay, they can be populated only by quickly growing forms with a short developmental cycle, such as the great multitude of green algae, diatoms, and similar forms. Stones and wood, on the other hand, are *imperishable*, or persist at least several years. They therefore provide an opportunity for permanent colonization, even to those *Aufwuchs* organisms which are slow growing and which, in contrast to the "annuals" mentioned above, form persisting colonies. But this does not fully explain the distinction between these two types of substrate. The one is non-living, the other is living, and the *Aufwuchs* attached to the latter is more or less under the influence of the metabolism of its bearer. Hence, the epiphytes and animals attached to a living leaf are exposed to great variations in concentrations of carbon dioxide and oxygen and in pH, since the metabolic processes previously described in the section on chemistry must prevail the strongest in the layers of water directly in contact with the leaf. There is scarcely a doubt that this factor is of great significance in the environmental selection of *Aufwuchs* organisms.

We must therefore distinguish between: (1) the *Aufwuchs* on a *persisting*, non-living substrate, and (2) the *Aufwuchs* on a *transitory* (usually living) substrate. (Even a portion of a dead plant undergoing decomposition must be regarded as a "living" substrate in a certain respect because of the metabolism of the saprophytes permeating it.) Both types of *Aufwuchs* occur under the very diverse environmental conditions of the littoral and are influenced by them, a fact that must be taken into account in a consideration of the *vertical distribution of this biocoenose*.[56]

Beginning on the uppermost shore we find first of all that in the *eulittoral*, as a consequence of the fluctuations in water level and of the influence of the waves (surf), very extensive changes in environmental conditions occur over a small vertical distance, and we can consequently expect equally sharp changes in the composition of its biocoenose. Yet the conditions necessary for the development of a typical *eulittoral Aufwuchs* do not prevail at all places on the beach. Most of the resistant species thriving under the extreme conditions of this biotope grow slowly and can develop only on a firm substrate of bedrock or large stones that cannot be rolled by the surf.

When it was stated (p. 179) that the eulittoral zone occupies that portion of the beach lying between the high and low water lines, it was not meant to imply that *only* the fluctuations in water level determine its extent. By means of the beating of waves (or the wetting of the shore and movement of the water resulting therefrom), the width of the eulittoral can be considerably enlarged both above and below. For this

[56]A community of a special type comprises those endolithic algae (blue-greens, heterokonts, and greens, as well as lichens) that penetrate into the substrate, and hence live *under* the surface of stones, particularly limestone, and in snail shells.

reason it is much more extensively developed on the surf shore of large lakes than in protected locations, a condition that has been specially emphasized by the Swedish investigators, particularly by Du Rietz (1939). How characteristic and sharp the zonation of the *Aufwuchs* can be under the influence of periodic desiccation and force of the waves is illustrated by means of the closely investigated relationships in Lunzer Untersee (cf. Kann, 1933).

In this lake the normal zone of the eulittoral is restricted to a vertical distance of 30 to 40 cm. The portion of the stony beach that is only occasionally washed over by waves bears a vegetation consisting solely of encrusting lichens and mosses. One can distinguish here, as in many other lakes, an uppermost *Verrucaria* zone. Then follows a zone of only 10 cm. vertical extent, lying dry about 170 to 280 days each year, with a noticeable dark-brown coloration, which on closer examination is seen to be produced by bushy growths of just *one* species, the blue-green alga *Tolypothrix distorta*. One cannot doubt that this severe selection results from the action of the *single* extreme factor, the temporarily complete desiccation. While dry the plants are exposed in addition to very high summer temperatures. Investigations have indicated that in this condition they can endure temperatures up to 70° C. without injury. Growth and metabolism naturally are possible only at the time of reflooding. In this *desiccation zone* or *emersion zone* there are also a few species of animals, which are just as resistant to water loss: several tardigrades, several bdelloid rotifers, and a few nematodes.

Below the *Tolypothrix* zone there follows almost without transition a very striking zone consisting of brownish to reddish-yellow pea-like crusts several millimetres thick, which extends in a vertical direction from about 20 cm. above to just below the low water line. Yet even the partially emergent portion scarcely ever dries up entirely, because the crusts suck up capillary water from below. The dominant forms of this zone are again blue-green algae, most important being *Rivularia haematites* with its hemispherical colonies interspersed with stratified layers of lime, and in addition, especially in the upper portion, the layer of *Calothrix parietina* resembling brown flat spots of chocolate and the curly tufts of *Scytonema myochrous*. Since moisture is not lacking, the number of species accompanying these dominant forms is very large. The outermost surface is populated above all by a host of diatoms; and living in the crusts, partly burrowing, are numerous kinds of animals, particularly nematodes, certain members of the Harpacticidae among the copepods, and insect larvae. Among the latter, in addition to numerous chironomids (which, to be sure, are fewer in Lunzersee than in other lakes) there should be mentioned in particular the larvae of the caddis fly *Tinodes*, to which is ascribed the formation of certain peculiar furrowed stones by dissolution of the lime substrate in the burrows. In its general aspect as well as in its plant and even animal communities this *Rivularia* zone resembles quite closely the *Aufwuchs* on stones in our mountain brooks. In both situations the same factor of strong water movement produces the characteristic life forms. Hence we may rightly designate this zone as the *surf zone*.

As the effect of the waves becomes weaker with increasing depth, which in Lunzer Untersee occurs barely 10 to 20 cm. below the lowest water level, the firm hard crusts of *Rivularia* give way to thick grey-green sediments permeated with loose precipitated lime, which cover over stones and wood with a fluffy spongy layer. As a regularly occurring dominant form, the blue-green alga *Schizothrix lacustris* can be mentioned, although numerous other algal species participate in the composition of this layer, which is also inhabited by a rich animal life. They are all, however, dependent on continual submergence; even a drying at room temperature results in their death. This *Schizothrix* zone, which in so far as it is not overgrown with filamentous algae can extend downward several metres, belongs essentially to the sublittoral. We also find animal *Aufwuchs* forms of impressive dimensions established here: the green, antler-like, branching colonies of the freshwater sponge *Euspongilla lacustris*, which is sessile by preference on wood-sticks and in the tissues of which the likewise green larvae of the neuropteran *Sisyra* are parasitic, as well as the spherical gelatinous colonies of the ciliate *Ophrydium*. The latter, which likewise are coloured green by zoochlorellae and which vary in size from a pea to a fist, are often sessile in large numbers on the encrusted stones and also on reed stems and similar substrates.

The zonation of the eulittoral *Aufwuchs* on bedrock and boulder shores as described for Lunzer Untersee is similarly developed although modified by local conditions in many lakes of the limestone Alps, as for example Traunsee in Upper Austria, which likewise has been studied in detail by Kann (1959). Corresponding to the large size of this lake and the resulting greater water movement, the zone of wave action exhibits a greater vertical extent and harbours the rhodophycean *Bangia atropurpurea* in addition to the species listed for Lunzer Untersee.

The lakes of Schleswig-Holstein, which likewise are rich in lime but are eutrophic lowland lakes, show a considerably different picture than the oligotrophic lakes of the limestone Alps (Kann, 1940). Although numerous species of Cyanophyceae occur, they do not produce a zonation like that in the alpine lakes, in part because of a lesser fluctuation in water level and consequently the lack of a well-developed desiccation zone. The zone of wave action is represented instead by a broad girdle of green algae, among which *Cladophora glomerata* and *Cladophora aegagropila* are the most important species. This luxuriant growth of *Cladophora* is probably controlled by eutrophy, since we also find *Cladophora* in those littoral portions of alpine lakes that are polluted by waste water. It is taken for granted that differences in the chemistry of the water likewise control variations in the appearance of the littoral *Aufwuchs*, as for example in the carbonate-poor lakes of the crystalline mountains.

A conspicuous depth arrangement is scarcely recognizable in the *upper sublittoral*, since here the substrate is most important for the development of the biocoenose. This is readily understandable when we consider that the water temperature within the epilimnion exhibits small or even negligible differences, and also that the intensity of CO_2 assimilation in these uppermost layers remains almost uniform, and indeed at times does not reach its optimum until a depth of several metres. It is needless to emphasize that the *lighted zone* of the sublittoral is characterized by an extremely rich and varied plant and animal life, which would take too much time to describe in detail.

The appearance of the *Aufwuchs* nevertheless begins to change considerably at depths usually coinciding with the thermocline, where life, and especially the metabolism of plants, is influenced in large degree by the decreasing light intensity and temperature. In this *twilight zone* of the *lower sublittoral* one notices first of all on microscopic examination of the plant *Aufwuchs* that there has been a quantitative decline of green types of algae in favour of brown, that is the diatoms, which here show especially deeply coloured chromatophores. But, in addition, there occur species and phenotypes that are lacking in the illuminated zone of the sublittoral. Thus, as first Lauterborn (1922) and later Geitler (1928) found, stones dredged from a considerable depth, perhaps between 10 and 20 metres, usually have a blackish covering, which on closer examination turns out to be a very distinctive mosaic of different genera of blue-green algae (for example *Chlorogloea*, *Pleurocapsa*, *Chroococcopsis*, *Oncobyrsa*, *Chamaesiphon*, *Lyngbya*), pervaded everywhere by disorderly growing filaments of the green alga *Gongrosira* and populated with numerous diatoms. Most noticeable, however, is the coloration of these blue-green algae. They are the same species that also occur in shaded brooks. But whereas in brooks they have a brown or a dirty grey-green colour, here they are usually bright red or violet. For this reason the mosaics of this *red-coloured deep water biocoenose* often give an over-all impression of magnificent colours. A last fact, which will serve to complete the description of this algal community, is that, in addition to the mosaics described above, beds of the red algae *Hildenbrandia*, *Chantransia*, and *Batrachospermum* also occur. Red-coloured Cyanophyceae of the genera mentioned also occur in the *Aufwuchs* of the *Fontinalis* meadows at depths of 10 to 15 metres.

How is this striking change in composition of the *Aufwuchs* to be explained? One might consider it to be the effect of temperature, since indeed many of the species listed also occur in cold mountain brooks. That this is not the correct explanation, however, is demonstrated by the fact that this red-coloured algal community in the same or even richer composition and in a more brilliant development occurs even at 26° in the tropical lakes (Geitler and Ruttner, 1935). Only *light*, therefore, can be the controlling factor, partly because of the quantitative, but most of all because of the qualitative, changes that occur with increasing depth. It is a known fact that plants are able to utilize for assimilation only those wave-lengths of radiation absorbed by the pigments of their chromatophores. These are, roughly speaking, the spectral colours that are complementary to their pigments. Accordingly, green chromatophores are able to utilize red light best, while the simultaneous occurrence of brown or red pigments in addition to chlorophyll increases the utilization of the short-waved green light. Hence, in the greater depths where green radiation predominates, as explained on page 18, brown or red algae are better adapted, or they are able to descend deeper than the majority of the algae with only green pigments. In addition, many blue-green algae have the ability to increase so greatly their content of the red pigment phycoerythrin under the living conditions existing in deep water that they assume an over-all bright red colour. It has been plausibly demonstrated by experiment that this colour results from the influence of green radiation, and consequently this phenomenon is called *chromatic adaptation.* Hence we see in the red-coloured deep water biocoenose a light-controlled selection or adaptation of the algae towards a better utilization of the radiation prevailing in the deeper layers of water. The fact that green forms likewise occur in this community (one might recall particularly the extensive carpets of *Fontinalis* at considerable depths, as previously described, as well as mats of *Cladophora aegagropila*) does not indicate anything contrary to this conception. Even among those species appearing green to our eyes, there are marked differences in the ability to utilize short-wave-length light in CO_2 assimilation, for example through an increased chlorophyll content (many of the Chlorophyceae occurring in deep water are strikingly dark green in colour) or through a change in the proportion of chlorophyll *a* to chlorophyll *b* in favour of *b*, whose limit of absorption, as compared with that of *a*, extends farther into the blue-green region.

The *Aufwuchs* on suitable substrates by no means stops at the lower limit of a positive assimilation balance, the depth of which naturally is dependent mainly on the transparency of the particular lake. However, here in the *profundal* it does not consist of assimilating algae, but rather of heterotrophic organisms—primarily sessile animals and less frequently fungi growing on wood.

In many lakes if one raises sunken logs or branches from a depth of 20 to 25 m., one sometimes finds them densely overgrown with mats several centimetres thick of the moss animalcule *Fredericella sultana.* Between them hydras also occur abundantly; the latter, however, are not confined to deep water but rather are especially numerous in the upper sub-littoral on the flexible stems of *Potamogeton natans* and similar species, preferring those places where these plants extend far into the open water and thus facilitate the capture of plankton crustacea. Even *Fredericella* is not confined entirely to the depths. Its colonies can be found, for example, on the under-side of floating planks and platforms, indicating that it is not temperature but light that they get away from in the depths. With the *Fredericella* zone of the alpine lakes, or similar animal biocoenoses in other waters, the *Aufwuchs* of the slope reaches its lower limit.

In the sea, especially in the great oceans, the area of the shore with its littoral vegetation is so small compared with that of the open water populated by plankton that it usually can be disregarded in considerations of production biology. It is different in the much smaller inland lakes. Here the *littoral production* increases in importance the more the longer the shoreline is in relation to the area of the lake and the shallower the slopes suitable for the growth of submersed littoral flora are. In such instances the production of the littoral vegetation can exceed that of the plankton. But methodological difficulties in the determination of the littoral biomass and its net production are so much greater than for the plankton that quantitative investigations of this kind are rare. In certain instances through the application of suitable methods success can be achieved in spite of these difficulties, as demonstrated by the study of Nygaard (1958) on a small Danish lake, which is only 11 metres deep, whose bottom is overgrown with macrophytes (phanerogams, mosses, Characeae).

4. *The Communities of the Ooze*

The common name "ooze" (*Schlamm*) refers to all *sediments* that become deposited within a body of water during the course of time. The original material of the lake basin outcrops only in scattered locations along steep shores, usually in the form of rocky walls and cliffs. Elsewhere it is completely covered by sediments, which often attain a thickness of many metres. According to composition these deposits are partly *inorganic*, partly *organic*; and according to origin they are either *autochthonous*, having been formed *in the lake itself* by life processes or physical-chemical processes separating them from the water, or *allochthonous*, having been introduced *from outside the lake* by inflowing water, falling of dust, etc. Whether the sediments deposited are primarily autochthonous or allochthonous is dependent mainly on the watershed of the lake. Lakes with a small inflow of surface water will contain primarily autochthonous sediments, whereas in river lakes and in mountain lakes with a large volume of water flowing through them, allochthonous sediments usually predominate.

Concerning the *allochthonous* sediments there is not much to say from a biological standpoint. Their quantity and composition depend on the ratio of the area of the lake to that of the watershed, as well as on the morphological and geological character of the watershed, the climate, and the plant covering of the watershed. The introduced particles undergo a sorting in the lake according to their size and their density. In the vicinity of affluents, boulders and gravel are deposited, and then follow concentric zones of coarser and finer sand. The finest clayey portion, which has a very slow rate of sinking, produces in the lake water a homogeneous turbidity. These clayey particles settle out quite uniformly over the entire lake bottom and contribute to the formation of the central plain. The *organic* component of allochthonous sediments can not be disregarded. A microscopical examination of these sediments reveals considerable quantities of humus particles coming from agricultural and forest land, portions of plant tissues, and to a lesser extent animal materials, such as hair, chitinous parts of insects, and the like. The contribution of *dust* is particularly impressive in spring when the conifers are in bloom, at which time the surface of lakes often is covered by a sulphur-yellow, mould-like film, and the characteristic pollen grains can be found in quantity in the fresh sediments.

Much more interesting to us are the *autochthonous* sediments arising in the lake itself. Here we must distinguish two different types: first, precipitations that take place *external to living organisms*, hence in the water, as a result of physical-chemical changes, which, to be sure, are brought about in most instances by life processes; second, the sedimentation of *plant and animal remains* from the community of the lake along with their inorganic and organic integuments and supporting materials. The precipitates of most importance belonging to the former group are those of calcite and iron.

As was explained on page 68, *calcite* is precipitated primarily through the assimilation activity of plants by the withdrawal of CO_2 or of the HCO_3-ion from dissolved bicarbonates. It either separates from the supersaturated solution in the open water, or else directly onto the outer surface of submersed macrophytes in the form of crystalline coverings, from which it likewise enters the circulation system of the lake on being washed off. These floating particles of calcite are deposited in the shore zone, particularly in backwaters behind prominences of the shore line, and there form the glistening whitish benches, which are very striking in an alpine lake when seen from above. This shore terrace sediment, *marl*, is of a greyish white coloration and can attain a thickness of many metres. Up to 90 to 95 per cent of it consists of carbonate of lime. Contributing to its composition are not only the $CaCO_3$ precipitated by plants but also to a considerable extent the shells and shell fragments of snails, especially the genera *Valvata* and *Limnaea* in the alpine lakes. In the Baltic lakes there is developed at a somewhat greater depth (7 to 12 m.) a very characteristic "shell zone," which owes its origin to the transportation of dead shells, predominantly those of the three-cornered mussel *Dreissensia.* Wasmund (1926) appropriately called these secondary accumulations "communities of the dead" (thanatocoenoses).

Why is marl not also deposited in deep water, or why is the calcite content of the central plain (as shown in the table on p. 194) so much less than that of the shore terrace? The explanation of this circumstance follows readily from the chemical stratification in the open water of a lake. We have seen that the CO_2 content of water increases with increasing depth and often exceeds the equilibrium value in deep water, so that as a result aggressive CO_2 is abundant, at least in the lowermost layers. This re-dissolves the settling particles of calcite, and hence there does not occur any significant enrichment of $CaCO_3$ in the mud but rather an increase in the $Ca(HCO_3)_2$ concentration in deep water (cf. p. 72). Thus the extensive marl benches furnish impressive evidence for the magnitude and course of the biochemical processes in lakes as well as for their geological significance.

The relationships concerning the precipitation and sedimentation of *iron* are fundamentally different. Organisms likewise participate in this process, at least in part, but less through altering solubility conditions in the open water than through their ability to store iron compounds in their outer coverings, as in the iron bacteria and numerous algae. The alteration of solubility conditions in the case of iron results primarily from the gradient of dissolved oxygen in stratified eutrophic lakes. As has already been pointed out on page 83, significant quantities of ferrous bicarbonate in the dissolved state can occur in our lakes only when oxygen is lacking, except at a strongly acid reaction. If oxygen is transported to the hypolimnion from above by eddy diffusion, there occurs a precipitation of ferric hydroxide (limonite), which settles upon the bottom of the lake. This process, as well as the sinking of iron organisms thriving at the oxygen boundary, leads to the accumulation of a considerable quantity of iron in the sediments of stratified eutrophic lakes (cf. p. 85). If the deep water contains H_2S, there occurs a precipitation of iron sulphide, deep black in colour, when the reaction becomes alkaline. A portion of the precipitated iron can be re-dissolved in the depths and returned to the cycle of materials in the open water, as has already been discussed (p. 85). Through the upwelling of iron-containing ground water in the shore zone there can be brought about the formation of littoral deposits of limonite (p. 84).

In contrast to that of lime and iron, the precipitation of *silicic acid* from water *always* occurs firmly bound to *living* cells. It is well known that the *diatoms*, whose siliceous shells in sinking to the bottom enrich the ooze with SiO_2, are most important in this process, but under certain conditions the siliceous cysts of the flagellate group, the *chrysomonads*, as well as the siliceous spicules of the freshwater sponges play an important role. The greatest quantities of diatoms are sedimented from the plankton of the pelagial region. In order to give some idea of the quantity of diatom frustules sedimented in a normal lake, it can be reported that in Lunzer Untersee in the course of a year 21 million diatom frustules were deposited per $cm.^2$ of lake bottom, of which only 2.4 to 18.6 per cent, with a mean of 8.7 per cent, originated in the littoral, whereas the major portion (predominantly *Cyclotella comensis*) originated from the plankton.[57] For this reason the deep water sediments on the whole are richer in SiO_2 than the littoral. In lakes with little inflow, in which allochthonous mineral sediments are negligible, there can remain after the destruction of the organic portion of the diatoms a more or less pure siliceous deposit, which is encountered as fossil *infusorial earth* or diatomaceous earth in the sediments of long-extinct lakes and seas.

This concludes the discussion of the most important inorganic components of the sediments, in so far as they are limno-autochthonous in origin, and there might now be cited two analyses (in round numbers) from Lunzer Untersee as an example of the magnitude of the differences between the shore sediments and deep water sediments arising from the conditions of sedimentation just described:

	Central plain (33 m.) (%)	*Shore terrace* (1 m.) (%)
SiO_2	34	5
CaO	11	51
MgO	6	1
Al_2O_3	12	3 (Al₂O₃ + Fe₂O₃)
Fe_2O_3	7	
Loss on ignition (carbonate CO_2 + organic matter)	31	43

In other lakes these relationships do not necessarily agree with those of Lunzer Untersee, especially with respect to the shore sediments. In mountain lakes with a small calcium content located in regions of primary rocks, less calcite is deposited, and the littoral sediments consist largely of allochthonous quartz sand. Or, particularly in shallow lake basins, there can occur a strong admixture of organic material, originating for example from a dense zone of reeds, which results in an increased CO_2 production and a re-dissolving of the calcite. A characteristic of a true marl bench is its sterility with respect to the growth of plants.

The earlier investigations of sediments were generally confined to analyses of structure and microfossils (pollen analysis) and to quantitative determination of the inorganic components, whereas the organic substances were usually summarized merely as loss on ignition. More recently special attention has been given to the organic portion of the sediments. As an example the detailed work of Kleerekoper (1957) can be cited, which is based on the investigation of cores of sediment from 37 lakes in the Canadian province of Ontario. The organic carbon content of the sediments in the majority of

[57] Züllig (1956) gives figures for the number of diatoms sedimented during a year in the Swiss lakes that are about 10 times smaller than those determined for Lunzer Untersee. This may possibly be related to the fact that *Cyclotella comensis*, which is dominant at Lunz, is very small.

these lakes lay between 40 and 60 per cent of the dry weight of the organic matter, and of this 40 to 60 per cent in turn consisted of *lignin*, which is attacked by bacteria with difficulty and is scarcely digestible as food for animals. In nine-tenths of the lakes the nitrogen content was below 3 per cent, and the phosphorus content 0.04 to 0.08 per cent of the organic carbon. In addition, ten amino acids could be demonstrated, and a series of other interesting determinations was made.

At any rate the most important and most interesting components of the sediments from a biological standpoint are the *organic substances*. Their presence transforms the lake bottom into an abode of life of very intense organic activity, the lively metabolism of which is able to affect significantly the processes in the free water. Even among the organic materials there are some that originally were dissolved in the water and were brought out of solution through physical-chemical processes. Such a material is the *humic matter* that is leached from the vegetation layer of the soil, and as a colloidal solution brings about the more or less brown coloration of many waters. These humic colloids on entering the lake are flocculated primarily through encountering dissolved salts (particularly calcium), and form a characteristic gelatinous sediment of dirty-brown colour, the *dy* of the Scandinavian investigators. Since waters with a high humic content ("brown water") are especially characteristic of bog (moor) regions, an intensive formation of *dy* occurs primarily in lakes that are imbedded in bogs or are fed by them. Hence, the humic material as such is allochthonous, but its precipitation as *dy* is mainly autochthonous.

Except for *dy*, which often predominates in Scandinavian lakes in particular, the organic components of the ooze are originally of a particulate nature. The leading role in their formation is played by the sedimentation of plankton, the magnitude of which increases with distance from shore and from affluents, and hence increases in importance the larger the lake is and the smaller the amount of water flowing through it. In addition, there are the organic remains washed in from the watershed and the littoral region, and captured as dust by the lake surface. All this, together with a greater or lesser component of inorganic matter, after being worked over by bottom animals and bacteria on the lake bottom, which is at least periodically *supplied with oxygen*, forms a very characteristic, finely divided sediment of grey to greyish-brown colour and at times of an elastic consistency, the *gyttja* (pronounced *Jüttja*) of the Scandinavian terminology. On the basis of the proportion of the mineral components (lime, clay, etc.), the intermixture with *dy* sediments, and other characteristics, a *gyttja* systematics has been established, which we need not consider here any further.

The quantity and the annual course of sedimentation in an alpine lake based on older hitherto unpublished data are illustrated in Figure 56. These data were obtained by lowering a bell jar, which was enclosed in a tripod with the opening directed upward, to the level bottom of Lunzer Untersee at a depth of 30 m., observing the necessary precautionary measures. At approximately monthly intervals the bell jar was again raised to the surface, whereupon the sediment that had accumulated in the previous interval could be investigated chemically and microscopically. The original hope of obtaining thereby an insight into

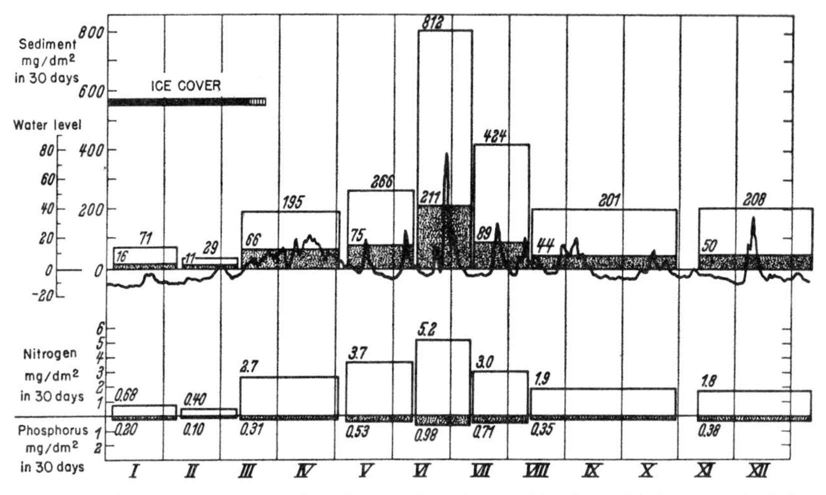

FIGURE 56. Annual course of sedimentation (in mg./dm.² per 30 days) and of the fluctuation in water level of Lunzer Untersee in 1934, according to unpublished analyses by F. Berger. *Above*: total quantity—inorganic portion (white) and organic portion (punctate). *Below*: nitrogen (white) and phosphorus (punctate).

the magnitude of production in a lake, specifically Lunzersee, was not realized by this investigation. It turned out that the seston suspended in the epilimnion was only incompletely sedimented because of the turbulence prevailing there, and that the largest part of it was removed from the lake via the outflow. Hence, the sediment originated chiefly in the metalimnion and hypolimnion. Although because of these large and irregular losses evaluation of the data for considerations of production biology is scarcely possible, nevertheless the data obtained give some idea of the quantity, the mineral or organic character, as well as the nitrogen and phosphorus content of the sediment deposited in each month of the year in this alpine lake. It should be emphasized that practically no sedimentation occurs under the ice during the winter months, whereas the influence of summer high water is already evident in the very great increase in sediment at this time, particularly the inorganic portion. The quantity of sediment deposited on a unit area of lake bottom in a year is considerable. In 1934 this amounted to 3056 kg. dry weight per hectare, for the most part allochthonous sediment, of which 784 kg. was organic matter, 20.4 kg. nitrogen, and 4.3 kg. phosphorus. These quantities, according to earlier investigations of Götzinger (1912), correspond to an annual sediment thickness of about 1 mm. Similar investigations with noteworthy results have been carried out more recently by Thomas (1950, 1955) on several Swiss lakes. He compared the monthly quantities of materials retained in the lakes (based on the difference between the analyses of the inflowing and outflowing water) with the sedimentation within the lakes and was able in this way to characterize clearly the eutrophic and oligotrophic types. (Cf. also the description on p. 166 of the investigations by Grim on Schleinsee.)

We shall now consider more closely the processes that take place in the *gyttja*, the transformations and decomposition which the organic matter undergoes. These transformations have been frequently touched upon in the course of this presentation, but in no aquatic biotope are they of such importance quantitatively as here. The changes occurring in the hypolimnion of lakes are similar, but the quantity of organic material per unit volume of water destined for decomposition there is indeed small in comparison with the accumulations of the same material in the sediments. Moreover, the former consists primarily of the easily decomposable substances which become mineralized even during sinking, whereas the substances attacked with difficulty, such as chitin and cellulose, to a large extent arrive unaltered on the bottom of the lake and there are gradually broken down. It is now proper to review briefly, so far as the present state of our knowledge permits, the most important groups of organic compounds and the fate that befalls them in the ooze. Kuznetsov (1959), in a book that has been translated into German, has recently summarized this information.

The most important building blocks of living matter, the *proteins*, fall victim to the attack of numerous oragnisms, among which the host of special "putrefaction bacteria" are most important. By means of enzymes, the compounds of high molecular weight are split into simpler ones. First are formed albuminoses and peptones, then amino acids, and from them nitrogen-free acids and ammonia. The end products of mineralization are: CO_2, H_2O, NH_3, and H_2S, the latter from the sulphur portion of the protein.

The decomposition of *cellulose*, the building material of the cell wall of plants, is accomplished by a chain of numerous independent reactions in which many mutually dependent organisms participate. These processes can be condensed schematically into the following two phases:

1. $C_6H_{10}O_5 + H_2O = C_6H_{12}O_6$ (hydration of cellulose, hexose formation).

2. The hexose is broken down further to organic acids, and ultimately to CH_4.

In view of the very great quantities of plant material that are annually sedimented in bodies of water and decomposed, and the great significance that the resulting decomposition processes have, both in the economy of nature and in the treatment of the waste materials of urban life in digestion tanks, we shall elaborate a bit on the methane fermentation of cellulose, following the presentation by Liebmann (1950). In his book the most important papers in the rapidly growing literature on this subject can be referred to (cf. also Kuznetsov, 1959).

The liberation of methane, which forms the last link in the chain of these decomposition processes, is brought about by a number of species of bacteria, only *four* of which have been grown in pure culture and are well known with regard to their physiology: *Methanosarcina methanica, Methanococcus mazei, Methanobacterium söhngenii*, and *M. omelianski*. The first-named species is of particular importance for the metabolic processes in lakes, because it is by far the most abundant in natural waters and in addition is capable of producing methane at low temperatures (around 5°), whereas the other species are active only at higher temperatures. All methane bacteria are strictly anaerobic and are especially sensitive to traces of oxygen. The consequence is that they can flourish only in biotopes that are completely anaerobic. Liebmann has demonstrated in Lunzer Untersee that in the uppermost layers of the *gyttja* there can be recognized a sharply marked stratification of organisms dependent on the O_2 content. Below a layer of fluffy ooze approximately 13 mm. thick populated principally by protozoa there is a "*Beggiatoa*-plate" only 2 mm. thick, in which the simultaneous occurrence of H_2S and small quantities of O_2 provides the necessary conditions for the development of sulphur bacteria (*Beggiatoa*). Methane bacteria do not yet occur. But 5 mm. deeper, where neither H_2S nor O_2 is present, *Methanosarcina* occurs in large numbers to the exclusion of all other organisms. (The methane bacteria are not capable of utilizing high molecular linkages. They require substances of low molecular weight, such as fatty acids, alcohols, and ketones. Cellulose and similar substances must therefore first be broken down by other bacteria to these simpler materials before methane formation can begin. It has been determined that even carbon dioxide in the presence of hydrogen can be reduced by methane bacteria.) Methane fermentation takes place only at an alkaline chemical reaction; consequently in acid biotopes, for example in bogs, cellulose remains unaltered.

Particularly in the shore zone where large quantities of cellulose are being broken down, the ooze is permeated with gas bubbles, which one can dislodge by probing with an oar and can collect in a glass cylinder and ignite. This demonstration becomes especially impressive when one taps the huge bubbles often formed under the ice of a lake by the rising of this gas, and then ignites the outflowing methane. At times a jet of flame 1 metre high is produced in this way.[58] In the closed digestion tanks of municipal sewage disposal systems the sludge provides favourable conditions for methane fermentation. Large

[58]The rising gas bubbles in Lake Beloye have been analysed by Rossolimo (1932). The chief constituents were 72–84 per cent CH_4, 5–18 per cent H_2, and 0.4–2.9 per cent CO_2.

quantities of a combustible gas mixture of high caloric value are liberated there and can be utilized technologically; 2.3 per cent of the total gas requirements of the city of Munich originates from sludge chambers.

Pectin plays an important role in the plant kingdom as a cementing substance of the cells. There are several aerobic and anaerobic species of bacteria (for example, *Bacillus amylobacter*) that have the special ability of "pectin fermentation," that is, the solution and decomposition of this carbohydrate. This process leads to the maceration of plant tissue and finds technical application in the "retting" of flax.

It is a well-known phenomenon that wood projecting into the ooze is only slightly affected. This might be attributed to the fact that the ability to break down *lignin* is possessed primarily by aerobic fungi (which only occasionally occur in water) and not by bacteria.

Starch and *fat*, the two most important reserve foods, are split by a great many organisms. The former is transformed into sugar by means of amylolytic enzymes and supplied to metabolism. Fat is broken down first to glycerol and organic acids, which in turn yield carbon dioxide and water on further break-down.

A relatively resistant substance is *chitin*, the skeletal material of the arthropods, which is deposited in great quantities as the cast-off integuments (exuviae) and dead bodies of the crustacea and insect larvae. In this case also there are specialized bacteria that are able to split this nitrogen-containing substance. For a long time the aerobe *Bacterium chitinovorum* has been known, through whose activity chitin is hydrolyzed, though only when oxygen is present. However, through the work of Steiner (1931) anaerobic chitin digesters have also become known, and it is these that complete the decomposition in the deeper layers of the ooze. The various chitinous parts of a given species of animal show a considerably different resistance. Most susceptible to destruction are the shells of *Daphnia*, and in freshly deposited sediment one can follow all stages in their hydrolysis. The same applies to the chitinous integuments of copepods. The filter combs, mandibles, and post-abdomens of the Cladocera and especially the "heads" (rostra) of *Bosmina* are particularly resistant. One can always find large numbers of these remains in the ooze. From the quantity of these remains in various layers of the sediments reliable conclusions can be drawn concerning the earlier plankton production of a lake, according to Frey (1960).

It should be strongly emphasized that these are only a few examples of the endless variety of bacterial processes that take place in water, most of which are still incompletely understood; that in addition there participate in these processes not only bacteria but also aquatic fungi (for example, the Saprolegniaceae), algae, flagellates, ciliates, and the higher animals of the ooze biocoenose, which nourish themselves on these organic materials, partly assimilating them and partly excreting them as simpler compounds.

As end products of this mineralization there are again set free those inorganic materials out of which the organic substances had previously been elaborated in the productive layer of the lake, hence primarily carbon dioxide, ammonium salts, phosphate, hydrogen sulphide, and water.

But only in the presence of oxygen does decomposition proceed to these inorganic end products. Under *anaerobic* conditions, as in the ooze of highly eutrophic lakes and especially in meromictic lakes, the course of decomposition is different and even less is known as to its details. By means of the anaerobic metabolism of the organisms living here, which must wrest from the organic linkages of their environment the oxygen necessary to their composition, a strongly reducing *milieu* is produced. A foul-smelling *sapropel* (sens. str. = *Faulschlamm*) results, deep black in colour from the presence of iron sulphide; this material is the habitat of the *sapropelic community* (Lauterborn, 1915), consisting of bacteria and colourless protista. Decomposition under these conditions does not proceed to complete mineralization, but stops short at *intermediate organic stages* (for example, methane; see above). That these materials are able to outlast even geological epochs is demonstrated to us by sapropelite, by sedimentary bituminous deposits, and not the least by petroleum. But even in well-aerated lake bottoms the deeper ooze layers are free of oxygen; thus here likewise there occurs a limited formation of sapropel, which is dispersed through the *gyttja*. This dependence of the sediments upon the chemical stratification, particularly with respect to oxygen, of the water in which they have originated has been taken notice of in geology and has opened up new points of view through evaluation of the results of limnological investigations (cf. Wasmund, 1930; H. Schmidt, 1935).

The mineral end products of decomposition do not by any means signify a cessation of chemical transformations. Rather these proceed further under favourable conditions and let scarcely a single substance be unaffected. The fate of *carbon dioxide* which in the free state or as bicarbonate diffuses out of the ooze into the layers of water overlying the bottom, and through eddy diffusion can be re-introduced into the metabolism of the lake, has already been described. *Ammonia* remains as such only in the interior of the ooze or in an oxygen-free hypolimnion. With access to oxygen, as at the surface of the sediments of an oligotrophic lake or even in a eutrophic lake at times of complete circulation, *nitrification* sets in as in agricultural land. This is well known as one of the most interesting *chemosyntheses* in nature, by means of which certain bacteria utilize the oxidation energy of ammonia for their anabolism and catabolism; they are not only able to do without organic substances, but these at times are even detrimental to them. *Nitrite bacteria* oxidize ammonia to nitrate according to the formula:

$$(NH_4)_2CO_3 + 3\ O_2 = 2\ HNO_2 + CO_2 + 3\ H_2O + 148 \text{ calories.}$$

Then the activity of the *nitrate bacteria* begins immediately, by which oxidation proceeds further to nitrate:

$$2\ HNO_2 + O_2 = 2\ HNO_3 + 44 \text{ calories.}$$

The liberated energy is utilized in carbon assimilation (out of CO_2), a process which in contrast to the photosynthesis of green plants goes on even in the dark. (The processes of nitrate reduction have already been described on p. 89).

Through the activity of the previously mentioned sulphur bacteria, *hydrogen sulphide* becomes involved in a similar very remarkable process, which is not yet clarified in all its details. It proceeds in two phases. In the first, hydrogen sulphide is oxidized to elemental sulphur, which is deposited within the cell in the form of tiny droplets. In the second phase this sulphur is further oxidized to sulphate, which is given up to the water. The energy relationships are the following:

$$\text{I. } 2\,H_2S + O_2 = 2\,H_2O + S_2 + 122 \text{ calories;}$$

$$\text{II. } 2\,S + 3\,O_2 + 2\,H_2O = 2\,H_2SO_4 + 282 \text{ calories.}$$

After the discovery of this chemosynthesis through the classic investigations of Vinogradsky (1888) it was assumed at first that this process ran essentially according to the above scheme for all species of organisms that utilize H_2S. Investigation in later years, however, indicated that this was valid in the strict sense only for the *colourless* sulphur bacteria of the type *Beggiatoa*, *Thiothrix*, *Achromatium*, *Thiospira*, etc., not for the other forms, and that in particular the red-*coloured* species, the purple bacteria or Thiorhodaceae, differed significantly from the colourless in their metabolism. It had already been demonstrated by the studies of Buder (1919), Bavendamm (1924), and others that these purple bacteria do not thrive in the dark. Van Niel (1936) showed on the basis of exact laboratory experiments carried out with pure cultures that the carbon assimilation of the Thiorhodaceae he worked with (mainly the genus *Chromatium*) is not a *chemosynthesis* but rather a special type of *photosynthesis*. Whereas in the reduction process yielding carbohydrate (CH_2O), which proceeds at the cost of light energy in the chlorophyll molecule, water functions as the hydrogen donor:

$$CO_2 + 2\,H_2O \rightarrow CH_2O + H_2O + O_2,$$

in the photosynthesis by the purple bacteria hydrogen sulphide takes over the role of hydrogen donor:

$$CO_2 + 2\,H_2S \rightarrow CH_2O + H_2O + S_2.$$

The correctness of the conclusions based on very many experiments according to the procedure of van Niel cannot be disputed. However, we might consider the extent to which the results agree with the ecological occurrence of the Thiorhodaceae. It is true that in great depths and hence in complete darkness usually only colourless sulphur bacteria are found, and the red species are lacking; likewise, that under suitable conditions the Thiorhodaceae can cover over the bottom ooze of shallow waters in such great numbers that the red colouration is visible from a considerable distance. On the other hand "bacterial plates" (Figure 47), consisting mainly of species of the genus *Chromatium* as described on page 147, sometimes occur at depths that lie far below the compensation point of chlorophyll-containing plants. Thus, Kuznetsov has already related for Belovod Lake that the extremely sharply developed *Chromatium* "plate" he found there was situated at a depth of 15 m. Although light measurements from this lake are not available to me, I believe I can assume on the basis of the other conditions in this highly eutrophic lake that this depth is located far below the compensation point of the photoautotrophic phytoplankton. Based on my own experiences I can report that the mass accumulations of *Chromatium* and *Lamprocystis* in Krottensee (Upper Austria) were located at a depth of 22 m. (Ruttner, 1937), and in the very large and deep Danau Manindjau (Sumatra) the accumulation of *Thiopedia rosea* occurred at 50 m. (Ruttner, 1952). In both instances the concentrations were located at the lower limit of oxygen. Moreover, it is striking that in both these lakes the maxima of the Thiorhodaceae occurred at the same depth as the mass accumulations of the colourless sulphur bacteria, such as *Achromatium mobile*. These observations suggest that at least some species or races of the Thiorhodaceae are capable of both photosynthesis and chemosynthesis, a conjecture already expressed by Buder and Bavendamm. It seems, however, that at least a certain small light intensity is necessary for the purple bacteria to thrive.[59] That the oxidation proceeds beyond the stage of depositing sulphur within the cells, as in *Chromatium*, is evident from the disappearance of the sulphur droplets when a water sample containing purple bacteria is allowed to stand without any addition of H_2S. Whether the energy liberated is used for chemosynthesis or only respiration is still questionable.

Even this does not exhaust the diversity of metabolism among the sulphur organisms. Thus there are *green sulphur bacteria*, which are characterized by having a green pigment called *bacteriochlorin*, and which behave like the

[59]Observations, however, that are completely contrary to this supposition and to all other previous findings have been reported from the Black Sea. As is well known, this sea is characterized by two noteworthy conditions: by an extensive freshening of the water (the salt content amounts to 15–18‰ in the uppermost layers and 20–22.3‰ from 100 to 2000 m. depth), and by an H_2S content below 200 m. increasing to almost 6 mg./l., which is associated with the disappearance of oxygen (cf. Caspers, 1957). Whereas in all other seas the biomass is compressed into the uppermost 100 to 200 m. as a consequence of the production-limiting effects of light and stratification, in the Black Sea this also applies for the photosynthetic plankton and for the zooplankton dependent on them, but in water at greater depths, which contains H_2S, the biomass after a brief decline increases again to very considerable amounts, which in fact even to depths of 2000 m. often exceed that of the upper layers in weight per $m.^3$ of water. This large biomass, according to the opinion of Kriss and Rukina (1953), results from the mass development of a filamentous organism, which attains a population density of several thousand per millilitre, and which the investigators consider to be a member of the Thiorhodaceae. Thus, if water samples from the depths concerned are allowed to stand 1–2 months out of contact with the air, red growths develop consisting of the same organism that can be cultured on nutrient substrates suitable for sulphur bacteria. Kriss and Rukina on the basis of physiological investigations of these cultures believe that the accumulation of biomass in the depths of the Black Sea results from the chemosynthesis of these organisms. The source of energy for this process, however, remains unexplained, since neither light nor O_2 can be involved. The two authors consider that the energy is gained from the decay of radioactive materials (which scarcely seems likely), or from the organic substances contained in the water. Still further investigations are needed to clarify this puzzling phenomenon. The idea should not be rejected that the organic substances raining down from the zone of production as soon as they reach the saltier, and hence denser, deep water are slowed so much in their rate of sinking that they accumulate at these levels, where their normal decomposition is obstructed by the toxic action of H_2S. If this is the case, then it would not be necessary to assume such a tremendously active chemosynthesis. (See also Kriss, 1954).

Thiorhodaceae with respect to their metabolism. Thus they are (photo)autotrophic, using the H_2S as a hydrogen donor, but in this instance the sulphur is deposited *outside* the cells rather than *in* them. In these forms oxidation apparently goes only to elemental S. Here belong *Chlorobium limicola* and the unique double organism *Chlorochromatium*, which consists of a colourless motile rod, surrounded by a short tubular colony of tiny green cells. Both species also occur in the plankton, and participate in the formation of the "bacterial plates" mentioned above.

Finally, there is still to be mentioned the group of the Thiobacteria (e.g. *Thiobacillus*), which like the colourless sulphur bacteria are chemoautotrophic (hence, they can thrive in complete darkness) and have the ability to oxidize thiosulphate. For further information about the metabolism and ecology of the sulphur organisms, see Kuznetsov, pp. 192–206, or concerning the systematics of the colourless and red species, see Bavendamm (1924).

The resulting sulphate is again brought into the metabolic cycle. It has already been stated (p. 92) that the major portion of the H_2S in the deep water of many lakes originates from *sulphate reduction.* This process occurs also in the sediments, and indeed there are a considerable number of bacteria (*Microspira desulfuricans* and others) that can bring it about in various ways. Of greatest interest is a process closely associated with cellulose fermentation or with the methane originating from it:

$$CaSO_4 + CH_4 = CaCO_3 + H_2S + H_2O.$$

Thus sulphur is involved in a continuous cycle between oxidation and reduction.

It has already been mentioned that sulphur bacteria, especially the red ones, require a certain low concentration of oxygen, keeping in mind that H_2S, which is also necessary, is not stable at a high oxygen level. Hence in oligotrophic lakes or in eutrophic lakes at the time of turnover, these organisms occur in the uppermost layer of the ooze but not in the free water. During the course of stagnation, as the oxygen content of the deep water is reduced from below upwards, a cloud of sulphur bacteria detach themselves from the bottom and continue to rise higher with progressive oxygen depletion (cf., however, what was described for Lunzer Mittersee on p. 147).

As already mentioned on page 147, the iron bacteria are also considered to be able to carry on chemosynthesis, although this matter has not yet been satisfactorily resolved. According to Kuznetsov (1959), in a more narrow sense the iron bacteria utilize the energy set free by the oxidation of ferrous to ferric compounds according to the following equation:

$$4\ FeCO_3 + 6\ H_2O + O_2 = 2\ Fe_2(OH)_6 + 4\ CO_2 + 58 \text{ calories.}$$

Presumably the interpretation of Starkey (1945) is more nearly correct, that only a portion of the bacteria that store iron in their sheaths are able to carry on chemosynthesis.

According to Sorokin (1958, cited by Kuznetsov) the oxidation of methane as a source of energy for chemosynthesis is of great significance in the ooze deposits.

Considering the total situation, chemosynthesis plays a minor role as compared with photosynthesis in the metabolism of lakes, according to Kuznetsov.

(*a*) *The redox potential of the sediments*

If we examine once more the condition in which the sediments of a lake exist as a result of the chemical transformations sketched roughly above, we find that there is a sharp and narrow transition from the more or less oxidized substances at the ooze-water interface to the strongly reduced substances in the deeper layers of the sediments. This gradient in the state of reduction is a consequence of the oxidative decomposition of organic matter, a process that does not cease even after complete utilization of the available free oxygen but proceeds further with the formation of reduced residual substances.

A few words might be inserted here concerning the concepts "oxidation" and "reduction." We know that not only the uptake or giving up of oxygen, but also changes in the content of hydrogen atoms (hydrogenation = reduction; dehydrogenation = oxidation) and even the mere uptake or giving up of electrons are understood to be oxidation or reduction processes. Thus, the transformation of bivalent (ferrous) iron into trivalent (ferric) iron is designated as oxidation and the reverse process as reduction, even though neither oxygen nor hydrogen enters into the reaction, but only an electron is given up by the bivalent atom or is taken on by the trivalent one: $Fe^{++} \rightleftharpoons Fe^{+++} + \epsilon$. *Each giving up of an electron is an oxidation, each taking on of one is a reduction.*

In a solution containing a mixture of a substance and its reduction product (which means that the oxidation-reduction process must be reversible) the reduced phase has the tendency to give up electrons and transform itself into the oxidized phase. The presence of free electrons, however, hinders this process through the development of large contrary electrostatic forces, and under such conditions any given quantities of Fe^{++} and Fe^{+++}, for example, can exist together without restriction. But if we draw off the electrons by means of a bright platinum electrode immersed in the solution, then the transformation of Fe^{++} into Fe^{+++} begins, and a current flows through the electrode. The electromotive force acting in this system can be determined potentiometrically by means of a calomel electrode. The potential of the solution determined in this manner is designated as the *oxidation-reduction potential* (shortened to *redox potential*); it is an expression of the oxidizing or reducing power of the solution. This power is dependent on the nature of the dissolved substances, as well as on the proportion of their oxidized and reduced components in the solution, but not, however, on their absolute quantities. Redox potential is greatly influenced by pH. It is necessary, therefore, to correct the potential (E) measured with the calomel electrode to a definite pH value (e.g. pH 7). Furthermore it is customary to select as the zero point the potential of a normal H_2 electrode, that is an electrode of gaseous H_2 at a pressure of 1 atmosphere (pH = 0).[60] A redox system of a certain potential (compared with an H_2 electrode) is able to oxidize another system of lower potential or to reduce a system of higher potential (cf. Michaelis, 1933, for details).

If we now attempt to relate these facts to the condition in the sediments of a lake, we must expect *a priori* that a stratification of the redox potential exists, decreasing in value downward from the surface of the sediments standing in contact with oxygen-containing water, because systems of oxidized and reduced substances occur here as is demanded by the theory for the origin of redox potentials. To be sure, not all these systems are directly reversible, considered from the purely chemical standpoint, but they become reversible under the influence of the catalysts of life processes (as was indicated previously in the examples NO_3–NO_2–NH_4 and SO_4–S–H_2S); we may assume, therefore, that the same principles prevail in these biological systems as in the purely chemical oxidation-reduction systems.

These assumptions were substantiated in their entirety by the thorough investigations of Mortimer (1941–42)[61] in the English Lake District. With the aid of an ingenious apparatus (shielded Pt electrodes, which permitted simultaneous measurements in numerous ooze levels) redox potentials were measured both in artificially established ooze-water systems (modified by the admission and exclusion of oxygen) and in natural sediment-water contacts in several lakes; and the microstratification of the redox potential in this boundary horizon, which is so significant for the metabolism of a lake, was determined. The essential facts and conclusions of this first comprehensive investigation of the subject under consideration will be discussed briefly in the following paragraphs.

At the water contact of the sediments in a lake whose hypolimnion contains sufficient quantities of oxygen (hence, in an oligotrophic lake,

[60]Redox potential is commonly expressed by the symbol E_h (r_H in Europe). Michaelis (1933) has this to say about it: "Another measure of the chemical oxidation-reduction potential is the *pressure* (in atmospheres) that gaseous hydrogen, which is in absorption equilibrium with a platinized platinum electrode in a buffered solution of the same pH as the redox system, must have in order to give the same potential as the indifferent electrode *without* hydrogen in contact with the redox system." (E_h is the negative logarithm of the hydrogen pressure in atmospheres.)

[61]Concerning the older literature, see Hutchinson (1957, pp. 693 ff.). Of the more recent studies one might refer to Hayes, Reid, and Cameron (1958), who tested on several Canadian lakes the methods developed by Mortimer.

or in a eutrophic lake at the time of complete circulation), redox potentials ($E_7 = E$ at pH 7) of approximately 0.6 volt are found. This is the same magnitude as occurs in the free water of the hypolimnion. Within the sediments, however, E_7 declines very rapidly, and as a

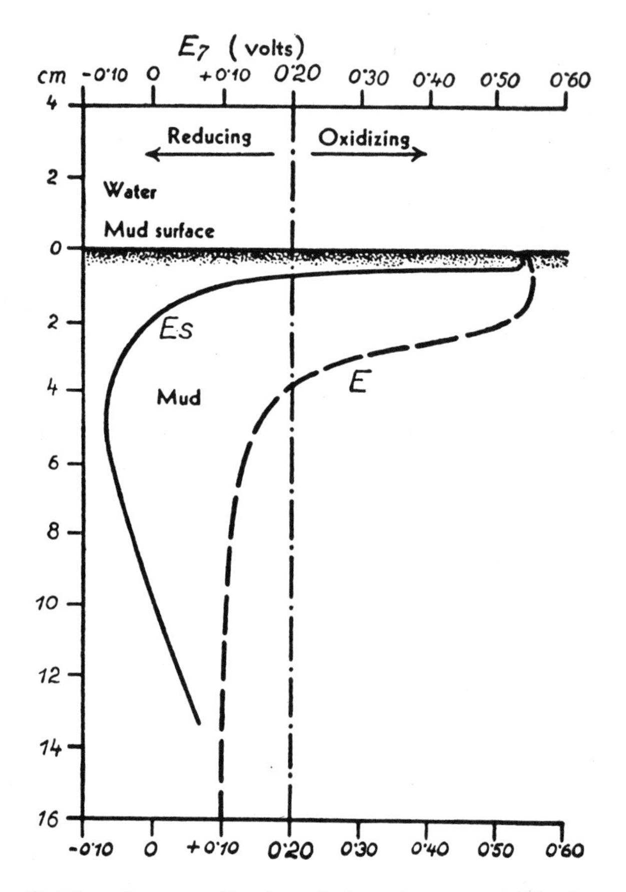

FIGURE 57. The winter stratification of the redox potential in the ooze of two lakes in northern England: *E*, Ennerdale Water (oligotrophic), and *Es*, Esthwaite Water (eutrophic). From Mortimer (1941).

general rule at a depth of 5 cm. reaches a minimum, the value of which is an expression of the reducing power of the ooze and hence is characteristic for the production type of the lake. At still greater depths in the ooze the redox potential can again exhibit an increase, indicating that the intensity of anaerobic decomposition has already diminished. Figure 57 shows the course of the E_7 curves in the sediments of two lakes of very different trophic level at the time of complete circulation (winter).

In this gradient of redox potential there could now be established the ranges of E_7 within which occurs the oxidation-reduction of certain systems important in metabolism: NO_3–NO_2, 0.45–0.40 volt; NO_2–NH_4, 0.40–0.35; Fe^{+++}–Fe^{++} (or their complexes), 0.30–0.20; SO_4–S, 0.10–0.06.

If the oxygen content of the water of the hypolimnion declines, and hence likewise its importation into the ooze, then the redox potential in the individual strata decreases, and the horizontal surfaces of equal E_7 values migrate upward towards the surface of the sediments. The range around $E_7 = 0.2$ volt, which is decisive for the reduction or oxidation of iron, has proved to be especially significant for the processes at the ooze-water contact. As long as a redox potential exceeding this value prevails at the surface of the sediments, and the curve $E_7 = 0.2$ volt runs within the sediments, the surface layer of sediments is infiltrated with a precipitated ferric hydroxide complex (in which humic acids and silicic acid can also be involved) and is distinguished by its brownish colour. With progressive decline in the oxygen content of the hypolimnion this "oxidation layer," which is often several millimetres thick, is destroyed from below upwards (through the reduction of Fe^{+++}), and it disappears as soon as $E_7 = 0.2$ volt is reached at the surface of the sediments. From this moment on, considerable quantities of Fe^{++}, Mn^{++}, NH_4, P, and similar substances occur in the deep water, and a stratification of dissolved substances proceeding from the bottom upwards is formed, which is identical to what we have come to associate with the hypolimnion of eutrophic and meromictic lakes. Mortimer (1941) advances the opinion that the ions mentioned above had been firmly bound in the iron gel of the oxidation layer (p. 217) and hence could not diffuse outward into the water so long as this layer was present. After destruction of the oxidation layer by reduction these firmly bound ions are released, and at the same time the blockade of the diffusion processes is ended. With further disappearance of oxygen, the isopleths of the low E_7 values rise into the free water, and bring about a macrostratification of the reduction potential in the hypolimnion similar to the compressed microstratification encountered beneath the ooze surface.

Figure 58 attempts to illustrate these processes on the basis of Mortimer's investigations in Esthwaite Water. For the sake of clarity, only the annual course of the "critical" 0.2-volt isopleth is shown, along with a number of the typical values for the layers lying above and below. It is noted that in May this line still runs within the ooze, about mid-June (after the beginning of stagnation) reaches the surface of the sediments, and from then on climbs steeply into the hypolimnetic water. During the summer, redox potentials between 0 and 0.1 volt prevail at the ooze surface and in the adjacent water. This is the range in which sulphate reduction takes place. Hence, under these conditions H_2S can occur in the water, and in the ooze this zone of E_7 (which in Esthwaite Water lies at a depth of 1 to 2 cm. during winter) is recognizable by the dark coloration resulting from the formation of iron sulphide. At the

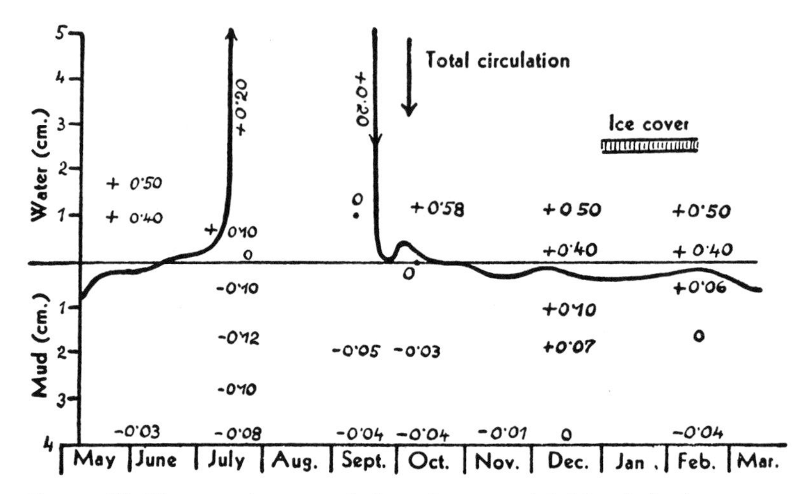

FIGURE 58. The annual course of the redox potential 0.2 volt in the ooze and adjacent water layers of the eutrophic Esthwaite Water in northern England. After Mortimer (1941).

time of stratification breakdown towards the end of September, the 0.2-volt isopleth appears again at the surface of the sediments, and, during winter, stays just beneath it, with small fluctuations controlled by differences in the oxygen supply. Naturally in an oligotrophic lake the 0.2-volt isopleth remains below the surface of the sediments throughout the year.

(*b*) *The fauna of the ooze*

The conditions of the sediments described above, brought about primarily by the metabolic processes of bacterial life, is very important in controlling the environmental conditions under which the *benthic animals* of the ooze exist. A limiting factor of major importance is the oxygen content of the stratum of water adjacent to the bottom. The interior of the ooze, as we have known since the time of Alsterberg's (1922) investigations, and further as the investigations on the vertical distribution of the redox potential have shown, is free of oxygen as a result of the oxidation processes occurring in the sediments and of the curtailed diffusional exchange. The consequence is a steep diffusion gradient of dissolved oxygen in the layers of water immediately overlying the surface of the ooze. Whether the uppermost layers of ooze are still "aerated" or whether the lack of oxygen extends across the surface of the sediments into the free water depends upon the quantity of reducing substances present in the ooze, which in part diffuse outward, and on the oxygen content of the overlying water and the magnitude of the eddy diffusion prevailing there. Because of these circumstances not only do the biochemical transformations in the uppermost layers of sediment proceed along entirely different lines, but also the living conditions for animals are basically altered. In oligotrophic lakes with an abundance of oxygen in the deep water and negligible quantities of reducing substances in the sediments, 100 to 200 species of animals, exclusive of protozoa, live in and upon the ooze. In eutrophic Furesee, on the other hand, Wesenberg-Lund (1917) could discover only 23 species. If the oxygen content of the hypolimnion disappears completely, and if in addition the water layers adjacent to the bottom are loaded with quantities of reducing substances (for example, H_2S), then any possibility of normal respiration ceases, and the ooze becomes completely azoic with respect to the higher types of animals. Only anaerobic protista are able to thrive here. Whenever more highly organized animals seek out this zone, they appear to do so only temporarily; for example the plankton midge larva *Chaoborus* (*Corethra*) during the day often burrows into the uppermost ooze layers but at night again rises into the epilimnion (Kaj Berg, 1937). Many species of animals are eliminated at a relatively high oxygen content; thus, according to Thienemann (1928), the schizopod crustacean *Mysis relicta*, which is abundant in the oligotrophic lakes of the Baltic region, occurs only at oxygen tensions of more than 50 per cent saturation. Noteworthy for entirely different reasons is the influence of oxygen content upon the distribution of midge larvae, which on the basis of numbers and weight form the most important component of the ooze community. As Thienemann (1918) has pointed out, the occurrence of the genus *Tanytarsus* characterizes the bottom fauna of the oxygen-rich oligotrophic lake type; whereas the absence of *Tanytarsus* and the predominance of the genus *Chironomus* characterizes the oxygen-poor eutrophic type. Between the two extremes of the *Tanytarsus* lakes and the *Chironomus* lakes additional types characterized by the occurrence of completely distinct midge larvae can be established, for example, the *Stictochironomus* lakes and the *Sergentia* lakes (cf. here the book by Lenz, 1928, as well as numerous other publications by the same author, and above all the large work *Chironomus* [1954] by Thienemann himself). Hence, investigation of the bottom fauna can yield important conclusions about the oxygen economy and the production type of a lake.

The profundal is populated by a *dependent* community, that is, the organic substances used in nutrition are produced not in this but in other biotopes, for example, in the trophogenic layer of the lake. Hence, except for predators such as the larvae of *Tanypus*, the animals depend for their nourishment on the organic content of the ooze, and it is this type of food intake that characterizes this community. Some of the species, such as the Hydracarina, the cyclopoids, and the turbellarian *Otomesostomum*, and others, swim or crawl upon the ooze surface and do not noticeably differ in their habits from the free-living species of the littoral region. Most species, however, burrow in the ooze, for example the ostracods and harpacticoids among the crustacea, the Tubificidae, Mermitidae, and nematodes among the worms, the chironomid larvae, and the tiny mussel *Pisidium*. Some are non-selective ooze eaters, and must pass large quantities through their intestines. According to the investigations of Alsterberg (1922), the Tubificidae in particular bring about a permanent re-arrangement of the sediments; their faecal pellets together with those of the chironomids form a characteristic component, which becomes evident on sifting of ooze samples.

Life in the deeper layers of ooze (according to more recent investigations, animal life is found to a depth of about 20 cm. in loose sediments) presupposes nevertheless that water containing dissolved oxygen is supplied to the animals for their respiratory needs. Numerous species construct long tubes, the walls of which in the chironomids are strengthened with silk, in which the animal by means of its sinuous movements produces a current of water and in which it can move rapidly up and down. The openings of these chironomid tubes project like tiny chimneys above the surface of the ooze. The tubificids protrude their posterior ends from their dwelling tubes and promote respiration by brisk undulations. An especially striking example of this means of providing the animal with water for respiration is the dwelling tubes of the magnificent mayfly larva *Ephemera*, which, to be sure, does not live in the central plain but rather in the ooze of the shore terrace. This animal, resembling a mole cricket, digs a U-shaped vertical burrow, with two openings clearly visible in calm water from a considerable distance, in which a lively stream of water is produced through the movement of the abdominal gills projecting over the back.

A further environmental condition that influences the composition of the community in the true profundal is the *uniformly low temperature*. The annual fluctuations in temperature here are unusually small (cf. p. 35), so that this biotope is an ideal abode for cold-stenothermal organisms. But only a portion of its inhabitants are strictly oligothermal; many can thrive at higher temperatures also and must therefore be considered eurythermal. A group of oligothermal species of great interest to animal geography is the so-called *glacial relicts*, which during the arctic climate of the Ice Age enjoyed a wide distribution in our region, but at the retreat of the glacier were forced to seek out certain refugia with constantly low temperatures. As Sven Ekman (1913) in particular has demonstrated, members of this group which as relicts of the glacial Baltic now live in its freshened border lakes are the crustaceans *Mysis relicta* and *Pontoporeia affinis*; in our alpine lakes, ostracods of the family Cytheridae and several turbellarians and mites are glacial relicts.

(*c*) *Stratified sediments*

In a lake with undisturbed sedimentation in which the deposits remain in the sequence in which they were laid down and are not mixed through the activity of ooze-dwelling animals, a stratification of the sediments is to be expected, because the composition of the organic and inorganic materials settling to the bottom varies with the time of year. These conditions occur especially in those lakes where a disappearance of oxygen from the deep water does not permit the development of a higher bottom fauna that continually works over the sediments in its feeding activities, and where, on the other hand, sapropel formation has not yet proceeded so far that the stratification could be destroyed by the formation of gas bubbles (methane).

The classic example of such an annual stratification, clearly defined, is furnished by Zürichsee (Switzerland) in which these relationships were discovered and investigated by Nipkow (1920). The remarkable thing about this lake is that the point in time at which conditions became favourable for a stratification of the sediments could be determined. This occurred, along with the incipient eutrophication of the lake (as a result of the dense colonization of its shore), about the year 1896, and was characterized by the first bloom of the diatom *Tabellaria fenestrata*. This phenomenon has been recorded historically, and is also demonstrable microscopically in the corresponding layer of sediment. Previous to that time, there had been deposited in Zürichsee an unstratified, bright grey, predominantly inorganic *gyttja* of the type characteristic of oligotrophic alpine lakes. Beginning in 1896 there was a disappearance of oxygen and a formation of sapropel, the profundal fauna disappeared, and the sedimentation showed an annual stratification (interrupted by occasional "catastrophes," that is, shore landslides). These layers consist of a summer portion coloured brightly through admixture of minerals, and a predominantly organic winter portion (arising from the winter blooms of *Oscillatoria rubescens*) coloured black by FeS. Nipkow (1920) has investigated all these layers on the basis of their content of organisms and has dated them exactly (Figure 59; cf. also Minder, 1943). Moreover, he was able to demonstrate the viability of quiescent forms of diatoms in sediment layers up to 12 years

cm.

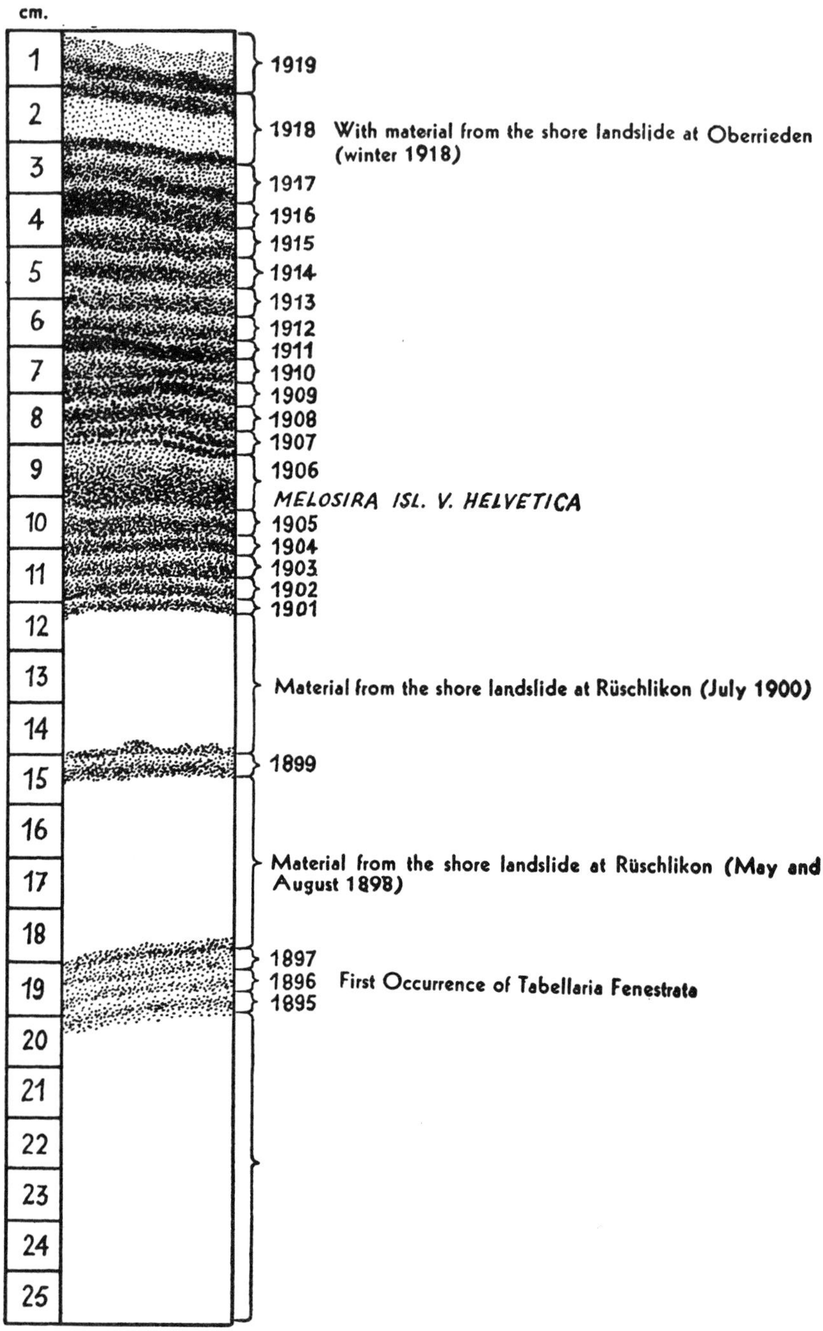

1 2 3 4 5 6 7 8 9 10 11 12 13 14 15 16 17 18 19 20 21 22 23 24 25

1919
1918 With material from the shore landslide at Oberrieden (winter 1918)
1917
1916
1915
1914
1913
1912
1911
1910
1909
1908
1907
1906
MELOSIRA ISL. V. HELVETICA
1905
1904
1903
1902
1901
Material from the shore landslide at Rüschlikon (July 1900)
1899
Material from the shore landslide at Rüschlikon (May and August 1898)
1897
1896 First Occurrence of Tabellaria Fenestrata
1895

FIGURE 59. Stratified sediments in Zürichsee. According to Nipkow, from Minder (1943).

old (Nipkow, 1950). The detailed structure of such stratifications has been studied especially by Perfiliev (1927) in Russian lakes. Annual calcareous laminae in lake sediments resulting from the biogenic precipitation of calcite by phytoplankton are described by Welten (1944).

But even where visible laminations are lacking, stratigraphic investigations of lake sediments can lead to valuable information concerning the ontogeny of our lakes. This has already been demonstrated by Lundqvist's investigations (cf. "Die Binnengewässer," 1927) on the sediments of Swedish lakes and the studies of Gams (1927) on the Lunzer lakes carried out according to the methods of *bog stratigraphy* and *pollen analysis*. All such studies are in agreement in showing that many lakes have undergone a gradual development towards eutrophy from their original condition of oligotrophy as a result of the growth of marginal vegetation, accumulation of sediments, and increased supply of nutrients. The detailed investigations of Deevey (1942) on sediment cores from Linsley Pond (Connecticut) have led to the same conclusion. The strata, which were dated by their pollen content, showed a development from a *Tanytarsus* lake into a *Chironomus* lake on the basis of the microfossils present, and also permitted other interesting conclusions concerning the physical-chemical relationships of this body of water in earlier periods.

Further progress in the field of sediments investigations has been made possible by the construction of an easily operated sediment corer (Livingstone, 1955). This apparatus enables sediment cores many metres long to be obtained from a boat anchored in water depths even exceeding 20 metres. By this means Frey has investigated a number of European lakes, especially in the Alps, with very good results. Thus one learns from his first publication up to now on the small meromictic Längsee in Carinthia (Frey, 1955), which is only 20 m. deep, that not only the point in time at which meromixis began but also its cause (the beginning of deforestation and of agriculture) could be determined from the sediments, and that it was also possible to determine the chitinous remains in the individual sediment layers the succession of animals during the colonization of the lake and its dependence on environmental conditions (trophic level and oxygen content).[62] Additional investigations concerned with palaeolimnology on the basis of the analysis of subfossil animal remains in lake sediments that might be

[62]After completion of the present manuscript there appeared the book by N. V. Korde "Biostratification and Typology of Russian Lake Sediments" (in Russian), published by the Academy of Sciences of the U.S.S.R., Moscow, 1960. Following a summary of the investigation and utilization of Russian lake sediments along with an apparently very complete reference to the literature, Korde describes some of her own investigations on the present condition and genesis of Russian lake sediments, based primarily on quantitative microscopic analysis.

mentioned are the already cited work by Deevey (1942) in Connecticut as well as a study by the same author in New Zealand (Deevey, 1955) and another by Frey (1958) in Germany.

Using other procedures Züllig (1956) has investigated the stratification in the sediments of a number of Swiss lakes. In the oligotrophic Walensee, in Bodensee, which although likewise still oligotrophic is already undergoing a slow transformation, and in the now eutrophic Zugersee, Züllig using an apparatus he constructed himself obtained short deepwater cores (50 to 120 cm.) representing the material deposited during the last 50 to 500 years. For sampling Zürichsee, which has become eutrophic, a Kullenberg piston corer was available, whereby cores up to 9 m. in length were obtained. The chemical investigation of these cores showed a clear dependence on trophic level. Particularly useful in this respect as indicators of the condition of the free water from which the sedimentation derives are the contents of xanthophyll (originating from the phytoplankton) and silicic acid arising from diatoms).[63]

(*d*) *The exchange between sediments and free water*

We should not conclude the section on the ooze biocoenoses without considering in retrospect the very interesting relationships between the sediments and the free water of a lake, which are difficult to investigate. During the course of our presentation there has been occasion a number of times to refer to the existence of such interrelationships, and it cannot be denied that they are of great significance in the metabolism of a body of water. Recently the interest of limnologists has been turned searchingly to these questions, and there now exist quite a few papers on this topic, of which only those of Alsterberg, Hutchinson, Mortimer, Thomas, and Hayes and his co-workers will be mentioned here. In them the chief concern is the extent to which the changes in and the distribution of dissolved substances in the free water of a lake, which we have summarized under the designation "biogenic chemical stratification," are influenced by the processes at the ooze contact, or in other words, the extent to which these stratifications are the result of the decomposition of sinking organic matter in the free water on the one hand and the transformations in the ooze on the other.

If we had a lake whose sediments arose entirely from its own organic

[63]It is very interesting that diatom shells, which indeed are known to occur as far back as the Upper Devonian and are very abundant in the Tertiary as well as the Quaternary, could be demonstrated microscopically only in the uppermost 20 to 40 cm. Possibly the delicate shells had been dissolved in the alkaline medium of the older lake sediments. According to the table presented by Korde (1960), the diatom layer in Russian lakes extends considerably deeper.

production and were therefore autochthonous, which showed in addition a sufficient supply of oxygen, and in which the decomposition products of metabolism in the sediments returned entirely to the free water, then this question would be of subordinate significance in production biology. In such a case, regardless of whether the major portion of the organic substance produced was already decomposed in the free water or not decomposed until after sedimentation upon the surface of the ooze, the changes in amounts of materials occurring in the hypolimnion would still correspond in magnitude to the total production. For that reason, only the temporal displacement would have to be considered, which arises from the fact that the decomposition of resistant organic components takes longer at the low temperatures in the sediments of our lakes than it does in tropical lakes (cf. p. 77).

We might encounter such relationships occasionally in oligotrophic lakes. But even in these (for example, in mountain lakes) the addition of allochthonous material by tributaries will play a considerable role in the sediments in the majority of cases. The changes brought about by the transformations in the sediments are caused not only by the mineralization of the material produced in the lake itself but also by the decomposition of allochthonous organic sediments. Further complications occur in eutrophic lakes, in which the hypolimnion exhibits a more or less complete disappearance of oxygen during stagnation periods. The anaerobic transformations prevailing in such a case do not result in a complete mineralization; there remain residual substances (cf. p. 200) the quantity of which can only be roughly estimated. If considerable quantities of iron occur in the deep water of such lakes, then the processes described on page 91 become active, so that this element and the phosphate associated with it are permanently stored up in the cycle of reduction and oxidative precipitation and are thus removed from the metabolism of the lake. In such a lake by far the greater part of the iron and phosphorus dissolved in the hypolimnion does not arise from current production but rather from the quantities accumulated in the sediments over many years.

For an understanding of the material economy of a lake under these conditions it is important to know how much the sediments influence the formation of the chemical stratifications, a goal from which we are still far removed, all the more so because individual lakes exhibit very great differences in this respect, dependent on many factors. We shall therefore have to confine ourselves to discussing briefly the basic principles on which the interaction of the sediments and free water is based, as well as the manner in which the changes resulting therefrom can be communicated to the water mass of the lake.

Like the plant-producing soils on land, the sediments of lakes in the temperate zone are an extremely complex system of crystalloid mineral components of the most diverse particle size, of inorganic and organic colloids, as well as of living organisms. It is possible only in exceptional instances to recognize the role played by the individual components; we usually perceive merely the results of the total metabolism in the system. The decomposition of organic substances in the interior of the sediments leads to an accumulation of decomposition products, which can be removed only by diffusion or by the strongly reduced turbulence in the hypolimnion. A diffusion gradient towards the ooze surface becomes established, over which CO_2, NH_4, P, Fe, Mn, etc. reach the water at the ooze contact.

A second exchange current proceeds in the opposite direction from the water into the ooze, for materials are also withdrawn from the water by the sediments. This withdrawal takes place in part through *chemical combination,* for example, of oxygen, and through exchange of ions in the zeolitic portions (permutites) of the sediments, and in part through *adsorption.*

By *adsorption* is commonly understood the uptake, or, better stated, concentration, of dissolved substances (or gases) at the surface of a body. By this means an equilibrium arises between the adsorbed portion and that remaining in solution determined by the so-called "adsorption isotherm," which can be expressed by the formula

$$\frac{C_a^n}{C_d} = k$$

in which C_a is the adsorbed portion, C_d the portion remaining in solution, and k and n are constants varying from one case to another. Out of this relationship arises an important difference compared with the processes that are based on chemical combination. Whereas in the case of the latter only the stoichiometric ratio of the two participants is significant and the resulting combination is *not* reversible, the quantity of adsorbed material with respect to the portion remaining in solution is determined only by the above-mentioned equilibrium, and the process is therefore indeed *reversible,* that is, with a decrease of the concentration in solution the quantity adsorbed also decreases. Hence, materials adsorbed from solutions can be completely washed out by treatment with pure water, and they are therefore also available to organisms through nutrient uptake.

The number of substances that function as adsorbents in the sediments of lakes is very large. Besides finely divided crystalloids, among which clay particles play the dominant role, there are chiefly the inorganic and organic colloids: gels of ferric hydroxide and silicic acid, humus colloids, the polymorphic inorganic and organic complexes, and finally the surfaces and integuments of living and dead organisms. Separation of the effects of these various adsorbents in the sediments is in general scarcely possible. Ohle (1935) in particular has pointed to the significance of colloid gels as nutrient regulators in lakes. There is also an investigation by Einsele (1938) concerning the role of ferric hydroxide, which will here be described briefly, serving at the same time as an example of the fundamentally different behaviour of chemical combination and adsorption and the effects they produce in the metabolism of a lake.

If oxygen is supplied to a body of water containing both ferrous and phosphate ions, a precipitation occurs. If the ratio of the two ions in the solution corresponds exactly to the stoichiometric ratio of Fe : P = 1 : 0.55 in the ferric phosphate molecule, the precipitate consists exclusively of ferric phosphate, and the overlying solution is practically free of iron and phosphorus. If, as is generally the case in lakes, an excess of ferrous ions was present, there is deposited in addition a corresponding quantity of ferric hydroxide, and the precipitate consists of a mixture of both compounds. The quantity and composition of these *chemical* compounds remains unaffected, regardless of the mineral content of the overlying water. Only when reduction occurs with the disappearance of oxygen are the ferrous and phosphate ions again set free. But since the precipitation is repeated in the next period of circulation, both elements remain continually withdrawn from the zone of production (p. 91).

However, when we suspend *precipitated* ferric hydroxide gel in water that contains phosphate, there also occurs a decrease in the phosphorus content of the solution, this time not through chemical combination but rather through adsorption on the ferric hydroxide gel. In accordance with the equilibrium described above, a portion of the phosphate remains in solution, and we can again bring the adsorbed portion completely into solution by washing the gel with phosphate-free water. This process is significant for the metabolism of lakes in those instances where the surface of the sediments is continually in contact with water containing oxygen, hence in oligotrophic lakes. Here the phosphate liberated in the decomposition of proteins is adsorbed by the film of ferric hydroxide that almost always covers the surface of the sediments. At the time of complete circulation this adsorbed phosphate is again washed out into the overlying currents of the phosphate-poor water and returned into the production of the lake.

Just as in the iron cycle, so in the formation and mode of action of other gels, chemical combinations and adsorption as well as transitions between these two processes are probably associated with one another, and it is the task of future investigations gradually to clarify these complicated relationships.

There still remains to be discussed the question of the means by which the changes in the composition of the solution originating at the ooze contact are transported, and how they can influence the stratification of the free water mass of the lake. That molecular diffusion can play only a very subordinate role in this process has already been impressively demonstrated by Alsterberg's (1935) calculations on the slowness of diffusion. Transport can therefore result only from water movements. In shallow bodies of water or in the epilimnion of deeper lakes this question requires no particular discussion. The currents set up by the wind and the resulting eddy diffusion provide for a rapid carrying away of the products of decomposition being liberated at the surface of the sediments, and supply these again immediately to the production going on in the trophogenic layer. The metalimnion and the hypolimnion, however, are withdrawn from the direct influence of the wind, and because of the stratification in density, eddy diffusion in a vertical direction is greatly curtailed. Are water movements here also, particularly at the ooze contact, effective in bringing about an appreciable transport of material? Large-scale water movements are produced even in these layers by *internal seiches* (p. 47), the significance of which in the production of vertical eddy diffusion Mortimer (1941–2) has pointed out. But the sediment contact itself is the site of processes that lead to water movements. First of all, the temperatures of the sediments and the water are scarcely ever exactly the same, as we know from the investigations of Birge, Juday, and March (1928). The sediments are colder in the summer and warmer in the winter than the overlying water. These *temperature differences* must lead to convectional streaming of the water at the sediment contact, which, insignificant as it might seem, can in the course of time result in a considerable transport of material. A further source of streaming not to be underestimated is the *movements of organisms*, not only those of the large ones like fish but also those of the small animals of the ooze and the overlying layers of water, whose small size is abundantly counterbalanced by their number and activity: these are the tubificids, the snake-like movements of which cause a lively turbulence in the layers near the bottom, the chironomids which, while remaining in their dwelling tubes, produce the same effect, the host of crawling and swimming organisms of all sizes down to the size of bacteria, whose energy of motion according to the laws of reaction must continually bring about a corresponding motion of the water. When one considers that the number of swimming micro-organisms—cilates, flagellates, and bacteria—amounts to thousands, or even hundreds of thousands, per ml., one can understand what great significance their movements can have for eddy diffusion.

In which direction will these movements arising at the ooze contact be propagated, and will the exchange of materials follow them? When a movement is started in a water mass stratified with respect to density (as is always the case in the metalimnion and in at least the upper portion of the hypolimnion), it encounters different resistance in different directions. Resistance is greatest along the stratification gradient, hence in a vertical direction, and least within the plane of stratification, hence in a horizontal direction. For this reason each movement, however it might be oriented initially, finally is turned aside in the direction of least resistance—the horizontal. What holds true for the mixing of large masses of water holds true also for turbulent eddy diffusion, which is effective only to a minor extent in a vertical direction, occurring predominantly within homothermal layers. It follows that transport of material from the walls of a lake basin likewise is mainly horizontal so long as a thermal stratification exists. This conclusion can be confirmed by direct observations, such as the intercalation of "plates" of local contaminations confined to a definite depth extending far out into the lake, or by the investigations of metalimnial oxygen maxima or minima (p. 79).

If the density is uniform in a water mass (as is nearly the case in the lower hypolimnion of deep lakes), then the influence described above which is directed horizontally ceases, and eddy diffusion, whether it is set up by internal seiches or by some other force, can spread unimpeded even in a vertical direction.

The direction of transport of materials from sediments into the free water is therefore determined by the distribution of temperature or density. For the extent of influence of this transport on the chemical stratification of a lake, however, the morphology or the profile of the lake basin is the main controlling factor, a relationship that had already been closely investigated by Alsterberg (1927) with respect to oxygen stratification and has been demonstrated by Hutchinson (1941) through exact investigations of the distribution of various qualities. The influence of the sediments is the greater the smaller the lake, or the larger the surface area of its sediments with respect to the volume of water. For the same reasons, level portions of the basin profile, because of their more extensive ooze contact, have a greater influence on this chemical stratification than do steep portions. In deep lakes with very steeply sloping bottoms, for example, in many mountain lakes and crater lakes, the transformations taking place in the sediments have a very minor influence upon the conditions in the free water. This can be inferred from the oxygen curves of our large oligotrophic lakes. However, in small oligotrophic lakes the influence of the bottom profile upon the oxygen curve is clearly recognizable, as already explained on page 80.

III. THE COMMUNITIES OF BOGS

A group of aquatic biotopes that cannot be inserted directly within the framework of our sketch of the ecological relations in standing inland waters is that of *bogs* (moors in European terminology). They are distinguished so markedly from other types of waters in their physiognomy, in the composition of their community, and in the physical-chemical peculiarities of the biotope, and are so uniform in their features over the entire earth, that we must devote special consideration to them, although too short, and restricted to the main features.

We shall best visualize the uniqueness of this community if, as a synoptic example of a primitive bog biotope, we consider in detail a small "raised bog" (*Hochmoor*) surrounded by forest. Seen from a distance it has the appearance of a clearing in the forest; more closely regarded, it seems to be a convex, watch-glass shaped portion of the forest floor, except that it does not consist of primary soil but exclusively of *peat*, which is plant material in the first stage of coal formation. This peat is saturated with water and imparts to the raised bog the characteristics of a gigantic, completely saturated sponge. The dome is surrounded by a swampy *moat*, the so-called *Lagg*.

The entire structure is distinguished by a highly characteristic concentric arrangement of vegetation zones. The spruce forest advancing through the moat decreases rapidly in height, and its pioneers, pressing forward to the edge of the bog, are stunted little trees of extremely slow growth. Specimens scarcely ½ m. high and as thick as one's thumb are revealed to be a hundred years old by a count of their annual rings. Adjacent to this "conflict zone" (ecotone) of the forest there is a girdle of crooked pines (*Pinus montana*), which here, as in the high mountains, replace the spruces. Beyond this zone there extend towards the bog the dwarf ericacean shrubs *Vaccinium myrtillus* (bilberry), *V. vitus idaea* (red whortleberry), *V. uliginosum* (bog whortleberry), *V. oxycoccus* (cranberry), *Andromeda polifolia* (rosmarin heather), and, especially in northern Germany, *Ledum palustre* (Labrador tea). The open bog surface occupies the almost level centre of the entire structure. It is taken over for the most part by the bread-loaf-shaped, often coalescing *hillocks* of peat mosses, and indeed there are just certain species of *Sphagnum* that are known to be producers of raised bogs. Between these hillocks lies a network of small, shallow pools, the *bog puddles* (*Schlenken*). The entire bog surface is swampy and difficult to walk upon; in places, where the surface is broken, deeper accumulations of water, the *bog pools* (*Blänken*) are formed. Relatively few species of vascular plants, but these occurring repeatedly in all bogs, extend into the peat moss, forming sparse stands:

several species of *Carex* (*C. limosa, C. pauciflora*), cotton grasses (*Eriophorum, Trichophorum*), the bladder rush (*Scheuchzeria palustris*), which is especially characteristic of this biotope, the club moss (*Lycopodium inundatum*), and in drier places the delicate rosettes of the "insect-eating" sundew (*Drosera*). In the hillocks grow certain liverworts (for example, *Cephalozia fluitans*), which kill the *Sphagnum* locally by entanglement and promote the formation of bog puddles in the scars arising arising in this manner.

With respect to *chemistry*, the *bog waters* (the bog puddles, bog pools, and lakes imbedded in the bog) are distinguished primarily by the following three characteristics: (1) an unusual *poverty of dissolved electrolytes*, especially lime: the total concentration corresponds approximately to that of rain water and amounts to scarcely a tenth of the quantity normally occurring in waters fed by springs; (2) a strongly *acid reaction*: the pH value in raised bogs usually lies between 3.5 and 4.5 and in flat bogs (*Flachmoor*) between 5.0 and 6.0; (3) a high content of *humic materials*, which impart a yellow to brownish colour to the water often in layers already of moderate thickness. In addition, especially in the bog puddles and bog pools, the *substrate*, which consists exclusively of organic material, exerts a strong influence. Vigorous decomposition processes that release much carbon dioxide are going on in the peat. For this reason these shallow pools continually have a pronounced excess of carbon dioxide, even at times of intense assimilation of the microflora. Moreover, reducing substances are given up to the water, which bring about an oxygen consumption. As a consequence we find in bog puddles that are scarcely 10 centimetres deep extremely pronounced stratifications of carbon dioxide and oxygen, which under the influence of diurnal fluctuations in temperature are subject to rapid change (Redinger, 1934).

The microscopic world of life of these aquatic bog biotopes, which often fills the shallow bog puddles in particular with a grey-green soup, is if possible even more characteristic than the surrounding macrophytic vegetation. Its features on the whole are so uniform wherever bogs occur, both in the arctic and temperate regions as well as in the mountainous regions of the tropics, that a quick look into a microscope is sufficient for recognition that a given sample belongs ecologically to these biotopes. The *microflora* of the bog puddles is characterized first of all by a complete dominance of the groups so rich in beautiful forms, the Desmidiaceae and Mesotaeniaceae, which can populate a *single* bog water with hundreds of varied species, and by a few blue-green algae (*Chroococcus turgidus, Stigonema, Hapalosiphon*, and others), which are constantly present and often very abundant. The other groups of algae for the most part contribute relatively few representatives to the bog flora, but these are often characteristic. This is true of the chrysomonads and heterokonts (for example, *Chlorobotrys*), the Dinophyceae (for example, *Gloedinium montanum*), and the diatoms (represented in bogs for the most part only by *Frustulia saxonica*, some *Eunotia*, and at times

Pinnularia, whereas in other biotopes this group is usually very rich in species). Of the Zygnemataceae there occurs *Zygogonium ericetorum,* usually distinguished by the possession of violet cell sap, a species which forms brownish-violet mats in the bog puddles; and of the Chlorococcales the strikingly large genus *Eremosphaera* is especially worthy of mention. The number of ubiquitous or generally distributed species is relatively small in bogs, and these scarcely ever occur as dominant forms.

The predominant role, which in the *microflora* of bogs is played by the desmids, in the *fauna* belongs to the testaceous rhizopods; these occur in a similarly great richness of species and include some forms that are restricted to this biotope. However, the number of ubiquitous species, or at least of those also occurring in other biotopes, is greater among the animals than among the plants. Several rotifers in particular can be designated as typical bog inhabitants, and in addition the cladocerans *Streblocerus* and *Acantholeberis* and also a few copepods. The bog fauna is more strikingly characterized by the lack or inconspicuousness of entire groups, for example the ostracods, the Hydracarina (which here are represented by the Oribatidae), the molluscs, and a number of groups of insects (for example, the Ephemeroptera).

Which environmental conditions bring about this rigid selection and give to the bog community its individuality? On the basis of the observation that all bog waters are poor in lime, the calcium content was considered to be the controlling factor, and the species were said to be "calciphobic." Precise investigations have demonstrated, however, that it is not the calcium ion that prevents specific bog organisms from conquering the lime-containing biotope, but rather the alkaline reaction constantly associated with a lime content in natural waters. The plants of the bog are therefore "acidophilic" or, better stated, "alkaliphobic." Whether this is true for animals or whether their behaviour is determined by some other factors has not yet been investigated. The research of Mevius (1924) has shown that the "alkaliphoby" of bog species is probably associated with changes in their plasma permeability, in the sense that at an alkaline or insufficiently acid reaction the permeability is increased and the cell becomes "flooded" with salts. For this reason many species can live even at pH 7 in solutions virtually free of salts, whereas they require pH 4 to 5 at somewhat higher concentrations. It follows that the *low electrolyte content* of bog waters is a very important ecological factor closely associated with the effect of the *acid reaction,* and, further, that certain narrow ranges of pH cannot be stated to be the range of toleration of particular species without reference to the concentration of electrolytes. Nevertheless, observations indicate that the number of species in a bog biotope decreases as soon as the reaction becomes *strongly* acid. Thus, approximately 180 species of desmids have been determined from the floating mat of Lunzer Obersee with a pH of 4.5 to 5.5, in contrast to only about 40 species from the adjacent more acid (pH 4 and less) raised bog, *Rotmoos*. Yet these differences are far less striking than those encountered in a comparison of alkaline and acid waters.

One can say in general that the *neutral point* forms a scarcely surmountable barrier for many plant organisms, and that from this circumstance the chemical reaction of the water becomes an *ecological factor of the first order.*

And yet the reaction *alone* is not able to control the specific peculiarity of the bog biocoenoses. In other acid biotopes, such as accumulations of rain water and similar situations, we find a biota which indeed has numerous clear-cut resemblances to that of the bog puddles, yet differs considerably both quantitatively and qualitatively from the latter. We can deduce with great probability that besides the acid reaction and the low salt content, the large quantity of *humic colloids* dissolved in the bog water is of decisive importance for the composition of this community. The humic substances can be effective both physically, as colloids through their adsorptive properties, and chemically (humic acids), and they produce a *milieu* of a special type, in which the bog flora finds its optimum conditions of existence.

Another question not always easily answered is why so many plants, even those which can thrive elsewhere in an alkaline *as well as* an acid environment, avoid the bog (the "zone of conflict" of spruce and other forest vegetation). An assumed toxicity of humic acids has been argued. This is not the important factor in all cases, however, as is evidenced by the fact that spruce seeds germinate abundantly on the raised bog and grow to a height of several centimetres, but then die. This condition is referable with certainty to the deficiency of oxygen in the saturated bog soil. In addition the scarcity of certain mineral nutrients (for example, calcium) may exclude many species.

How does the acid reaction of the bog waters originate? There is as yet no consistent interpretation for this; first one possibility and then another is pushed into the foreground. The cause of this uncertainty lies primarily in the fact that in these unusually dilute and therefore almost unbuffered solutions, minimal quantities of an H^+ ion dissociating substance, hence an acid, suffice to bring about a considerable lowering of the pH, and consequently it is difficult to determine these acids chemically. It is possible that the prevailing condition of pH in a bog water is never determined by *one* cause but rather by the combined action of several physical-chemical processes that liberate H^+ ions. Even the high *CO_2 content* of an unbuffered water must produce an acid reaction, as is indeed the case with distilled water. As a matter of fact the pH value increases when water from a bog puddle is subjected to a vacuum and the CO_2 is driven out. In addition (especially according to the interpretation of Sven Odén, 1922) the acid nature of the dissolved humic materials (*humic acids*) can be effective. Furthermore, through the activity of the *sulphur bacteria* always present in bogs, traces of free H_2SO_4 are produced. (Everyone who at any time has passed over a quaking bog has noticed the odour of H_2S.)

The *adsorption of bases*[64] (Baumann and Gully, 1909 to 1913) by the peat colloids should be mentioned as an especially effective factor and one which continually ameliorates disturbances. Both the living *Sphagnum* membrane and the peat itself have the remarkable property of adsorbing the base (for example, calcium) from dissolved salts, and of setting acids free. If tap water is filtered through a sufficiently thick layer of peat or *Sphagnum*, it becomes reduced in lime and acid in reaction. This property, which is similar to that of a buffer, is especially important for the bog biocoenose, because it can within certain limits so alter, and to a certain extent so assimilate, the alkaline water that gets into the bog by one means or another, that this water becomes similar to the bog water with respect to its reaction and its salt content, and can therefore cause no disturbance to the community.

The *origin of a bog* can be understood forthwith from a knowledge of the physical-chemical conditions of its existence, or at least the fundamental processes. Three conditions must be present: an abundant supply of (atmospheric) water, a high atmospheric humidity (whence the widespread occurrence of bogs during the Atlantic Climate), and a production of plant substance exceeding decomposition. Wherever these conditions occur, bogs can originate: hence where springs emerge (spring bogs) or on moist slopes (hanging bogs), but most abundantly where there are accumulations of water in valleys and lowlands. Most instructive indeed are those cases in which a raised bog represents the *final stage of senescence of a lake*, and such an example will be considered briefly (Figure 60). Through deposition of sediments a lake becomes continually shallower. With progressive filling of the basin, the shore flora at the same time continually advances further towards the centre of the lake and finally covers its entire surface. Filling in is accelerated by the large quantities of new plant substance formed annually, and when the sediments finally reach the original surface of the water, a *marsh* has been formed from the lake. At the same time, however, there also occur marked changes in the physical-chemical conditions of the environment. Through the great production of CO_2 by the decomposition processes and through the effect of the resulting

[64]The question whether this is a true adsorption or a chemical process in the nature of an "exchange" has been answered in favour of the latter by Anschütz and Gessner (1954).

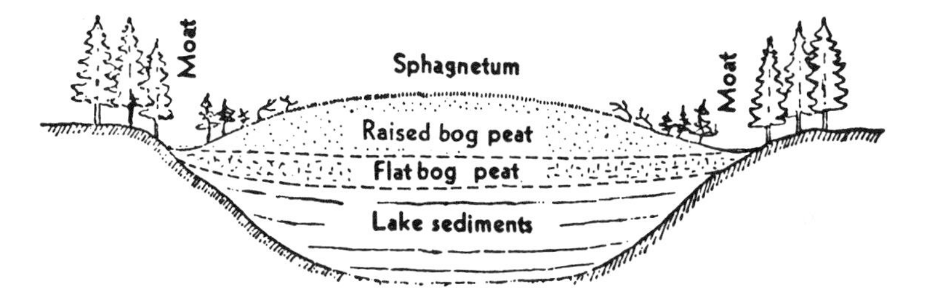

FIGURE 60. Cross section through a raised bog that has arisen from a small lake.

humic colloids, the pH of the originally alkaline water is lowered. It sinks below neutrality, and plants of the acid environment become established: certain sedges, the cotton grasses, the marsh trefoil (*Menyanthes*), and others, as well as characteristic mosses (for example, species of *Calliergon* and *Drepanocladus*). A *flat bog* is formed with a water-saturated bottom of cyperacean peat and pH values about 6. With increasing acidity the preliminary conditions for the colonization by peat mosses are provided. First appear the less specialized species, such as *Sphagnum teres, subsecundum, platyphyllum, cymbifolium,* and others. These, by means of the previously described base adsorption, bring about a further lowering of the pH, and now the typical hillock formers become established—*Sphagnum magellanicum, papillosum, fuscum,* and *rubellum.* These, in closely crowded cushions growing vertically upwards and dying at the base, gradually bring about a convexity of the entire surface through anastomosis of their hillocks, and build up the characteristic shape of a *Sphagnum* peat *raised bog.* Transitional stages with raised bog plants but as yet without the convexity are designated as *intermediate bogs* or *transition bogs.*

A particular development of bogs, the *quaking bog,* deserves special mention. It is a type of marginal vegetation of lakes in which the plant cover, which is only partially rooted in the bottom and elsewhere floats like a raft, gradually pushes out from shore towards the middle of the lake. It consists of a disc of peat held together by a network of roots, which is wedged against the shore and which participates in the fluctuations in water level of the lake, hence can never be flooded over. The vegetation of the lakeward-extending edge, for example in the extensive quaking bogs of Lunzer Obersee, consists of the vigorously growing, hillock-forming sedge (*Carex diandra*), as well as of marsh trefoil (*Menyanthes trifoliata*) and the finger fern (*Comarum palustre*), whose horizontal runners extend into the open water and like raptorial arms hold secure any drifting material. *Only* the floating portion of the mat bears a typical raised bog vegetation, solely because an inundation with alkaline lake water can never occur, and the water penetrating into the quaking bog from below is decalcified by the peat. It is a peculiar sight to see floating on a lake of high lime content a decidedly "calciphobic" flora with many species of *Sphagnum*, *Trichophorum alpinum*, *Carex limosa*, sundew, etc. The firmly attached portion near shore, which becomes flooded over at high water, is, in contrast, free of raised bog plants and exhibits a normal marsh vegetation. It is obvious that quaking bogs form primarily on small lakes with weak wave action, and it seems that they find favourable conditions for growth only over shallow benches and in protected bays.

Even larger *lakes* that are located in bog regions and are not maintained by alkaline affluents can assume bog characteristics. Such instances are especially common among the Scandinavian lakes. Their water exhibits a very low salt content and an acid reaction, and is coloured yellow to brown. Because the production of these lakes in Sweden is usually very low, Naumann (1921), in his system of lake types, regarded these waters as an extreme case of oligotrophy, and arranged them under the name "dystrophic type," in a certain sense as the final stage of the trophic series. Yet Thienemann (1925, in *Die Binnengewässer*) recognized early that this is a *fundamentally* different type, which is not subordinate to but co-ordinate with the production series of generally alkaline lakes. He differentiated *clear-water lakes* and *brown-water lakes*. Both groups can exhibit different degrees of trophy. Although the majority of brown-water lakes must be designated as poor, yet the total production of algae in a bog puddle proves that bog water can very well be eutrophic. The *deciding difference* between the two series is the presence or lack of the characteristic properties of bog waters, primarily the acid reaction. In the majority of cases it is correct to designate the *alkaline type* as clear-water lakes and the *acid type* as brown-water lakes, although there are also acid (usually weak) clear-water lakes. With respect to chemical stratification, the *brown-water lakes*, although low in production, continually exhibit an *oxygen depletion*, which is not attributable to production within the lake but rather to the utilization of oxygen by organic substances originating in the surrounding bog. In conjunction with the oxygen depletion there is also an enrichment of the hypolimnion with the products of decomposition (CO_2, NH_3, P, and usually Fe and Mn in addition) (Müller, 1937, 1938).

IV. THE COMMUNITIES OF RUNNING WATER

Up to this point only the eulittoral surf zone of lakes has provided the occasion to discuss the influence of water movement upon the composition and configuration of a community. Yet the phenomena encountered in running water[65] are much more homogeneous and pronounced. In addition there is a well-known and pronounced difference between the relationships in brooks and streams on the one hand and in the eulittoral of lakes on the other. In the eulittoral zone the water movements are produced by wave action, which to be sure is able to elicit physiological effects similar to those observed in flowing waters, but by which the water particles oscillating back and forth remain essentially in *one* place. In a brook or river, on the other hand, the entire water mass independent of all secondary movements taking place in it is flowing more or less rapidly, so that the shore and bottom are continually bathed by new water. Hence, when we carry out a time series of observations at one place along the shore on the physical and chemical properties of the water, we are measuring not the same but always a new water mass, which can exhibit completely different characteristics brought about by a variety of influences. The previously measured water has already been transported downstream. These conditions must be considered above all in any study of the temperature relations and the metabolism of a watercourse (Schmitz, 1961).[66] This is also the basis for the marked difference between the stream plankton on the one hand and the littoral or benthic community of flowing water on the other. Whereas the plankton is carried downstream in the same water mass, the benthic community is stationary and is thus continually being bathed by new water and exposed to the mechanical effects of the current. To be sure, the influence of water movement varies greatly in importance depending upon whether we consider the upper, middle, or lower course of a river. But the degree of turbulence in a stream hardly declines so far as to permit the formation of stratifications such as we find in lakes, a characteristic that especially distinguishes flowing from standing waters. Even a deep river is, to some extent, continually in a condition of complete circulation.

A second characteristic of incisive biological significance is the transport of water masses over very great distances, even through contrasting climatic provinces, finally ending after a longer or shorter time with the emptying into the sea and the resulting death of most organisms carried

[65]In his major communication before the XIV International Limnological Congress in Vienna (1959), Schmitz (1961) reported that the discharge volume of all the rivers of the world amounts to 37,000 km.^{3} per year, as compared with 200,000 km.^{3} for the volume of the lakes. This discharge volume would cover the surface of the earth with a layer of water 25 cm. thick.

[66]Since, indeed, currents also arise in standing waters, similar phenomena can likewise occur there, especially in the larger lakes. They are limited, however, by the size of the basin and hence are much less effective than in flowing waters.

along by the current. The duration of this transport, which is dependent on the length of the river and on the current velocity, determines whether or not there is the development of a true plankton biocoenose—a *potamoplankton*. We can, indeed, capture plankton organisms with a net in many streams, particularly those containing lakes, ponds, or floodplain waters in their watersheds; but we do not know whether these plankton organisms are able to grow and reproduce under the altered conditions, or whether they are merely "tychoplankton" unable to adjust to river conditions. It has already been demonstrated by older investigations on the Rhine, Oder, Havel, and Mississippi rivers that the plankton entering the river course from the lakes or floodplain waters in general decreases quantitatively downstream, and the more rapidly the stronger the current is. Chandler (1937) and Reif (1939) obtained very clear-cut results in the outflowing streams of small plankton-rich lakes in the states of Michigan and Minnesota, respectively, in which they demonstrated a surprisingly rapid decrease in the net plankton. It could also be demonstrated that the origin of this rapid decrease is associated with the biotal film (Bewuchs) of the river bed containing submersed vegetation.

Investigations in the outflowing streams of Lunzer Obersee and Untersee (Ruttner, 1956) led to essentially the same results for the nannoplankton. Already a couple of hundred metres below the lakes the nannoplankton had completely disappeared. In these instances the filtration effect of submersed phanerogams was lacking, although it is likely that even here a surface effect of the brook bed, which is provided with large stones overgrown with moss, was effective, in the manner that suspended particles are retained at the substrate interface where there is scarcely any current, as explained on page 229. It is evident that this effect was influenced to a large degree by the stage of the stream: at high water there was a much smaller decrease of the plankton.

Associated with this removal of plankton from the outflowing waters of lakes there is a phenomenon that is very important for production biology: it is well known that lake effluents are especially rich in animal life, which nourishes itself on the plankton retained here (Illies, 1956). On the other hand, the flowing water always contains benthic algae and also animals that have been wrested loose from their substrate (Butcher, 1932).

We can speak of a true potamoplankton only when the prerequisite conditions for it have been provided, so that a special biocoenose adapted to these particular conditions can develop through selection of the species that are washed in. As a general rule this is true only in slowly flowing streams of great length, for example, the Volga (treated by Behning in a monograph, 1928), in which even the high water of spring requires almost two months to reach the sea. Within such periods of time, which are even longer in summer, the majority of plankton organisms find a sufficient range for the completion of their generations. Even in rivers that are relatively short there can be a rich development of true plankton forms with a rapid rate of reproduction, as indicated by the studies of Schmitz (1961) on the Werra, which is only about 300 km. long. Schmitz could clearly distinguish three zones: an upper zone free of plankton, a middle zone populated only by tychoplankton, and a lower zone populated in summer by planktonic *Cyclotella* and by *Thalassiosira* at densities of 10^4 to 10^6 cells per ml.

With respect to the biotas of the shore and bottom of streams, the slower the current is the more the composition and configuration of these communities approach those of standing water. Hence, if we wish to become acquainted with the influence of water movement in moulding life, we shall study it best in the rapidly flowing mountain brooks.

It has already been explained on page 50 that in moving water the current can be either laminar or turbulent, depending on the current velocity as well as on the diameter and the roughness of the channel. At the current velocities in our brooks and rivers, which usually lie between 0.5 and 3 metres per second, and at the prevailing width and roughness of their channels, we can expect only turbulent flow under natural conditions. This consists of eddies and unordered currents at right angles to the direction of flow produced by movements within a water mass as it is displaced downstream with a certain velocity. The velocity of flow is not uniform over the cross section of a channel: it decreases especially in the direction of the bottom and the shores. Directly over the bottom of the channel in the so-called *boundary layer* (Grenzschicht) there is a very rapid decrease in velocity, which approaches zero at progressively smaller distances from the bottom. The thickness of this boundary layer with its reduced current velocity, which generally on the average amounts to a few millimetres, is dependent not only on the current velocity (the faster the water flows, the thinner the boundary layer is) but also on the viscosity, and hence is thicker in cold water than in warm, according to page 11.

The boundary layer with its almost non-moving film of water closely appressed to the substrate is of great significance for the microscopic biota, for the attachments of algal spores, and so on (p. 239). Likewise, the *dead water spaces* arising downstream from obstructions are important for the life of many of the larger animals (insect larvae, fishes) (Ambühl, 1959; Schmitz, 1961). As is apparent in Figure 61, there is

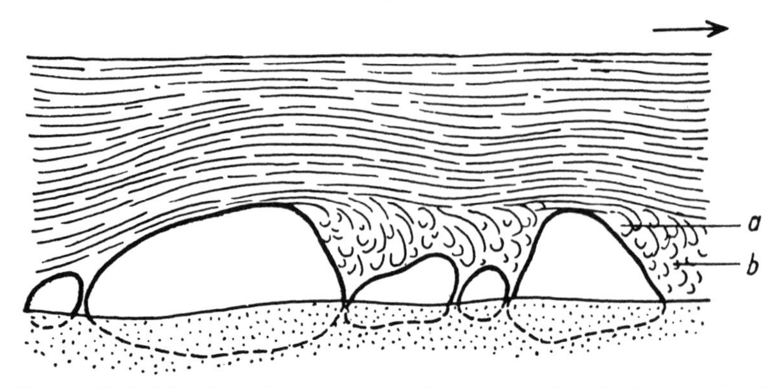

FIGURE 61. Origin of dead water spaces behind obstructions in the current. After Ambühl (1959), somewhat modified. *a*, leeward margin; *b*, dead water space.

formed downstream from a submerged stone in the angle between the current flowing over it and the substrate a circumscribed region affected only by turbulence, from which the main stream flow is almost excluded, and in which therefore the animals find protection against being washed away.

The *leeward edge* of a submerged obstruction provides an especially favourable biotope against being washed away. The zone where the eddy that fills the dead water space separates from the main stream flowing down the valley is a zone that is almost free of current processes, according to Ambühl (1959). This zone is often clearly marked by a row of animals on the substrate there (Figure 62), and in it likewise the first moss coverings are able to establish a foothold.

According to the velocity of the water movements hydrologists distinguish a *streaming* and a *rushing*. The quantitative difference between the two can be made clear by the circumstance that "rushing" is faster whereas "streaming" is slower than the speed of propagation of a wave formed on the surface, such as that produced by throwing a stone into the water. Current velocity in natural waters depends not only on the stream gradient but also on the stream width and water depth, as evident in the following table from the publication of Einsele (1960), in which v is the current velocity per second, and Q is the volume of flow per second:

Gradient	Stream width, 200 m. Stream depth, 4 m.		Stream width, 20 m. Stream depth, 0.5 m.		Stream width, 2 m. Stream depth, 0.25 m.	
(m./km.)	v(m.)	Q(m.^{3})	v (m.)	Q (m.^{3})	v (m.)	Q (m.^{3})
0.5	1.5	1200	0.5	5	0.25	0.125
1.0	(2.5	2000)	0.6	6	0.35	0.175
2.0	—	—	0.8	8	0.50	0.250
5.0	—	—	1.3	13	0.70	0.350
10.0	—	—	1.8	18	1.00	0.500

The measurement of current velocity is generally accomplished by means of a *current meter*, consisting of a propeller set in motion by the water and provided with a revolution counter. For certain purposes such as the measurement of currents in the biologically important layer near the bottom the use of an L-shaped *pitot tube* is recommended, especially the improvement by Prandtl (1931), by means of which the current velocity is determined from the pressure exerted in the direction of the current. Also recommended is the use of a hot wire current meter,

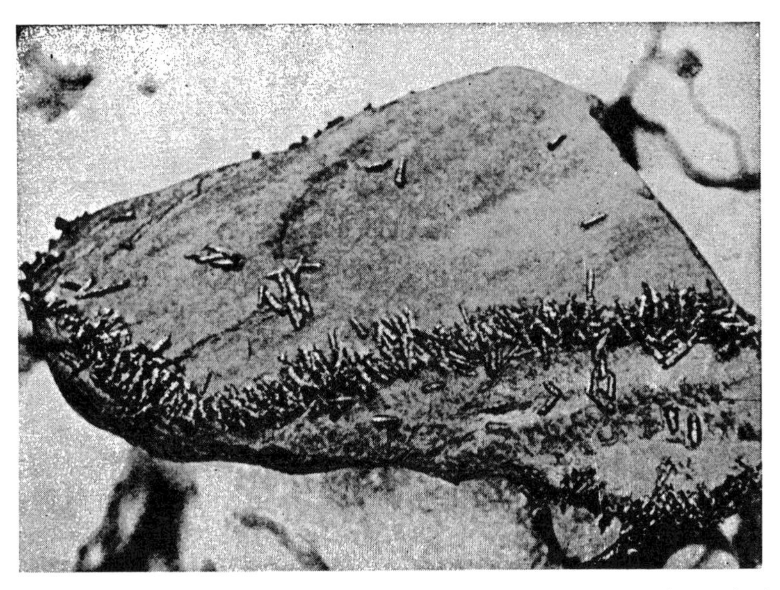

FIGURE 62. Concentration of trichopteran larvae (*Halesus*) at the leeward edge of a stone located in the current. After Ambühl (1959).

in which a heated thermal probe is cooled more or less strongly according to the current velocity, and this in turn produces corresponding changes in the thermal current recorded by a galvanometer. For the study of small current velocities at the bottom a method developed by Pomeisl (Pleskot, 1953) gives good results, whereby the direction and speed of propagation of a small cloud of colour introduced by a pipette is observed.[67] The current velocities determined with a current meter at different depths in a measured cross section of a stream serve above all for the calculation of the volume of flow per second.

[67]A very interesting method for measuring the intensity of weak and especially turbulent currents is that recommended by McConnell and Sigler (1959), in which the loss of weight of standard salt tablets is determined after they have been exposed to the current for a certain period of time.

The influences of a current of water are manifested not only in the *morphology* of the brook organisms but also in the *quantity of organic production* per unit area. The latter is immediately apparent when one inspects a mountain brook, in which rapid stretches alternate with slowly flowing *lenitic* stretches having the same substrate. In the rushing water of the rapids the stones are thickly overgrown with mosses and algae, and in addition there is a richly developed animal life, such as one would not expect in an oligotrophic mountain water. The stones of the lenitic regions, on the other hand, exhibit a much smaller *Aufwuchs* and usually fewer animals as well.[68] If representatives of this *torrential* fauna are transplanted into standing water, many of them perish with symptoms of suffocation even after a few hours.

Early workers were inclined to attribute these specific effects of swiftly flowing water to its higher oxygen content. It is easy to demonstrate, however, that even cascading water seldom has an oxygen content higher than that corresponding to the momentary saturation equilibrium with respect to the air,[69] whereas in standing water supersaturations occur commonly. The effect of strongly agitated water in promoting growth and respiration must therefore have some other basis. In quiet or in weakly agitated water the organisms are surrounded by a closely adhering film of liquid (analogous to the formation of the "boundary layer" described above), which speedily produces around the animal or plant a cloak impoverished of substances important for life. In a rapid

[68]Yount (1956) studied the influence of current velocity and light intensity on the number of species and productivity of diatoms in Silver Springs (Florida) by determining the number of cells and chlorophyll content of the *Aufwuchs*. In slowly flowing shaded locations he found a small production but a greater number of species, whereas in locations with faster current and good light there was a greater productivity but fewer species. This phenomenon of a larger population density of a smaller number of species is also encountered in other biotopes in which an extreme factor severely restricts the composition of the community, whereby, however, the development of those species adapted to the special conditions is not curtailed but rather is favoured (cf., e.g., the communities of thermal waters).

[69]This readily follows from the definition of saturation equilibrium, which is determined by vigorously agitating the water with air until saturation results. The generally small supersaturations or undersaturations often observed in flowing waters can be explained by the fact that equilibrium is not established instantaneously (cf. Järnefelt, 1949; Russ [in Pleskot, 1953]; Lindroth, 1957). Supersaturations can also arise from the condition that cascading water (as in a waterfall) is placed under pressure in its onrushing. The oxygen content of flowing waters that are exposed to organic pollution, especially those with small current velocities, can remain continually below the saturation level in spite of constant addition of oxygen from the atmosphere and in spite of photosynthesis under suitable conditions, a state that is almost always the rule in densely populated industrial regions. Schmassmann (1951) has established types of oxygen regimes for flowing water existing under such influences. On the other hand, under the influence of vigorous photosynthesis sizable O_2 supersaturations can occur around noon and in the afternoon, at times amounting to several hundred per cent saturation.

current, however, the formation of such exchange-hindering investitures is strongly curtailed, and the absorbing surfaces are continually brought into close contact with new portions of water as yet unutilized. In this manner, moving water promotes respiration and the uptake of nutrients much more than quiet water of the same content; it is not absolutely but rather *physiologically* richer in oxygen and nutrients. *A current consequently promotes respiration as well as eutrophication* (Ruttner, 1926).

These favourable effects of strong current are evident, however, only where the floor of the brook or river is composed of bedrock or of coarse stones that cannot be moved by the current. Rubble that can be rolled by the current forms an unfavourable substrate for colonization by algal *Aufwuchs* and an unsuitable dwelling place for animal life, as is likewise sand that is continually being shifted about (Lauterborn, 1916–18; Einsele, 1957).

Concerning the dependence of the particle size of the sediments and hence of the composition of the river channel on the current velocity, Einsele (1960) has presented an instructive table, from which a few data important for our consideration are given below:

Velocity range (cm./sec.)	*Bottom composition*
3–20	Mineral organic mud; large quantities of organic detritus
20–40	Fine sand
40–60	Coarse sand to fine gravel
60–120	Small, medium, to fist-size gravel
120–200	Large stones to boulders

Continuing current velocities of considerably more than 2 metres per second occur under natural conditions only in the bedrock channels of mountain brooks, where at higher stream flows even large stones and boulders are moved along. In large rivers high current velocities of longer duration bring about damage to the banks and scouring of the channel. The latter have resulted in serious consequences (lowering of the ground water level by several metres!) in the artificially regulated streams (Rhine, Danube), where through the narrowing of the channel an increase in current velocity has been brought about. The percentage of the total mass of material being transported that is deposited at a given current velocity Einsele calls the relative sedimentation.

But even a slowing of the current can under suitable circumstances bring about a richer life, although on entirely different grounds than the production increase just considered in stronger current. If the current velocity sinks below a certain value—somewhat less than 20 cm./sec.—then not only do stones and sand lie unmoved but also the lighter organic particles suspended in the water are to a large extent deposited (Einsele, 1960) regardless of whether they arise from the upper course of a stream (e.g. as algal tufts) or are washed into the watercourse from outside. In this sediment enriched with organic materials an abundant animal life can now develop, which lives on this material without being excluded by the disappearance of oxygen, since this is prevented by the high degree of turbulence. Hence, this enrichment of animal life does not come about as in the above-described fast currents by an increase of autotrophic production within the biocoenose of the lenitic region in question, but rather it owes its origin to a production that has occurred at other locations in the water course or even outside of this. Thus, similar relationships are present as in the profundal zone of lakes, where the community thrives only at the cost of those organic substances that reach it from the autoproduction of the trophogenic zone ("Nährschicht").

An especially striking characteristic of the torrential community is the close adaptation of the *life forms* to the current. It is obvious that only those species are able to exist in this biotope which in some manner can offer resistance to the mechanical forces of the current. Among the *plant Aufwuchs* the most important basic forms occurring in moving water have already been mentioned on page 184 (cf. also Jaag, 1938). Especially widespread is the occurrence of a flatly developed thallus applied tightly to the substrate. We find this principle of construction developed very nicely in the chrysophycean *Phaeodermatium*, in several blue-green algae (for example, *Chamaesiphon fuscus* and *Ch. polonicus*), among the green algae for example in the remarkable *Rhodoplax schinzii* whose blood-red layer is characteristic of the torrential flora of the Rhine Falls at Schaffhausen studied in detail by Jaag, in *Gongrosira*, and above all in the beautiful red alga *Hildenbrandia rivularis*. The yellow, brown, red, and green blankets, which frequently cover large areas, impart to the channel of the mountain brook *vegetational colorations* often visible from a distance (Geitler, 1927). A firm gelatinous layer of hemispherical shapes, like that we have found in the surf zone, is likewise formed here by *Rivularia* and species of *Nostoc* and *Schizothrix*. In calcium-containing water these gelatinous growths are often held together by means of lime. Especially remarkable examples are the wart-like or hemispherical growths composed of the united calcareous tubes of the desmid *Oocardium stratum*, in the mouths of which the green cells rest like corks. Floating growths or turfs, which must be particularly strong, are characterized by especially powerful hold-fast organs, for example, the turfs of moss and *Cladophora* firmly attached to stones, or the diatoms of the genus *Gomphonema*, whose gelatinous stalks are strikingly short and thick in comparison with those of their relatives living in standing water.

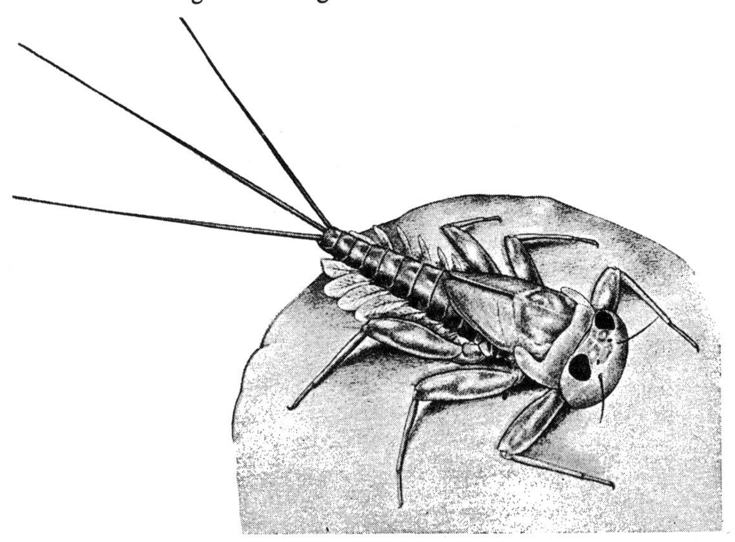

FIGURE 63. *Ecdyonurus austriacus*, seen from above. From Pleskot (1949).

FIGURE 64. *Ecdyonurus* in the current, seen from the side. From Ambühl (1959).

The adaptations that *animals* have developed to life in flowing water are greatly varied. Naturally, the stronger the current is, the less prominent are the free swimming forms. Only fishes (for example, the salmonids with their muscular, almost round, stream-lined bodies) are able to swim upstream in rushing water and to overcome rapids. In weaker currents swimmers also occur among the insects, such as the nymphs of the mayflies *Baëtis rhodani* and *Centroptilum luteolum* which have "fish-shaped" bodies. The number of forms continually associated with a solid substrate is much greater. The dominant principle of construction among these is an extensive *flattening*. This along with a very

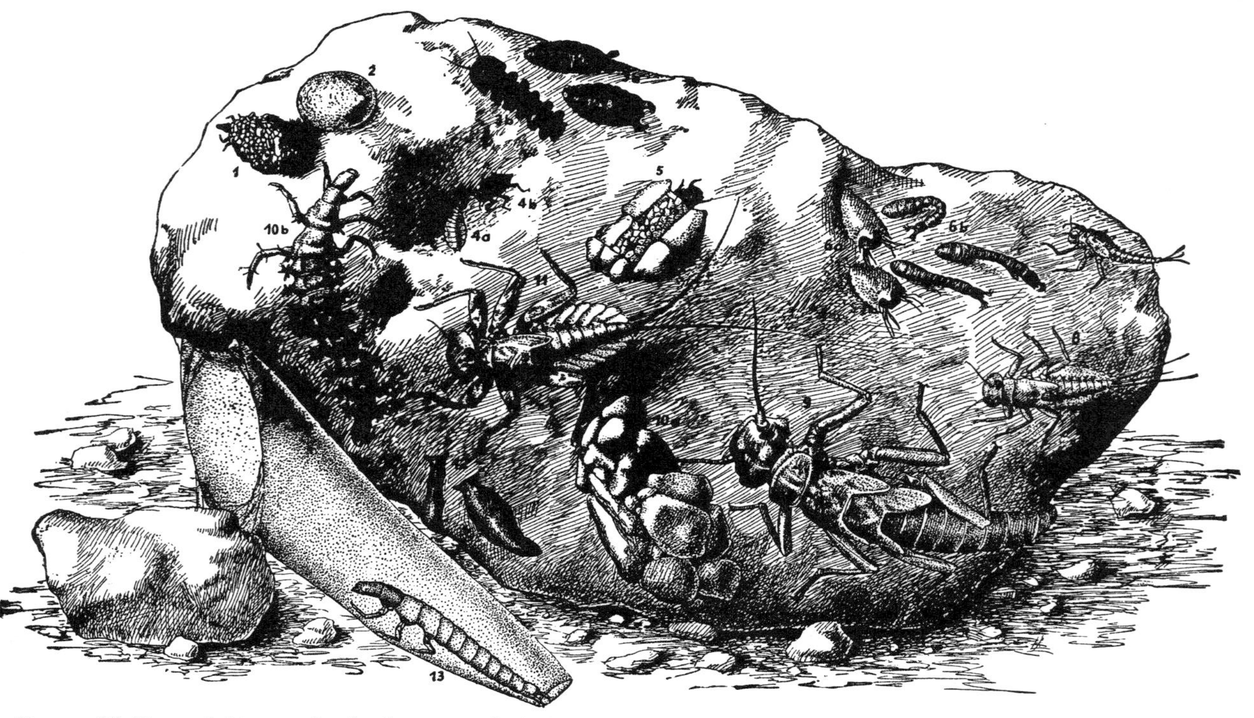

FIGURE 65. Torrential types of animals on a rock. 1, *Synagapetus* (Trichoptera), larva in a case of sand grains; 2, *Ancylus* (Gastropoda); 3, *Liponeura* (Diptera): *a*, pupa, *b*, larva; 4, *Helmis* (Coleoptera): *a*, larva, *b*, imago; 5, *Silo* (Trichoptera), larva; 6, *Simulium* (Diptera): *a*, pupa, *b*, larva; 7, *Baëtis* (Ephemeroptera), nymph; 8, *Rhithrogena* (Ephemeroptera), nymph; 9, *Perlodes* (Plecoptera), nymph; 10, *Rhyacophila* (Trichoptera): *a*, pupal case, *b*, larva; 11, *Epeorus* (Ephemeroptera), nymph; 12, *Planaria alpina* (Turbellaria); 13, *Philopotamus* (Trichoptera), larva in its catching net.

well developed marginal contact, which to a certain extent makes them a part of the substrate, permits the current to pass unimpeded over their flatly-arched bodies. Thereby the vertical component of the current presses their bodies all the more tightly against the substrate. These adaptations are immediately apparent when we compare the morphological appearance of related genera, such as the long-legged, terete nymph of the mayfly *Cloëon* living in standing water, and the larva of *Ecdyonurus* (Figures 63, 64) of mountain brooks belonging to the same order and resembling a flattened crab. A similar efficacious "half stream-line form" is present in the most diverse groups (cf. Figure 65), such as *Planaria alpina*, the larva of the beetle *Helmis*, and especially the nymph of the mayfly *Prosopistoma*, as well as the snail *Ancylus*, which resembles a Phrygian cap, and finally the pupal cocoons of *Simulium*. Even many caddisfly cases composed of foreign materials are arranged in this form: the genera *Synagapetus* and *Thremma* accomplish the adaptation by fastening securely to the edge of their somewhat hemispherical cases a margin of the finest sand grains, whereas *Silo*, *Goëra*, and *Lithax*, which live in cases round in cross section, bring about this adaptive flattened form and at the same time make their cases heavier through the lateral application of little stone wings. In species living in the strongest current there is often developed in addition a special mechanism of attachment, most beautifully illustrated in the dipteran *Liponeura* (Figure 66), belonging to the Blepharoceridae, whose flattened body, resembling that of a woodlouse, bears six powerful suction cups on the under-side, one per segment. These animals occur, often by the hundreds, in cold mountain brooks on the upper surface of smooth flat rocks in rushing water. Frequently associated with them are the larvae of the black flies *Simulium* (*Melusina*), whose blood-sucking adults can be dangerous to cattle. These larvae, easily recognized by their enlarged posterior ends, attach themselves securely to the substrate by means of a terminal corona of innumerable fine hooklets, and are able to withstand a very strong current. A second corona of hooks located on the first thoracic segment assists along with the other in the looping locomotion of the larva. In addition the stream-lined larva is secured by a secreted thread, by means of which it is able to abandon the substrate and float in the current.

The nymph of the mayfly *Epeorus alpicola* (Figure 67) occurring in the cascades of the Alps has developed a special attachment device in the form of a "suction cup" occupying the entire ventral surface, consisting of the overlapping gill membranes. Similar structures occur in the related genus *Rhithrogena* (Ephemeroptera). In addition there are often

FIGURE 66. Suckers on the under side of *Liponeura cinerascens* (Diptera).

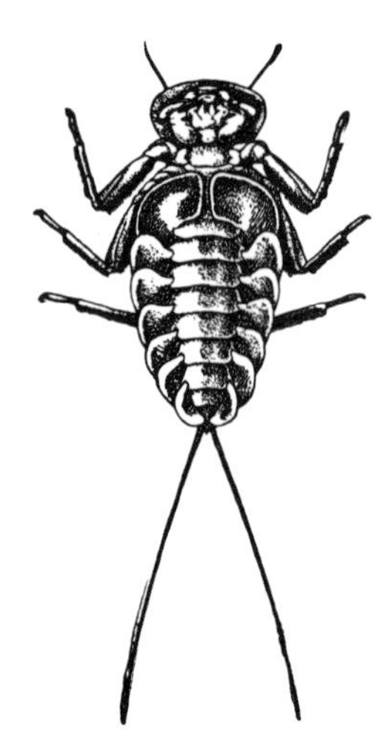

FIGURE 67. Under surface of the mayfly nymph *Epeorus alpicola*.

powerful claws on the extremities for clinging to cracks in rocks (*Rhyacophila*—Trichoptera) and to turfs of moss and algae (*Helmis*—Coleoptera). Instructive in this connection is a comparison of the legs of the lake-dwelling water mites, which are adapted for swimming, with the unguiculate legs of the brook inhabitants of this group.

In general one finds that the *smaller* the brook inhabitants are the less they exhibit adaptations for living in the current. Thus the flagellates and ciliates occurring here, the rotifers and nematodes, and even the small crustaceans, do not show a life form noticeably different from that of those living in standing water. This lack of difference is related to the fact that the velocity of the current in a channel rapidly decreases as it approaches the boundary layer at the bottom, and at submicroscopic distances becomes almost zero. It is thus apparent that a larger organism, which raises itself several millimetres above the substrate, extends into the region of stronger current and therefore must be better adapted than a microscopic organism, which moves about permanently in the slowly flowing film of water covering the bottom. Under other circumstances it would be inconceivable how the delicate swarm cells of algae could remain adherent to a smooth substrate even in the strongest current and contribute to its algal covering.

A biotope similarly protected from the current is that represented by the cracks and small fissures in rocks as well as the spaces between the individual stones of the substrate. These are the domiciles of the "cavity dwellers," the morphology and behaviour of which Pleskot (1953) has studied in detail. Their adaptations (e.g. in the Leptophlebiidae, Figure 68) are typically undulating movements and the perception of contact stimuli (thigmotaxis) (Ambühl, 1959). Obviously the under surfaces of stones, which are often densely populated, also offer an excellent protection.

FIGURE 68. *Habroleptoides modesta* (Ephemeroptera), a cavity dweller. From Pleskot (1949).

This habitat, although considerably withdrawn from the current, is often surpassed in this respect by the interior of moss and algal turfs. It is amazing the wealth of free-living protozoa, algae, and small animals of the most diverse groups that can live here under these conditions.

The influence of current upon the morphology of the respiratory organs is great. Aerial respiration is even more difficult for insects in flowing water than in standing water; they are usually forced to extract oxygen from the water. Their larvae ordinarily have a closed tracheal system and usually develop blood gills or tracheal gills. In the imagoes of the beetle *Helmis*, which have an open tracheal system, special modifications are present to facilitate gaseous respiration under water. In the pupae of *Simulium* and *Liponeura*, which usually are attached at shallow depths, cuticular gills provide for alternating aerial and aquatic respiration with fluctuations in water level. Since, as was specified above, flowing water is physiologically richer in oxygen than standing water, a reduction of respiratory surface is commonly observed in stream animals.

A number of forms, for example, the nymphs of the majority of Plecoptera (stoneflies), do not, generally speaking, exhibit any respiratory appendages. Still-water forms seek to promote respiration by setting up a current of water by means of undulating body movements, ciliary apparatus, or movements of the respiratory organs. In moving water such provision for ventilation can be progressively reduced as the current becomes stronger. Thus many species in mountain streams (*Epeorus, Rhithrogena, Baëtis, Rhyacophila, Liponeura*, and others) possess erect immobile gills, especially if they are living in "rushing" water.

Currents of water are also commonly utilized in the *feeding of animals*. The most impressive example is the funnel-like or sac-like catching nets directed towards the current which several trichopteran larvae spin (*Plectrocnemia, Polycentropus, Philopotamus, Hydropsyche*). The larvae of *Simulium* by means of their large comb-shaped premandibles filter the water flowing past them. The majority of the brook animals are phytophagous, nourishing themselves by grazing on the algal *Aufwuchs* of stones and mosses through the use of varied and sometimes highly specialized scraping and collecting organs.

The nature of the *substrate* is also of great importance for the character of the brook biocoenose. Yet its influence is not specific for this biotope, as is already known to us from our consideration of the littoral of lakes. It is to be emphasized, however, that a firm, stony, or bouldery substrate with its characteristic, often mosaic-like formations of plant *Aufwuchs* predominates under all circumstances in rapidly flowing water, because the looser sediments are washed away. In lenitic sections (but also in dead water and below moss and algal turfs), on the other hand, sand and ooze deposits occur. The biocoenose of these deposits closely resembles that of the corresponding biotopes of lakes according to the temperature of the littoral or profundal ooze facies. In the water-filled capillary spaces of sand bars there is an interesting animal community, the *psammon*, which also occurs in the sandy shores of lakes, consisting of rotifers, turbellarians, nematodes, insect larvae, and others.

This unusual community, which was first described by Russian and Polish investigators and later was studied more closely in other regions, particularly the United States (cf. especially the papers of Pennak, 1940, and Neel, 1948, in which the older literature is reviewed), occupies a position intermediate between the "edaphon" of the soil and the communities of the aquatic biotope, as well as between the open water of the eulittoral and the ground water. It is especially well developed in lake and stream beaches consisting mainly of quartz sand, and forms a zone as much as 2 to 3 m. wide above the water line to the upper limit of capillary suction.[70] The environmental

[70]In the concept of psammon Neel also includes the sandy beaches lying beneath the surface of the water.

conditions, which in recent times have been studied especially by Ruttner-Kolisko (1956), are somewhat similar to those in lake sediments, in that there is a stratification of oxygen, carbon dioxide, pH, and temperature. Yet these conditions are strongly influenced by meteorological factors (precipitation, insolation, evaporation) and frequently by the washing over of waves running onto the sand beaches, all of which circumstances expedite the exchange of substances. At the surface of the *psammal* often a luxuriant algal flora can develop, which can colour the sand green even at a depth of 5 mm., and which together with organic detritus makes possible the existence of a fauna. The latter is usually restricted in its occurrence to the upper 4 to 5 cm. At greater depths the oxygen content is already too small, as a rule, to permit animal life. The quantities of organisms that can populate this biotope under favourable conditions are amazingly large. Thus Pennak (1940), in a surface sand sample of 10 ml. (which contained 2 to 3 ml. of water), taken 150 cm. above the water line of a lake, found 4,000,000 bacteria, 10,000 protozoa, 400 rotifers, 40 copepods, 20 tardigrades, and smaller numbers of other metazoa.

The numerous investigations of Russian rivers have furnished evidence that the benthic colonization of a stream is particularly strongly dependent on the character of the bottom substrate (which in turn is controlled by the current velocity, as explained on page 234). According to Zhadin (1950) the following benthic biocoenoses can be distinguished in streams: *lithorheophile* biocoenoses on stones or other solid substrates; *phytorheophile* on water plants; *argillorheophile* in clay deposits; *psammorheophile* in sand deposits lying within the current region; *pelorheophile* in soft muddy substrate in the current region; *psammophile* in sand deposits in still water; and *pelophile* in mud deposits in still water. These biocoenoses can attain a considerable areal extent where conditions for deposition of sediments are homogeneous, such as occurs over broad stretches in the larger streams. Thus here these biocoenoses are well demarcated from one another.[71]

Of critical importance for the life of flowing water is the light intensity prevailing in the water or at the bottom. As already explained on page 13, the light climate of bodies of water is dependent on the extinction of radiation. This is controlled by two factors—absorption by water

[71]In brooks and small mountain streams the current relationships are extremely complex, so that the character of the bottom can change greatly over very small distances. The consequence is a dense mosaic of colonizers of different substrates (stone, sand, moss, etc.), which no longer can so easily and unequivocally be grouped according to their substrate preferences. The colonization relationships in brooks, therefore, are different than in streams and not directly comparable. For this reason Illies (1961) proposed that this habitat—the *rhitron*—be distinguished from that of the stream—the *potamon*. Within the habitat of the rhitron the vegetation and fauna change strikingly in a downstream direction from the source. A succession of zones can be recognized with characteristic colonization relationships. Up to the present this zonation has been closely investigated in middle European brooks (Illies, 1952; Schmitz, 1957), but we now know that brooks in other geographical regions have a similar zonation. Illies (1961) distinguishes three zones adjoining the source region of a brook in a typical case—the *epirhitron*, *metarhitron*, and *hyporhitron* (cf. the old classification on p. 244).

itself and by the substances dissolved in it on the one hand, and scattering by suspensoids on the other. In flowing waters the second factor—turbidity—understandably is more important. The decrease in light intensity by turbidity can be so extensive under certain conditions that the compensation point of photosynthesis is already reached at a shallow depth, and an assimilating plant life cannot develop on the bottom of the stream. In streams of shallow depth turbidity is generally not so effective in bringing about extinction. In clear water streams the light intensity at the bottom is often greater than the light optimum of many water plants (Schmitz, 1960). These relationships are different in the large and deep streams. The magnitude of the differences in light transmission that can occur even in mountain streams is shown in Figure 69. The injurious effect of turbidity is manifested also in the deposition of a sediment on the surface of macrophytes and algal *Aufwuchs*.

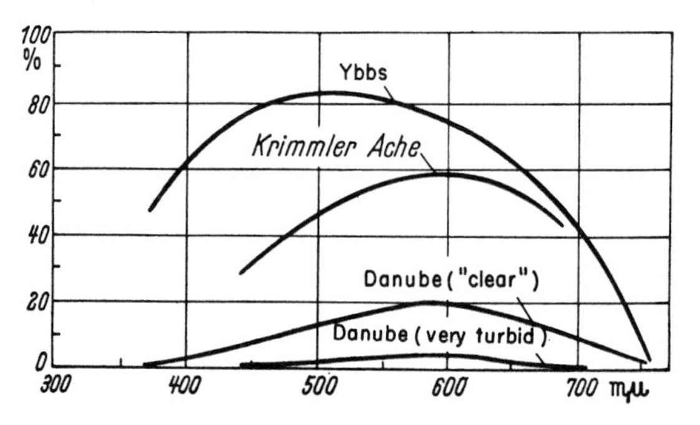

FIGURE 69. Light transmission in several streams of the alpine region, in per cent of the total incident radiation. From Dirmhirn (1953).

A factor that brings about a regular change in the composition of the biocoenoses of a river course from its source to its lower course and also brings about zonations along the shore bottom is *temperature*. As already indicated earlier, it makes a considerable difference in investigating the temperatures of a water course whether one makes measurements in a water body that is being carried downstream by the current (in the "flowing wave"), or at a particular place along the river bed. If we follow a single mass of water flowing from the source (which might have a constant temperature of about 5°) to the lower course of the stream, there occurs at first a rapid and later a slower increase in temperature, on which the daily fluctuation is impressed (Figure 70). The amplitude of this fluctuation in summer becomes smaller from day to day and approaches a constant equilibrium value as soon as the source temperature no longer is effective (assuming uniform meteorological conditions). From this it is evident that for an understanding of the thermal relations of a stream, continuing observations at a considerable number of stations from source to the lower course are required, since measurements in a progressively moving water mass can scarcely be carried out.

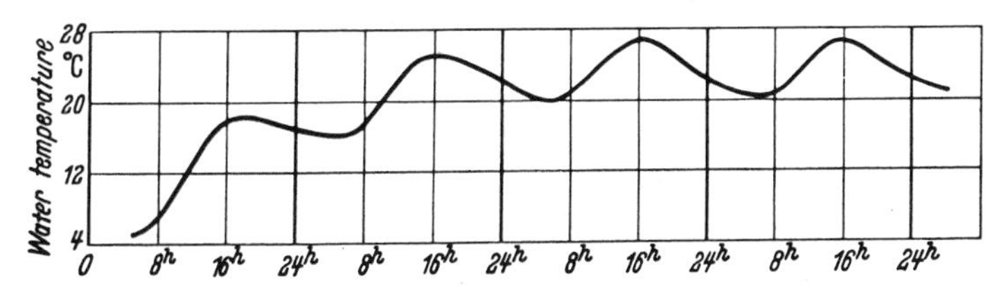

FIGURE 70. The course of temperature in a given water mass ("flowing wave") of a brook over a 4-day period, beginning at the source. After Eckel and Reuter (1950).

In the upper course the temperature at fixed stations, which is decisive for the ecology of the brook bed, depends (exclusive of meteorological, geographical, and topographic factors) above all on the *distance from the source*, until usually in the middle or lower course the equilibrium temperature with an approximately constant daily fluctuation is attained (Eckel and Reuter, 1950; Pleskot, 1953).

For ecological considerations it is important not only to have individual observations available but also to obtain insight concerning the thermal climate of a biotope, the thermal sum reached during the course of a vegetation period, and the mean temperature during this time interval. Such a procedure, which was introduced by Schmitz and Volkert (1959) as a limnological method and has been used with success, is the measurement of the temperature-dependent rate of inversion of a sucrose solution, which is enclosed in sterile tubes and exposed for any desired length of time at various places in the water course.

A mountain brook for a longer or shorter section below its origin is characterized by a relatively constant condition of temperature corresponding to that of the source. In temperate latitudes (but also in high mountains in the tropics) this is the habitat of cold-stenothermal organisms, which in this biotope (including the source itself) are represented by a much larger number of species than among the deep-water fauna of lakes. I shall cite only two especially striking examples: the chysophycean *Hydrurus foetidus*, whose magnificent turfs consisting of brush-shaped branched gelatinous threads colour the rapids of cold brooks deep brown, and whose temperature maximum (hence the temperature at which the plants perish) has been experimentally determined to be 16°; and the well-known animal *Planaria alpina*. The latter inhabits the cold upper course of brooks and is replaced in the middle course by *Polycelis cornuta,* whereas the warm lower course is populated by *Planaria gonocephala.* Additional examples from the animal kingdom are contained in the work of Pleskot (1951). The old division of a river course by the fishermen into regions of trout (Forelle), grayling (Äsche), barbel (Barbe), and bream (Brachsen or Blei) is also based primarily on thermal conditions. The interposition of a lake into a brook course brings about a sudden alteration of the summer water temperature and thoroughly changes the composition of the biocoenose. The cold-stenothermal forms disappear, and in their place there occurs a mass development of eurythermal species, such as the caddisfly larva *Hydropsyche angustipennis.* Similar relationships are likewise exhibited by spring-fed brooks having a less constant temperature as the result of a superficially located place of origin. One can therefore adopt an ecological division into *summer-cold* and *summer-warm* water courses.

A few words might be said about the sources, or *springs,* themselves. They are the sites of emergence of ground water flowing along an impervious rock stratum. According to the type of emergence there are distinguished *rheokrenes* (flowing springs), the water of which flows away from the spring mouth with a gradient; *limnokrenes* (pool springs), which come out at the bottom of a basin and the overflow of which forms the beginning of the spring brook; and *helokrenes* (marsh springs), whose water, oozing out of the ground diffusely, produces a marshy place ("Nassgalle"). It is not surprising that these relationships favour the formation of biocoenoses differing mainly in their zoological characteristics; these will not be elaborated on here.

With respect to *chemistry,* springs are often differentiated from the brooks arising from them. The reason for this lies in the fact that the water at the mouth of the spring comes out of the solubility equilibrium prevailing in the ground into sudden contact with the atmosphere, and a new equilibrium with respect to the gas content is only gradually restored. Because of the lower oxygen content of the soil atmosphere, springs as a rule have a lower oxygen content than surface waters at the same temperature. Exceptions occur, for instance when the spring water can saturate itself before emergence by flowing through broken rocks or gravel. In addition, the CO_2 content of spring water, and in limestone regions the associated content of $Ca(HCO_3)_2$, is often several times greater than the level corresponding to atmospheric equilibrium. With the escape of carbon dioxide the excess lime precipitates out and gives rise to deposits of tufa around the source of the spring, which can attain a considerable thickness and can build up a spring mound. This phenomenon is most imposing in the *travertine* deposited by strongly supersaturated volcanic springs.

The deposition of lime, however, as already explained on page 63, does not proceed to equilibrium, which is the reason most brooks evidence a greater or lesser supersaturation with lime even after stabilization of the solubility equilibrium with the atmosphere has been brought about below the spring. This is similar to the supersaturation encountered in standing waters. Concerning the content of other dissolved substances the same can be said for streams as already described for lakes. Fluctuations in total salt content (determined by the measurement of electrolytic conductivity) brought about during the course of a year by precipitation and by the melting of snow are considerable (cf. Ruttner, 1914), and yet as a general rule they do not appreciably change the chemical character of the water. However, this chemical character is being changed continually along the course of the stream by various affluents. Particularly the waste matters, and especially those from industries, are significant, since they can completely change the chemistry of the water and the character of the colonization, and indeed under certain conditions can destroy all life.

What has been said concerning the temperature constancy of summer-cold brooks and its ecological consequences applies even more to those springs that arise from deep rock strata. Their *temperature* is known to be similar to the *mean annual temperature* of the site, and is subject to unusually small fluctuations in the course of time. Such springs in our region are therefore preferred refugia of cold-stenothermal species and of glacial relicts. On the other hand, springs that gather a stream of ground water near the surface (as a consequence of the shallow location of the stratum impervious to water) exhibit considerable fluctuations in temperature in the course of a year.

The animal life of springs is not confined to sites of emergence, but continues as *ground water fauna* (Chappuis, 1927, in *Die Binnengewässer*) into the subterranean cavities, ranging from the capillary spaces of layers of gravel and sand to fissures and larger caves in the formations. This fauna is completely atypical and peculiar in composition as a result of the modified external conditions. The factors that have shaped the aspect of this community are primarily the complete darkness and the absolute protection against climatic and atmospheric influences. It is well known that cave animals tend to be blind, that is, they possess atrophied eye anlagen. In addition these ground water dwellers are distinguished by a far-reaching loss of pigment. Under the protection and the seclusion of the subterranean spaces many ancient forms have been preserved, and endemic species and genera evolved. We cannot here consider this interesting condition any further except merely to mention that particularly in the cave waters of the warmer regions (for example, the Karst caves in the Adriatic region) a wealth of special forms has become known. One need only recall the cave *Proteus* of the Adelsberg Grotto or the freshwater serpulid *Marifuga* discovered by Absolon and Hrabě (1930). In central Europe the ground water fauna is relatively poor in species, a fact which perhaps is associated with the Pleistocene cover of ice over the mountains.[72] For, although the ground water domain was not directly affected by this change in climate, nevertheless the latter destroyed the basis for nourishment of the subterranean animal world, which is dependent on the importation of organic materials suspended in the ground water. In spite of this one of the most remarkable finds was made in our region of the crustacean *Bathynella* about 2 mm. in size, which is related to Palaeozoic forms. The most abundant representative of the indigenous ground water fauna of central Europe is the splendid cave amphipod *Niphargus*, blind and pure white, which can be found in most wells.

Springs having a higher mean annual temperature than the ground from which they arise must in a hydrographic sense be designated as *thermal waters*. They, like other springs, originate either from the *vadose* water derived from the earth's surface, which as a result of geological stratification reaches great depths and there becomes warmed, or from *juvenile* water, which originates in the interior of the earth and comes to the surface in volcanic regions or along thermal fissures. Juvenile water is usually laden with gases and salts. *Balneology* designates as thermal waters those springs with a temperature greater than 20°. This holds good, however, only for a temperate climate; in the tropics springs normally have temperatures of 24° to 25°. From an *ecological standpoint* we include as thermal waters those biotopes in which the limiting and selecting influence of high water temperature is clearly apparent on the composition of the biocoenose. As a rule this is the case at temperatures near 30°.

The influence of increasing temperature on the composition of the community manifests itself in two different ways: in the progressive exclusion of species, genera, and entire families, and in the occurrence of specific thermal forms. Considering the *plant kingdom* first of all, at the lowest thermal temperatures (30° to 35°) almost all groups of the algae as well as mosses and phanerogams are still present, and the aspect of the community scarcely differs from that encountered in

[72]There was recently discovered in the ground water of the Upper Rhine lowland and of the Main River valley (hence in a formerly *unglaciated* region) a considerable number of remarkable new species and even genera of small crustacea.

normal tropical waters. Up to about 38° the green algae survive, and up to 41°, or at the highest 45°, a few diatoms survive. Above 45° only blue-green algae are present, but these still are represented by a considerable number of species. The German Limnological Sunda Expedition (Geitler and Ruttner, 1935) observed 14 species above 45°, 10 above 50°, 7 above 55°, and 3 species of this group of algae above 60°; above 66° algae were no longer encountered. The number of specific thermal algae, hence those that develop luxuriantly *only* at high temperatures, is not very large. Among the best known are *Mastiglocladus laminosus* and *Phormidium laminosum*, which can occur in unusual abundance at temperatures around 50°, and *Synechococcus elongatus*, which in the instances observed thus far reaches at 66° to 69° the highest limit of algal life.[73] The considerably higher limits of life in thermal waters that are often quoted are doubtless referable to inaccurate measurements of temperature, for the temperature of thermal waters fluctuates very strongly from place to place, and only a measurement taken in the algal turf itself can be significant.

The previous statements apply to alkaline thermal waters. In *solfataric vents*, whose water contains free mineral acid (pH below 3), the picture is considerably different. At temperatures above 45° only *one* species has been observed in this biotope, but this in abundance—*Cyanidium caldariorum*, a single-celled bluish-green alga belonging to the Bangiales, which according to the investigations of F. Berger can thrive even at a pH of 0.4 (corresponding to a normal H_2SO_4 solution). Around 40° there occur in such biotopes, likewise abundantly, two diatoms, *Pinnularia acoricola* and *Eunotia fastigiata*, and at still lower temperature (37°) the genus *Zygogonium*, which we have already become acquainted with in acid bog puddles. These facts likewise are an indication of the great ecological significance of the hydrogen ion concentration (Hustedt, 1938; Geitler and Ruttner, 1935).

The *animal* component of the thermal community, especially at higher temperatures, is qualitatively and particularly quantitatively less rich than the plant component. There are only a few authentic instances of animals having been found at temperatures in the neighbourhood of 50°, for example the midge larva, *Dasyhelea tersa*, discovered by Thienemann (Johannsen, 1932) during the German Limnological Sunda Expedition living in great numbers on the turfs of *Phormidium laminosum* on a spring wall washed over with 51° water, in the Gedeh region of west Java. The same *Dasyhelea* colonized even extremely acid solfataric vent waters (pH 2.7) at a temperature of 38° The

[73]On the other hand *bacteria* were discovered by Molisch (1926) living even at 77.5° in Japanese thermal waters.

thermal fauna is composed of protozoa, rotifers, nematodes, a few crustaceans, midge larvae, and especially numerous beetles and snails. For the most part they are widespread eurythermal forms which are also found in other biotopes. However, there are also apparently specific thermal animals, among which by far the most remarkable is the crustacean found in Tunis at 45°, *Thermosbaena mirabilis*, occupying a completely isolated position in systematics.

The influence of man's technology and industrialization is even more evident in flowing waters than in lakes. Whereas in lakes the effects are largely confined to changes (usually unfavourable) in water economy and metabolism, in flowing waters as a result of engineering control structures often new types of waters are produced that otherwise do not occur in nature or are very imperfectly developed there.

When a dam or barricade is constructed in the valley of a brook or river it can serve the purpose of providing a storage of water for low water periods, or of making use of a section of a valley for the continuous production of water power or for irrigation of cultivated land. In the former case the dams involved are as a general rule quite high, and they transform the stream valley above the dam into a body of standing water often of considerable length and depth. Such a *reservoir* (Stausee) differs from a natural lake with respect to its water economy in that the outflow of water normally occurs at a considerable depth below the surface and only at times of high water flows over the spillway. This brings about a modification of the thermal economy that only seldom occurs under natural condition, namely, a stronger warming of the epilimnion, since the warm surface water is prevented from leaving the lake, and the removal of water taking place at greater depths results in the warmed surface water being sucked below.

In the first years after the flooding of land it is often observed that as a result of the decay of the meadow top soil and other vegetation a disappearance of oxygen develops in the hypolimnion, even though organic materials may not have been supplied to the lake from outside via industrial effluents or domestic pollution.

In other respects a storage reservoir behaves similar to a natural lake in its biogenic stratification and plankton distribution. A eutrophication (with disappearance of oxygen, etc.) as a rule occurs only when cities or industries located upstream dump their wastes into the stream. As long as this water is moving the organic substances introduced (within certain limits) become oxidized over a relatively short distance, whereas their decomposition in the stratified water of a reservoir can lead to familiar calamities (see Bleiloch Reservoir, p. 92).

In the case of a *mainstream dam* (Flusstau or Laufstau) the cross section of the dam is small in relation to the volume of stream flow, and as a result the current is not entirely stopped but only slowed down, whereby turbulence is maintained in the entire water mass although weakened, and thermal and chemical stratifications cannot develop. Hence, in a mainstream reservoir scarcely any effects injurious to life can result from decomposition processes, especially not to animal life on the bottom of the reservoir. Einsele (1957, 1960), who has concerned himself closely with the problems of mainstream reservoirs and current velocities in their waters, has reached the conclusion that the bottom of a mainstream reservoir because of the quiescence of rolling gravel and sand, as already explained on page 234, and the accumulation of various organic materials deposited there as a result of the reduced current, forms biotopes especially rich in animal life, since the turbulence is sufficient to prevent a disappearance of oxygen.

In recent times a large number of investigations have been carried out on reservoirs, as for example in the Soviet Union where the Academy of Sciences has established a special institute for this purpose at Borok (Gouv. Jaroslavl.) which publishes its own journal.

The difficulties already described for making quantitative investigations of the littoral of lakes apply even more strongly to flowing waters. Hence, our knowledge of the production of flowing waters is based mainly on approximations. A very noteworthy attempt to comprehend the biomass of a very large spring run (Silver Springs, Florida) in numerical terms was successfully undertaken by Odum (1956). His investigations were based on the biogenic changes in the O_2 content of the water. McConnell and Sigler (1959) on the other hand attempted to determine the production in a rapidly flowing brook, in which the O_2 method was not applicable, by measuring the chlorophyll content of the epilithic algal *Aufwuchs* of entire stones taken along the profile of the brook, and to infer the magnitude of assimilation from this parameter. The methods used up to this time for quantitative biological investigations (especially of the bottom fauna) have been reviewed in gratifying completeness by Macan (1958) and Albrecht (1959).

GLOSSARY OF TECHNICAL EXPRESSIONS

absorption coefficient. The volume of a gas (referred to 0° C. and 760 mm. pressure) which is dissolved at a certain temperature by 1 ml. of liquid. In optics "absorption" means the reduction of radiation on passing through a material (Lat. *ab* away, *sorbere* to suck in).

acidophilic. Acid loving (Lat. *acidus* sour, Gr. *philein* to love).

adiabatic. Referring to a temperature change that does not involve an exchange of heat with the surroundings (warming by compression, cooling by expansion (Gr. *adiabatos* not passable).

adsorption. The concentration of gases, dissolved materials, or ions on the surface of solid particles (Lat. *ad* to, *sorbere* to suck in).

aerobe (aerobic). Organisms that can prosper only when oxygen is present more or less abundantly (Gr. *aer* air, *bios* life).

aggressive carbon dioxide. The quantity of free carbon dioxide exceeding the equilibrium relationship of a bicarbonate solution, which is capable of "attacking" (i.e. dissolving) calcium carbonate and other substances (Lat. *aggredi* to attack).

alkalinity. The acid combining capacity (SBV in German terminology) of a (carbonate) solution, expressed in milliequivalents (the number of ml. of tenth normal HCl required to bring 100 ml. of the solution being investigated to the methyl orange end point).

alkaliphobic. Avoiding an alkaline reaction (Ar. *al-qili* ashes of saltwort, Gr. *phobein* to fear).

allochthonous. Arising in another biotope (Gr. *allos* other, *chthon* land).

amplitude. The difference in height between the highest and lowest part of a wave; ordinarily in English "wave height" (Lat. *amplitudo* extent).

anaerobe (anaerobic). Organisms which either by obligation or facultatively thrive in the absence of oxygen (Gr. *an* without, *aer* air, *bios* life).

anomaly of water. This arises from the fact that water has its greatest density at 4° C. Hence, at each point within the temperature range from 0° to 4° there is a density value that also occurs above 4° (Gr. *anomalos* irregular).

anorgoxydants. Organisms that meet their energy needs by the oxidation of inorganic substances.

apatite. Calcium phosphate with chloride, hydroxyl, or fluoride Ca(Cl, F, OH)$Ca_4(PO_4)_3$; forms hexagonal crystals; earlier was often confused with fluorite (Gr. *apati* deceit).

assimilation. The transformation of absorbed nutrient substances into body substances (Lat. *assimilare* to make like).

autochthonous. Arising in the biotope under consideration (Gr. *autos* self, same, *chthon* land).

autolysis. Self-dissolving. The breakdown of dead organisms by their own enzymes without the intervention of bacteria (Gr. *autos* self. *lysis* loosening).

autotrophic. The nutrition of those plants that are able to construct organic matter from inorganic (Gr. *autos* self, *trophein* to nourish).

benthal. The region of the shore and the bottom of waters, benthal (as noun) or benthal zone (Gr. *benthos* depth).

biochemical oxygen demand (BOD = BSB in German terminology). The decrease in oxygen content in milligrams per litre of a sample of water in the dark at a certain temperature over a certain period of time, which therefore is brought about by the bacterial breakdown of organic matter. Usually the decomposition has proceeded so far after 20 days that no further change occurs. As a rule the oxygen demand is measured after 5 days (BSB_5), at which time 70 per cent of the final value has usually been reached.

biocoenose. A community of organisms whose composition and aspect is determined by the properties of the environment and by the relations of the organisms to each other (Gr. *bios* life, *koinos* common, *osis* condition).

biogenic. Arising as a result of life processes of organisms (Gr. *bios* life, *genos* origin).

biomass. The total particulate organic matter present beneath a unit surface area in a body of water (Gr. *bios* life).

biotope. A place of life; the totality of the environmental conditions under which a biocoenose exists (Gr. *bios* life, *topos* place).

buffer. A mixture of weak acids and their salts which (in solution) is able to greatly minimize changes in the hydrogen-ion concentration.

calomel. Mercury chloride, Hg_2Cl_2. A *calomel electrode* is a galvanic element of a constant potential, consisting of mercury, calomel, and potassium chloride solution. The latter can be brought into contact with a fluid through a porous substance, the mercury by means of a metal wire. In this manner a "lead-off electrode" is formed, so that electrochemical potentials arise only at a second electrode (Gr. *kalos* beautiful, *melas* black).

calorie. The quantity of heat which when added to a unit weight of water raises the temperature 1° C. Units: gram calorie, kilogram calorie, ton calorie (Lat. *calor* heat).

carbon dioxide of equilibrium. That quantity of free carbon dioxide necessary to prevent the precipitation of calcium carbonate from a solution of calcium bicarbonate.

chemocline. A stratum of stronger concentration gradient of dissolved substances (Gr. *klinein* to incline).

chemosynthesis. The synthesis of organic matter from mineral substances with the aid of chemical energy (in contrast to photosynthesis) (Gr. *synthesis* placing together).

chlorophyll. The green pigments of plants (Gr. *chloros* green, *phyllon* leaf).

chorology. The study of the geographic distribution of organisms (Gr. *chora* place, *logos* discourse).

chromatic adaptation. The ability of a blue-green alga to modify its coloration so as to be complementary to the quality of the light reaching the organism (Gr. *chroma* colour, Lat. *adaptare* to fit).

chromatophores. The variously formed bodies in plant cells consisting of protoplasmic material that bear the assimilation pigments (Gr. *chroma* colour, *pherein* to bear).

chromatoplasm. The peripheral plasma layer of the cells of blue-green algae that contains the assimilation pigment (Gr. *chroma* colour, *plasma* formed).

circulation period. The interval of time in which the density stratification of a lake is destroyed by the equalization of temperature, as a result of which the entire water mass becomes mixed (Lat. *circulus* a small circle).

clinograde. The stratification curve of temperature or of a chemical substance in water that exhibits a uniform slope from the surface into deep water (Gr. *klinein* to slope, Lat. *gradi* to step, walk).

colloids. Substances that, in contrast to crystalloids, are not distributed as individual molecules or ions in a liquid but rather as larger aggregates of molecules; hence, they are intermediate between true solutions and suspensions (Gr. *kolla* glue, *eides* resembling).

colorimetry. Determination of the concentration of a dissolved substance by comparison of its colour intensity with that of a corresponding solution of known concentration (Lat. *color* colour, *metrum* measure).

compensation point. The depth at which assimilation and dissimilation are equal (Lat. *compensare* to weigh several things with one another).

condensation. Transformation of a substance from the gaseous to the liquid state (Lat. *condensare* to make dense).

consumers. Organisms that nourish themselves on particulate organic matter (Lat. *consumere* to take wholly).

convection. Movements of particles of a fluid as a result of changes in density (Lat. *convehere* to bring together).

Coriolis effect (Coriolis force). The diverting force of the earth's rotation which causes horizontally moving water or air particles to be diverted towards the right in the Northern Hemisphere and towards the left in the Southern (after the French physicist, G. Coriolis).

cosmopolitan. See *ubiquitous.*

cyclomorphosis. Periodically repeated changes in the body form of successive generations of plankton animals (Gr. *kyklos* ring, cycle, *morphe* form).

denitrification. Reduction from nitrate to nitrite and further to elemental nitrogen (Lat. *de* from, *nitrum* saltpeter, *facere* to make).

density. Weight in grams of a unit volume (1 ml.) of a substance.

detritus. Finely divided settleable material suspended in the water: organic detritus, from the decomposition of the broken down remains of organisms; inorganic detritus = settleable mineral materials.

diffusion. The gradual reciprocal penetration of two substances in contact with each other as a result of molecular thermal movement. *Diffuse radiation* is scattered, moving out towards all directions in space, or radiation that is reaching a point from all directions (Lat. *diffundere* to pour out).

dissimilation. Metabolic processes by means of which simpler substances (down to the inorganic end products of decomposition) arise from complex organic compounds (physiological combustion) (Lat. *dissimilis* unlike).

dissociation. The splitting of a molecule of a substance (a salt, acid, or base) existing in solution into electrically charged particles (*ions*). The positively charged hydrogen ion and metallic ions are called *cations*, the negatively charged hydroxyl ion and acid ions are called *anions* (Lat. *dissociare* to separate; Gr. *ion* going; Gr. *kata* down, *ana* up).

dy. A bottom deposit of precipitated humic colloids (Swedish).

dystrophic lakes. Brown-water lakes with a very low lime content and a very high humus content, often characterized by a severe poverty of nutrients (Gr. *dys* badly, *trophein* to nourish).

ecology. The study of the relationships of organisms to their environment (Gr. *oikos* house, *logos* discourse).

ecotypes. Races of a species that are distinguished on the basis of physiological characteristics (reactions to the environment) (Gr. *oikos* house, *typos* the mark of a blow, general form).

electrolyte content. The quantity of ion-forming materials (acids, bases, salts) contained in the water (Gr. *lyein* to loose, to undo).

electrolytic conductivity. The unit is the electrical conductivity, expressed in "reciprocal ohms," of a column of liquid 1 $cm.^2$ in cross section and 1 cm. high possessing a resistance of 1 ohm. In dilute solutions the conductivity is approximately proportional to the concentration.

emersion zone. The uppermost portion of the eulittoral, which lies above the water level most of the year (Lat. *emergere*, from *e* out, and *mergere* to dip).

epilimnion. The turbulent superficial layer of a lake lying above the metalimnion which does not have a permanent thermal stratification. (Gr. *epi* on, *limne* lake).

epiphytes. Plants that are not rooted in the bottom but rather use other plants as a substrate without penetrating into them and without withdrawing nutrient substances from them (pseudoparasites) (Gr. *epi* on, *phyton* plant).

equivalent (weight). That quantity of a chemical element or compound that can combine with or replace 1 gram atom of hydrogen (1.008 g. = 1 gram equivalent). If the combining weight of the compound is expressed in milligrams, then the term milliequivalents is used (Lat. *aequus* equal, *valere* to be strong).

equivalent conductivity. The electrical conductivity of a solution divided by the number of gram equivalents of the dissolved substance contained in 1 ml. of the solution.

estuary. The mouth region of a river that is affected by tides and marine salt water (Lat. *aestuarium* arm of the sea.).

eulittoral. The shore zone of a body of water lying between the limits of annual fluctuation in water level (Gr. *eu* well, Lat. *littus*, prop. *litus* shore).

eurytopic. An organism that has a wide range of tolerance, e.g. *eurythermal* with respect to temperature, *euryphotic* with respect to light (Gr. *eurys* broad, *topos* place).

eutrophic. Waters with a good supply of nutrients and hence a rich organic production (Gr. *eu* well, *trophein* to nourish).

galvanometer. A precision instrument for measuring weak direct currents (after the Italian physician, L. Galvani).

Geiger counter. A tube-shaped device for counting the incidence of charged particles (alpha particles, single electrons, beta particles, gas ions, gamma rays) based on impact ionizations.

gel. A colloid system of readily mouldable shape consisting of solid and liquid components (Lat. *gelare* to freeze).

glacial relicts. Survivors of the Pleistocene biota that are restricted to particular localities (Lat. *glacies* ice, *relictus* left behind).

glass electrode. If two solutions of different pH are separated by a thin glass membrane, a reciprocal action through this membrane is brought about which leads to an electric potential that can be measured. If the pH of one solution is known, that of the second can be calculated from the potential.

global radiation. The total radiation from the sun and the sky reaching the earth's surface (Lat. *globus* a ball, sphere).

gradient. A change in a physical property related to unit of length, e.g. temperature per metre (Lat. *gradi* to step, go).

gross production. The total amount of new organic matter formed in a given time, including that which is simultaneously utilized in metabolic processes (Lat. *grossus* thick, *producere* to bring forth).

gyttja. The kind of sediment typical of aerated lake bottoms (Swedish).

heat conduction. The transfer of heat within a substance or from one substance to another without radiation, current, or mixing.

helokrene. A marsh spring (Gr. *helos* marsh, *krene* spring).

heterograde. A curve for temperature or a chemical factor in a body of water that exhibits a non-uniform slope from the surface downward into deep water (Gr. *heteros* other, Lat. *gradi* to step, walk).

heterotrophic. The nutrition of plants and animals that are dependent on organic matter for food (Gr. *heteros* other, *trophein* to nourish).

hexose. A simple sugar (monosaccharide), $C_6H_{12}O_6$ (Gr. *hex* six).

holomictic. Lakes that are completely circulated to the bottom at the time of winter cooling (Gr. *holos* entire, *miktos* mixed).

humus substances. Organic substances only partially broken down, which occur in water mainly in a colloidal state (*humus colloids*). *Humic acids* are large-molecule organic acids that dissolve in water (Lat. *humus* soil).

hydric. Attributable to water (Gr. *hydor* water).

hydrobiology. The study of life in water (Gr. *hydor* water, *bios* life, *logos* discourse).

hydrogenation. The addition of hydrogen to organic compounds. *Dehydrogenation*, the removal of hydrogen.

hydrolysis. The partial splitting caused by water of a neutral salt into its component free acid and base; according to the strength of these products the solution reacts acidic or alkaline (Gr. *hydor* water, *lyein* to loose, to undo).

hypolimnion. The deep layer of a lake lying below the metalimnion and removed from surface influences. (Gr. *hypo* under, *limne* lake).

infra-red. The region of long wave-length radiation beyond the visible red (Lat. *infra* below).

interference. Superposition of the same kind of waves having a difference in phase, whereby the energy of the resulting wave can be increased, reduced, or extinguished (Lat. *inter* between, *ferire* to strike).

interference filter. By means of phase differences these filters bring about the extinction of all transmitted light except in a certain selected spectral band. They are used for the separation of particularly narrow spectral bands. In *interference gradient filters* the transmitted spectral band changes continuously along the filter.

ions. See *dissociation.*

isobath. A line of equal depth (Gr. *isos* equal, *bathys* deep).

isopleth. A line for the same numerical value of a given quantity (Gr. *isos* equal, *plethos* quantity).

isotonic. Solutions of the same osmotic pressure (Gr. *isos* equal, *tonos* tension).

isotopes. Elements that occupy the same place in the periodic table but have a different atomic weight (Gr. *isos* equal, *topos* place).

juvenile water. Spring water originating from the interior of the earth (Lat. *juvenilis* young).

katharobes. The organisms of "pure" water poor in organic matter (Gr. *katharos* pure, *bios* life).

laminar flow. The organized unidirectional movement of a liquid or a gas (Lat. *lamina* leaf, thin layer).

lenitic. Slowly flowing (Lat. *lenis* mild, soft).

light climate. The optical conditions in a body of water resulting from special absorption characteristics, scattering, and turbidity (Gr. *klima* slope).

limnokrene. A pool spring (Gr. *limne* lake, pond, *krene* spring).

limnology. The study of inland waters (Gr. *limne* lake, *logos*, discourse).

limonite. Brown iron ore, colloidal ferric hydroxide (Gr. *leimon* meadow).

littoral. The shoreward region of a body of water (Lat. *littus*, prop. *litus* shore).

longitudinal waves. These arise when the direction of oscillation of each individual point of the wave coincides with its direction of propagation (Lat. *longitudo* length).

μ (Gr. *mu*) = 0.001 mm.; mμ = 0.000,001 mm.

Maare. Volcanic explosion funnels, usually in the form of circular lakes.

macrophytes. Large plants (Gr. *makros* great, *phyton* plant).

membrane filter. A specially prepared filter of cellulose ester with a controlled pore diameter. For limnological purposes "Group 3" filters (Fabrik "Membranfilter," Göttingen) with a pore diameter of 0.3–0.5 μ are usually used (Lat. *membrana* skin covering, parchment).

meromictic lakes. Those lakes that at the time of winter cooling undergo only a partial circulation down to a depth determined by a density stratification (Gr. *meros* part, *miktos* mixed).

metalimnion. The layer of water in a lake between the epilimnion and hypolimnion in which the temperature exhibits the greatest difference in a vertical direction (Gr. *meta* between, *limne* lake).

microphytes. Small plants (Gr. *mikros* small, *phyton* plant).

mixotrophic. Applied to plants that have the ability to assimilate CO_2 but in addition depend in part on organic substances for their nutrition (Gr. *mixis* mixing, *trophein* to nourish).

monimolimnion. The deep water of a meromictic lake that is not involved in the annual circulation (Gr. *monimos* continuing).

morphology. Study of configuration or form (Gr. *morphe* form, *logos* discourse).

nannoplankton. Those organisms suspended in open water which because of their small size cannot be collected by nets. They can be recovered by sedimentation or centrifugation (Gr. *nannos* dwarf).

nekton. The powerful swimmers among the freshwater animals that to a large degree are capable of moving about voluntarily from place to place (Gr. *nektos* swimming).

net production. The assimilation surplus in a given period of time after subtracting the amount of dissimilation in the same time interval (Fr. *net*, fr. L. Lat. *nitidus* clean, pure, Lat. *producere* to bring forward).

neuston. The community of the surface film of water: *epineustic*, living in the air on the upper surface; *hyponeustic*, living in the water on the under surface (Gr. *epi* on, *hypo* under, *nein* to swim).

normal hydrogen electrode. A platinized (covered with Pt-black) platinum electrode which is immersed in a solution of pH 0 and is surrounded by hydrogen at atmospheric pressure.

normal solution. A solution containing one gram equivalent of a substance per litre.

ohm. The unit of electrical resistance: one ohm is the resistance offered by a column of mercury 1 $mm.^2$ in cross section and 106.3 cm. long at 0° (after the physicist G. Simon Ohm).

oligoaerobe (*oligoaerobic*). Organisms that thrive at small tensions of oxygen (Gr. *oligos* small, *aer* air, *bios* life).

oligothermal. Confined to a range of low temperatures (Gr. *oligos* small, *therme* heat).

oligotrophic. Waters with a small supply of nutrients and hence a small organic production (Gr. *oligos* small, *trophein* to nourish).

orthograde. A stratification curve for temperature or a chemical factor in a body of water which has a straight uniform course (Gr. *orthos* straight, Lat. *gradi* to step, walk).

osmosis. The passing of liquids through membranes. *Osmotic pressure* is the tendency of a solution to reduce its concentration by attracting more of the solvent to it (Gr. *osmos* impulse, *osis* process).

oxidation. A chemical process that can occur: (1) in the uptake of oxygen (combustion); (2) in the removal of hydrogen (H_2S—S); (3) in the increase of the valence (for example, from ferrous to ferric compounds). *Reduction* is the reverse process (Gr. *oxys* sharp; Lat. *reducere* to lead back).

pelagial. The region of the free water in seas and inland lakes (Gr. *pelagos* the high sea).

permutite. A substance that exchanges its cations or anions for those of a solution with which it is in contact (Lat. *permutare* to change completely, to exchange).

pH. The negative logarithm of the hydrogen ion concentration expressed in gram equivalents.

photoelement. An element consisting of a metallic plate (iron, copper), a thin layer of a crystalline semi-conductor (silver selenide, cuprous oxide), and a transparent metallic film. On illumination a photoelectric current arises from the surface of the metallic film and passes through the conductor enclosed between this and the metallic plate.

photosynthesis. Elaboration of organic matter (carbohydrate) from CO_2 and H_2O with the aid of the energy of light (Gr. *phos, photos* light, *synthesis* placing together).

phototaxis. Orientation of movement of organisms on the basis of light stimuli (Gr. *photos* light, *taxis* arrangement).

phototropism. Orientation in response to the stimulus of a light gradient (Gr. *photos* light, *trope* a turning).

phylogeny. Race history (Gr. *phylon* race, *geneia* origin of).

phytoplankton. The plant portion of the plankton (Gr. *phyton* plant).

planimetry. Measurement of the surface area of plane figures by tracing their circumference with a mechanical-mathematical instrument (*planimeter*) (Lat. *planus* flat, level, *metrum* measure).

plankton. The community of the free water (Gr. n. of *planktos* wandering).

pleuston. The community of organisms floating on the surface of a lake (Gr. *plein* to sail, float).

poikilohaline. Of changing salt content (Gr. *poikilos* varied, *hals* salt).

polymerization. The formation of cómpound molecules out of homogeneous units (Gr. *polys* many, *meros* part).

polythermal. Confined to a range of high temperature. (Gr. *polys* many, *therme* heat).

potamoplankton. The "true" stream plankton in large rivers, which forms its own community independent of tributaries and floodplain waters (Gr. *potamos* river).

primary production. The production of organic matter from inorganic materials within a certain period of time by autotrophic organisms with the help of radiant energy (Lat. *primus* first, *producere* to bring forward).

producers. Organisms that are able to build up their body substance from inorganic materials (Lat. *producere* to bring forward).

profundal. The deep region of a body of water below the light-controlled limit of plant growth (Lat. *profundus* deep).

proteins. Organic materials that are broken down by enzymes or acids to amino acids (Gr. *proteios* primary, holding first place).

psammon. The community of the interstices of sand deposits of lake and river shores (Gr. *psammos* sand).

pyrite. FeS_2, crystallized mainly as shining golden cubes or octahedrons (Gr. *pyrites* pertaining to a mineral that strikes fire).

reaction of a solution. The acid, neutral, or alkaline condition determined by the ratio of the H^+ and OH^- ions.

redox potential (oxidation-reduction potential). The electrical potential of a bright platinum electrode immersed in a solution containing a mixture of the oxidized and reduced states of a substance, compared with a normal hydrogen electrode (cf. p. 204).

reduction. See *oxidation.*

reflection depth. That depth in a natural body of water from which a reflected light ray is still able to reach the surface (Lat. *reflectere* to bend back).

resistance thermometer. An arrangement for measuring temperature on the basis of thermally induced changes in the electrical resistance of thin wires.

RGT-rule. See van't Hoff's Law.

rheokrene. A flowing spring (Gr. *rhein* to flow, *krene* spring).

rheotropism. Orientation in response to the stimulus of a current gradient (Gr. *rhein* to flow, *trope* a turning).

saprobes. Organisms living in water polluted with organic materials (Gr. *sapros* rotten, *bios* life).

sapropel. Foul-smelling ooze (= *Faulschlamm*) (Gr. *sapros* rotten, *pelos* clay, mud).

seiche. A standing wave in a lake (perhaps from Fr. *sèche* dry, since part of the shore is laid bare by the recession of the water).

seston. All the particulate matter suspended in water (Gr. *sestos* strained, strainable).

Solfataras. Volcanic fumaroles rich in sulphur compounds (from the "Solfatara" in the Phlegraean Fields; Ital. *solfo* sulphur).

specific gravity. A ratio that denotes how many times heavier a body is than the same volume of water at 4°C.

specific heat. The quantity of heat in calories that must be added to a unit weight of a substance in order to raise its temperature 1°C.

spectrum (light spectrum). The sequence of all wave-lengths from red to violet that arises from the resolution of the light they comprise (Lat. *spectrum* image).

stability of stratification. The work that must be done to destroy or equalize the density stratification existing in a lake.

stagnation period. The period of time in which through warming (or cooling) from above a density stratification is formed that prevents a mixing of the water mass (Lat. *stagnum* a piece of standing water).

standing crop. The biomass present in a body of water at a particular time.

stenotopic. An organism with a narrow range of tolerance, e.g. *stenothermal* in relation to temperature, *stenophotic* in relation to light (Gr. *stenos* narrow, *topos* place).

stoichiometric. The relative quantities of elements in a chemical compound according to their combining weights (Gr. *stoicheion* a first principle, element).

sublittoral. The shore zone from the lowest water level to the lower boundary of plant growth (Lat. *sub* under, *littus*, prop. *litus* shore).

sulphate reduction. Formation of hydrogen sulphide from sulphates (Lat. *sulphur* sulphur, *reducere* to lead back).

suspension. Very finely divided particles of an insoluble solid material dispersed in a liquid (Lat. *suspendere* to suspend below).

tectonic. Brought about by the formation or the movement of the earth's crust (Gr. *tektonike* architecture).

thermal gradient. The temperature difference within a layer 1 m. thick.

thermistor. A name applied to metal oxides (semi-conductors), which have a high negative temperature coefficient of electrical resistance. Hence, the resistance *decreases* with rising temperature. (In the platinum wire of the older style resistance thermometer, on the other hand, the resistance *increases.*) The sensitivity of a thermistor is about ten times greater than that of a platinum wire. Since the resistance is very large, the thermistor element can respond to very small temperature changes. The designation thermistor is a contraction for "thermal sensitive resistor."

thermocline. (Gr. *therme* heat, *klinein* to slope.) See *metalimnion.*

thermocouple. A pair of wires of different metals soldered together at their ends, which yields an electrical current (thermoelectric current) when the temperatures of the junctions are different (for example, as a result of radiation). *Thermopile*, several thermocouples connected in series (Gr. *therme* heat).

thermotaxis. Orientation of the direction of movements of organisms on the basis of thermal stimuli (Gr. *therme* heat, *taxis* arrangement).

thigmotaxis. Orientation on the basis of touch stimuli arising from contact, as with the substrate. As a result many animals do not become quiescent until the largest possible portion of their body is in contact with the substrate (Gr. *thigma* touch, *taxis* arrangement).

torrential. Living in a rushing stream (Lat. *torrens* roaring).

transverse waves. These arise when the direction of oscillation of each individual point is at right angles to the direction of propagation of the wave (Lat. *transversus* turned or directed across).

travertine. A mineral formed in lime-rich springs supersaturated with CO_2, usually through the activity of plants (Ital. *travertino* a stone from Tibur).

tripton. The non-living suspended matter (detritus) in water (Gr. *tribein* to rub, pulverize).

trophogenic layer. The superficial layer of a lake in which organic production from mineral substances takes place on the basis of light energy (Gr. *trophe* nourishment, *gennan* to produce).

tropholytic layer. The deep layer of a lake where organic dissimilation predominates because of light deficiency (Gr. *trophe* nourishment, *lyein* to loose, undo).

turbulence. Unorganized movement in liquids and gases resulting from eddy formation (Lat. *turba* disorder).

tychoplankton. Forms of the littoral community occurring in the plankton "accidentally" (Gr. *tyche* fortune, chance).

ubiquitous. Species of plants or animals that are able to thrive under very different conditions (in different biotopes); in contrast, species are *cosmopolitan* that are distributed over the entire earth in their particular biotope (Lat. *ubique* everywhere; Gr. *kosmos* world, *polites* citizen).

ultraviolet. The region of short-wave radiation beyond the visible violet (Lat. *ultra* beyond).

uninodal oscillation. Having only one point where the oscillating body of water remains at rest. A *binodal oscillation* has two such points or nodes.

vadose water. Spring water originating from the surface of the earth (rain) (Lat. *vadosus* shallow).

van't Hoff's Law. (RGT-rule = Reaktiongeschwindigkeit-Temperatur-Regel). States that a chemical reaction proceeds approximately twice as fast at a temperature increase of 10° C. (after the Dutch chemist J. van't Hoff).

vegetational coloration. Change in the colour of a body of water or of its substrate by plants (Lat. *vegetare* to be active).

viscosity. Resistance to flow in a liquid (Lat. *viscum* birdlime).

volt. The unit of electromotive force: 1 volt is the force that produces an electrical current of 1 ampere in a conductor with a resistance of 1 ohm (after the physicist A. Volta).

Werfener strata. The sandstones and schistose clays of the lowermost part of the alpine Triassic (corresponding to the New Red Sandstone of regions outside the Alps). This contains gypsum (and rock salt) in abundance, so that the springs arising from these strata can be rich in sulphate (and chloride) (named after the place Werfen in Austria).

zooplankton. The animal portion of the plankton (Gr. *zoion* animal).

BIBLIOGRAPHY

A. COMPREHENSIVE WORKS

BREHM, V. 1930. Einführung in die Limnologie. Biologische Studienbücher X. Berlin.

CLEGG, J. 1952. The Freshwater Life of the British Isles. London.

DEFANT, A. 1929. Dynamische Oceanographie. Einführung in die Geophysik III. Berlin.

EDMONDSON, W. T. 1959. Ward and Whipple's Fresh-water Biology. 2nd ed. New York.

FOREL, F. A. (1) 1892–1904. Le Léman. Monographie limnologique. 3 vols. Lausanne.—(2) 1901. Handbuch der Seenkunde. Stuttgart.

FRITSCH, F. E. The Structure and Reproduction of the Algae. Vol. 1 (1935), Vol. 2 (1952). London.

GESSNER, F. Hydrobotanik I (1955), II (1959). Berlin. (Also considers the ocean.)

HALBFASS, W. 1923. Grundzüge einer vergleichenden Seenkunde. Berlin.

HENTSCHEL, E. 1923. Grundzüge der Hydrobiologie. Jena. (Also considers the ocean.)

HUTCHINSON, G. E. 1957. Treatise on Limnology. Vol. I. New York and London.

HYNES, H. B. 1960. The Biology of Polluted Waters. Liverpool Univ. Press.

KALLE, K. 1943. Der Stoffhaushalt des Meeres. Akad. Verl. Ges., Leipzig.

KLEEREKOPER, H. 1944. Introducao ao Estudo da Limnologia I. Rio de Janeiro.

KUZNETSOV, S. J. 1959. Die Rolle der Mikroorganismen im Stoffkreislauf der Seen. Translated from the Russian by A. POCHMANN. VEB Dtsch. Verlag d. Wissensch., Berlin.

LAMPERT, K. 1925. Das Leben der Binnengewässer. 3. Aufl. Leipzig.

LENZ, F. 1928. Biologie der Süsswasserseen. Biologische Studienbücher IX. Berlin.

LIEBMANN, H. 1951. Handbuch der Frisch- und Abwasserbiologie. Munich.

MACAN, T. T., and E. B. WORTHINGTON. 1951. Life in Lakes and Rivers. London.

MITSCHERLICH, E. A. 1923. Bodenkunde. 4. Aufl. Berlin.

OLTMANNS, F. 1923. Morphologie und Biologie der Algen. 2. Aufl. III Bd. Jena. (Also considers the ocean.)

PAVLOVSKY, E. N., and V. I. ZHADIN. 1950. Das Leben des Süsswassers der U.d.S.S.R. Leningrad.

PENNAK, R. W. 1953. Fresh-water Invertebrates of the United States. New York.

PRANDTL, L. 1931. Abriss der Strömungslehre. Braunschweig.

SAUBERER, F., and O. HÄRTEL. 1959. Pflanze und Strahlung. Leipzig.

SAUBERER, F., and F. RUTTNER. 1941. Die Strahlungsverhältnisse der Binnengewässer. Leipzig.

SERNOV, S. A. 1958. Allgemeine Hydrobiologie. Berlin.

STANDARD METHODS for the Examination of Water and Wastewater. 1960. 11th ed. Amer. Publ. Health Assoc., New York.

STEUER, A. 1910. Planktonkunde. Leipzig. (Also considers the ocean.)

THIENEMANN, A. 1925. Die Binnengewässer. Stuttgart. Reference has been made in our book particularly to the following volumes: BEHNING. 1928. Das Leben der Wolga.—CHAPPUIS. 1927. Die Tierwelt der unterirdischen Gewässer.— GROTE. 1934. Der Sauerstoffhaushalt der Seen.—HAEMPEL. 1930. Fischereibiologie der Alpenseen.—HARNISCH. 1939. Die Biologie der Moore.—HUBER-PESTALOZZI. Das Phytoplankton des Süsswassers (five parts have appeared since 1938).—LUNDQVIST. 1927. Bodenablagerungen und Entwicklungstypen der Seen.—NAUMANN: (1) 1932. Grundzüge der regionalen Limnologie; (2) 1930. Einführung in die Bodenkunde der Seen.—MAUCHA. 1932. Hydrochemische Methoden in der Limnologie.—PESTA. 1929. Der Hochgebirgssee der Alpen.—PIA. 1933. Kohlensäure und Kalk.—REMANE and SCHLIEPER. 1958. Die Biologie des Brackwassers.—RYLOV. 1935. Das Zooplankton der Binnengewässer.—SEIDEL. 1955. Die Flechtbinse *Scirpus lacustris*.—THIENEMANN: (1) 1925. Die Binnengewässer Mitteleuropas; (2) 1928. Der Sauerstoff in eutrophen und oligotrophen Seen; (3) 1950. Verbreitungsgeschichte der Süsswassertierwelt Europas; (4) 1954. "Chironomus."

WELCH, P. S. (1) 1952. Limnology. 2nd ed. New York and London.—(2) 1948. Limnological Methods. Philadelphia and Toronto.

WESENBERG-LUND, C. (1) 1939. Biologie der Süsswassertiere. Translated by O. STORCH. Berlin.—(2) 1943. Biologie der Süsswasserinsekten. Berlin.

B. JOURNALS

Archiv für Hydrobiologie. Founded by A. THIENEMANN, edited by H. J. ELSTER and W. OHLE. Published by E. Schweizerbart'sche Verlagsbuchhandlung, Stuttgart. (In the following literature summary abbreviated as: Arch. Hydrobiol.)

Hydrobiologia. Acta hydrobiologica, limnologica et protistologica. Edited by P. VAN OYE and collaborators. Published by Dr. W. Junk, The Hague.

Internationale Revue der gesamten Hydrobiologie. Founded by R. WOLTERECK, edited by H. CASPERS and F. GESSNER. Published by the Akademie-Verlag, Berlin. (Abbreviated: Int. Rev.)

Limnology and Oceanography. American Society of Limnology and Oceanography. Lawrence, Kansas, U.S.A.

Memorie del Istituto Italiano di Idrobiologia. Pallanza, Italia.

Polskie Archiwum Hydrobiologii. Polska Akad. Nauk, Warszawa.

Schweizerische Zeitschrift für Hydrologie. Edited by O. JAAG. Published by Verlag Birkhäuser, Basel.

Verhandlungen der Internationalen Vereinigung für theoretische und angewandte Limnologie. Edited by the Gen. Sekr. d. I. V. L. Published by E. Schweizerbart'sche Verlagsbuchhandlung, Stuttgart.

Vie et milieu. Bulletin du Laboratoire Arago. Published by Herrmann, Paris.

Wetter und Leben. Zeitschrift für Bioklimatologie. Founded by F. Sauberer, edited by O. Eckel, F. Lauscher, and I. Dirmhirn. Published by Verlag der Oesterreichischen Gesellschaft für Meteorologie.

C. SELECTED REFERENCES REFERRED TO IN THE TEXT[1]

ÅBERG, B., and W. RODHE. 1942. Ueber die Milieufaktoren in einigen südschwedischen Seen. Symb. Bot. Upsal. 3.

ABSOLON, K., and S. HRABĚ. 1930. Ueber einen neuen Süsswasserpolychäten aus den Höhlengewässern der Herzegowina. Zool. Anz. 88.

ALBRECHT, M.-L. 1959. Die quantitative Untersuchung der Bodenfauna fliessender Gewässer. (Untersuchungsmethoden und Arbeitsergebnisse.) Zschr. Fisch. N. F. 8.

ALSTERBERG, G. (1) 1922. Die respiratorischen Mechanismen der Tubificiden. Lunds Universitets Årsskrift, N. F. 2/18.—(2) 1926. Die Winklersche Bestimmungsmethode für in Wasser gelösten, elementaren Sauerstoff sowie ihre Anwendung bei Anwesenheit oxydierbarer Substanzen. Biochem. Zschr. 170.—(3) 1927. Die Sauerstoffschichtung der Seen. Botan. Notiser.—(4) 1929. Ueber das aktuelle und absolute O_2-Defizit der Seen im Sommer. Botan. Notiser.—(5) 1935. Die Dynamik des Stoffwechsels der Seen im Sommer. Lund.

AMBÜHL, H. (1) 1955–60. Die praktische Anwendung der elektrochemischen Sauerstoffbestimmung im Wasser. Schweiz. Zschr. Hydrol. 17, 20, 22.—(2) 1959. Die Bedeutung der Strömung als ökologischer Faktor. Schweiz. Zschr. Hydrol. 21.

ANDERSON, G. C. 1958. Some limnological features of a shallow saline meromictic lake. Limnol. and Oceanogr. 3.

ANSCHÜTZ, J., and F. GESSNER. 1954. Der Ionenaustausch bei Torfmoosen (*Sphagnum*). "Flora," 141.

APSTEIN, C. 1896. Das Süsswasserplankton. Kiel and Leipzig.

ARENS, K. Physiologisch polarisierter Massenaustausch und Photosynthese bei submersen Wasserpflanzen. I. Planta, 20, 1933; II. Jahrb. wiss. Bot. 83, 1936.

AUERBACH, M., W. MAERKER, and O. SCHMALZ. 1926. Hydrographisch-biologische Bodensee-Untersuchungen. Verh. naturw. Ver., Karlsruhe.

AUFSESS, O. VON. (1) 1903. Die Farbe der Seen. Diss., München.—(2) 1905. Die physikalischen Eigenschaften der Seen. "Die Wissenschaft," 4, Braunschweig.

BÄCKSTRÖM, H. L. J. 1921. Ueber die Affinität der Aragonit-Calcit-Umwandlung. Zschr. physikal. Chemie, 97.

BAUMANN, A., and E. GULLY. 1909–13. Untersuchungen über die Humussäuren. Mitt. d. K. Bayer. Moorkulturanstalt, 3, 4, 5.

[1]More extensive compilations of literature can be found in the comprehensive works referred to above as well as in some of the works listed in the present section.

BAVANDAMM, W. 1924. Die farblosen und roten Schwefelbakterien. "Pflanzenforschung," 2.

BEHNING, A. L. 1924. Zur Erforschung der am Flussboden der Wolga lebenden Organismen. Monographien der Biologischen Wolga-Station zu Saratow, No. 1. (Russian.)

BELING, A., and HOLGER W. JANNASCH. 1955. Hydrobakteriologische Untersuchungen der Fulda unter Anwendung der Membranfiltermethode. Hydrobiologia, 7.

BERE, R. 1933. Numbers of bacteria in inland lakes of Wisconsin, as shown by the direct count microscopic method. Int. Rev. 29.

BERG, KAJ. (1) 1937. Contributions to the biology of *Corethra* Meigen (*Chaoborus* Lichtenstein). Det Kgl. Danske Videnskabernes Selskab. Biol. Medd. 13.—(2) 1943. Physiographical studies on the River Susaa. Fol. Limnol. Scand. 1.—(3) 1948. Biological studies on the River Susaa. Fol. Limnol. Scand. 4.

BERGER, F. (1) 1955. Die Dichte natürlicher Wässer und die Konzentrations-Stabilität in Seen. Arch. Hydrobiol. Suppl. 22.—(2) 1958. Ueber die Ursache des Oberflächeneffekts bei der Lichtmessung unter Wasser. Wetter u. Leben 10. —(3) Der Austritt von Gasblasen aus der Schnittfläche von Nymphaeaceen-Blättern. Oesterr. Bot. Zschr. (in press).

BIRGE, E. A. (1) 1916. The work of wind in warming a lake. Trans. Wis. Acad. Sci. 18.—(2) 1922. A second report on limnological apparatus. Trans. Wis. Acad. Sci. 20.

BIRGE, E. A., and C. JUDAY. (1) 1911. The inland lakes of Wisconsin: The dissolved gases of the water and their biological significance. Wis. Geol. and Nat. Hist. Surv. Bull. 22.—(2) 1922. The inland lakes of Wisconsin: The plankton, its quantity and chemical composition. Wis. Geol. and Nat. Hist. Surv. Bull. 64.—(3) 1929–32. Transmission of solar radiation by the waters of inland lakes. Trans. Wis. Acad. Sci., four papers in vols. 24 to 27.—(4) 1934. Particulate and dissolved organic matter in Wisconsin lakes. Ecol. Monogr. 4.—(5) 1919. Further limnological observations on the Finger Lakes of New York. Bull. U.S. Bur. Fish. 37. (Seneca Lake.)

BIRGE, E. A., C. JUDAY, and H. W. MARCH. 1928. The temperature of the bottom deposits of Lake Mendota. Trans. Wis. Acad. Sci. 23.

BOURRELLY, P. 1954. Recherches sur les Chrysophycées. Paris.

BREHM, V., and F. RUTTNER. 1926. Die Biozönosen der Lunzer Gewässer. Int. Rev. 16.

BRÖNSTED, I. N., and C. WESENBERG-LUND. 1912. Chemisch-physikalische Untersuchungen der dänischen Gewässer nebst Bemerkungen über ihre Bedeutung für unsere Auffassung der Temporalvariationen. Int. Rev. 4.

BROOKS, J. L. (1) 1946. Cyclomorphosis in *Daphnia*. Ecol. Monogr. 16.—(2) 1947. Turbulence as an environmental determinant of relative growth in *Daphnia*. Proc. Nat. Acad. Sci. 33.

BRÜCKNER, E. 1909. Zur Thermik der Alpenseen und einiger Seen Nordeuropas. Geogr. Zschr. 15.

BRYSON, R. A., and C. R. STEARNS. 1959. A mechanism for the mixing of the waters of Lake Huron and South Bay, Manitoulin Island. Limnol. and Oceanogr. 4.

BUDER, J. 1919. Zur Biologie des Bakteriopurpurins und der Purpurbakterien. Jahrb. wiss. Bot. 58.

BURKHARD, R. 1955. Zur Ermittlung der relativen Sauerstoffsättigung von Wasser. "Vom Wasser," 22.

BUTCHER, R. W. 1932. Studies in the ecology of rivers. II. The microflora of rivers with special reference to the algae in the river-bed. Ann. Bot. 46.

CANTER, H. M., and J. W. LUND. 1951. Studies on Plankton parasites III. Ann. Bot. 15.

CASPERS, H. (1) 1961. Biologie des Elbe-Aestuars; Vorwort. Arch. Hydrobiol. Suppl. 26/I.—(2) 1957. Black Sea and Sea of Azov. Geol. Soc. America, Mem. 67.—(3) 1959. Vorschläge einer Brackwasserzonen-nomenklatur ("The Venice System"). Int. Rev. 44.

CHANDLER, D. C. 1937. Fate of typical lake plankton in streams. Ecol. Monogr. 7.

CHOLODNY, N. 1929. Zur Methodik der quantitativen Erforschung des bakteriellen Planktons. Zentralbl. f. Bakteriol. II/77.

CHU, P. S. (1) 1942. The influence of the mineral composition of the medium on the growth of planktonic algae I. Methods and culture media. Jour. Ecol. 30.—(2) 1943. The influence of the concentration of inorganic nitrogen and phosphate phosphorus. Jour. Ecol. 31.

CLARKE, F. W. 1908–24. The Data of Geochemistry. Washington Dept. of the Interior, U. S. Geol. Survey Bulletin, 5 editions.

COFFIN, C. C., F. R. HAYES, L. H. JODREY, and S. G. WHITEWAY. 1949. Exchange of materials in a lake as studied by the addition of radioactive phosphorus. Canad. Jour. Res. D 27.

COKER, R. E., and H. H. ADDLESTONE. 1938. Influence of temperature on cyclomorphosis in *Daphnia longispina*. Jour. Elisha Mitchell Scient. Soc. 54.

COLLET, L. W. 1925. Les Lacs. Paris.

COMITA, G. W., and J. J. COMITA. 1957. The internal distribution patterns of a calanoid copepod and a description of a modified Clarke-Bumpus plankton sampler. Limnol. and Oceanogr. 2.

CUSHING, D. H. 1951. The vertical migration of planktonic crustacea. Biol. Rev. Cambridge Philos. Soc. 26.

DEEVEY, E. S. (1) 1940. Limnological studies in Connecticut. V. A contribution to regional limnology. Amer. Jour. Sci. 238.—(2) 1942. Studies on Connecticut Lake Sediments. III. The Biostratonomy of Linsley Pond. Amer. Jour. Sci. 240.—(3) 1955. Paleolimnology of the upper swamp deposit, Pyramid Valley. Rec. Canterbury Mus. VI/4.—(4) 1955. Some biogeographic implications of paleolimnology. IVL-Verh. 12.—(5) 1957. Limnological studies in Middle America. Trans. Connecticut Acad. Arts Sci. 39.

DEFANT, A. 1929. Dynamische Ozeanographie. Einführung in die Geophysik III. Berlin.

DEMOLL, R. (1) 1922. Temperaturwellen und Planktonwellen. Arch. Hydrobiol. 13.—(2) 1925. "Teichdüngung" in Handb. d. Binnenfischerei Mitteleuropas, Bd. 4.

DIRMHIRN, I. 1953. Ueber die Strahlungsvorgänge in Fliessgewässern. Wetter u. Leben, Sonderheft II.

DOMOGALLA, B. P., and E. B. FRED. 1926. Ammonia and nitrate studies of lakes near Madison, Wisconsin. Jour. Amer. Soc. Agron. 18.

Du Rietz, E. 1939. In: Führer für die Exkursion des IX. Internationalen Limnologenkongresses (Zur Kenntnis der Vegetation des Sees Tåkern). Acta phytogeographica Suecica XII.

Eckel, O. (1) 1935. Strahlungsuntersuchungen in einigen österr. Seen. Sitz. Ber. Akad. d. Wiss. Wien, math. naturw. Kl. IIa, 144.—(2) 1950. Ueber die numerische und graphische Ermittlung der Stabilität der Gewässer nach W. Schmidt. Schweiz. Zschr. Hydrol. 12.

Eckel, O., and H. Reuter. 1950. Zur Berechnung des sommerlichen Wärmeumsatzes in Flussläufen. Geographiska Annaler.

Edmondson, W. T. (1) 1946. Factors in the dynamics of rotifer populations. Ecol. Monogr. 16.—(2) 1956. Measurements of conductivity of lake water in situ. Ecology, 37.—(3) 1956. The relation of photosynthesis by phytoplankton to light in lakes. Ecology, 37.—(4) 1957. Trophic relations of the zooplankton. Trans. Amer. Micros. Soc. 76.—(5) 1960. Reproductive rates of rotifers in natural populations. Mem. Ist. Ital. Idrobiol. 12.

Einsele, W. (1) 1936. Ueber die Beziehungen des Eisenkreislaufes zum Phosphorkreislauf im eutrophen See. Arch. Hydrobiol. 29.—(2) 1938. Ueber chemische und kolloidchemische Vorgänge in Eisen-Phosphatsystemen unter limnochemischen und limnogeologischen Gesichtspunkten. Arch. Hydrobiol. 33.—(3) 1940. Versuch einer Theorie der Dynamik der Mangan- und Eisenschichtung im eutrophen See. Naturwissenschaften.—(4) 1941. Die Umsetzung von zugeführtem, anorganischem Phosphat im eutrophen See und ihre Rückwirkungen auf seinen Gesamthaushalt. Zschr. Fisch. 39.—(5) 1944. Der Zeller See, ein lehrreicher Fall extremer limnochemischer Verhältnisse. Zschr. Fisch. 42.—(6) 1957. Flussbiologie, Kraftwerke und Fischerei. Schr. d. Oesterr. Fischereiverb. 1.—(7) 1960. Die Strömungsgeschwindigkeit als beherrschender Faktor bei der limnologischen Gestaltung der Gewässer. Oesterr. Fisch. Suppl. 1.

Einsele, W., and J. Grim. 1938. Ueber den Kieselsäuregehalt planktischer Diatomeen und dessen Bedeutung für einige Fragen ihrer Oekologie. Zschr. Bot. 32.

Einsele, W., and H. Vetter. 1938. Untersuchung über die Entwicklung der physikalischen und chemischen Verhältnisse im Jahreszyklus in einem mässig eutrophen See (Schleinsee bei Langenargen). Int. Rev. 36.

Ekman, S. (1) 1915. Die Bodenfauna des Vättern, qualitativ und quantitativ untersucht. Int. Rev. 7.—(2) 1913–14. Studien über die marinen Relikte der nordeuropäischen Binnengewässer I–V.

Elster, H. J. (1) 1939. Beobachtungen über das Verhalten der Schichtgrenzen nebst einigen Bemerkungen über die Austauschverhältnisse im Bodensee (Obersee). Arch. Hydrobiol. 35.—(2) 1955. Limnologische Untersuchungen im Hypolimnion verschiedener Seetypen. Mem. Ist. Ital. Idrobiol. Suppl. 8.—(3) 1958. Das limnologische Seetypensystem, Rückblick und Ausblick. IVL—Verh. 13.

Elster, H. J., and W. Einsele. 1937. Beiträge zur Hydrographie des Bodensees (Obersees). Int. Rev. 35.

Exner, F. M. 1928. Ueber Temperaturseiches im Lunzer Untersee. Ann. der Hydrographie und maritimen Meteorologie.

Ferling, E. 1957. Die Wirkungen des erhöhten hydrostatischen Druckes auf Wachstum und Differenzierung submerser Blütenpflanzen. Planta, 49.

FINDENEGG, I. (1) 1935. Limnologische Untersuchungen im Kärntner Seengebiete. Arch. Hydrobiol. 28.—(2) 1942. Die Bedeutung des Nährstoffgehaltes der Seen für die Menge und Art ihres Planktons. "Der Biologe," 11.—(3) 1947. Ueber die Lichtansprüche planktischer Süsswasseralgen. Sitz. Ber. Akad. d. Wiss. Wien, math.-naturw. Kl. I, 155.

FOX, H. MUNRO. 1925. The effect of light on the vertical movement of aquatic organisms. Proc. Cambridge Philos. Soc. (Biol. Sci.) 1.

FREY, D. G. (1) 1955. Längsee: A history of meromixis. Mem. Ist. Ital. Idrobiol. Suppl. 8 (Colloque IUBS, n. 19).—(2) 1960. The ecological significance of cladoceran remains in lake sediments. Ecology, 41.—(3) 1958. The late-glacial cladoceran fauna of a small lake. Arch. Hydrobiol. 54.

FREY, D. G., and J. B. STAHL. 1958. Measurements of primary production on Southampton Island in the Canadian Arctic. Limnol. and Oceanogr. 3.

FRITSCH, F. E., and P. K. DE. 1938. Nitrogen fixation by blue-green algae. Nature, 142.

FRY, F. E. J. 1947. Effects of the environment on animal activity. Univ. Toronto Studies, Biol. Ser. 55.

FRYER, G. 1956. A cladoceran *Dadaya macrops* (Daday) and an ostracod *Oncocypris Mülleri* (Daday) associated with the surface film of water. Ann. Mag. Nat. Hist. 12/IX.

GAMS, H. 1927. Die Geschichte der Lunzer Seen, Moore und Wälder. Int. Rev. 18.

GEITLER, L. (1) 1928. Ueber die Tiefenflora an Felsen im Lunzer Untersee. Arch. Protistenk. 62.—(2) 1927. Ueber Vegetationsfärbungen in Bächen. Biologia Generalis, 3.—(3) 1942. Zur Kenntnis der Bewohner des Oberflächenhäutchens einheimischer Gewässer. Biologia Generalis, 16.—(4) 1922. Die Mikrophytenbiozönose der *Fontinalis*-Bestände des Lunzer Untersees und ihre Abhängigkeit vom Licht. Int. Rev. 10.

GEITLER, L., and F. RUTTNER. 1935. Die Cyanophyceen der Deutschen Limnologischen Sunda-Expedition, 3. Teil. Arch. Hydrobiol. Suppl. 14.

GERLOFF, G. C., and FOLKE SKOOG. 1957. Nitrogen as a limiting factor for the growth of *Microcystis aeruginosa* in southern Wisconsin lakes. Ecology, 38 (4).

GESSNER, F. (1) 1935. Phosphat und Nitrat als Produktionsfaktoren der Gewässer. IVL-Verh. 7.—(2) 1944. Der Chlorophyllgehalt der Seen als Ausdruck ihrer Productivität. Arch. Hydrobiol. 40.—(3) 1949. Der Chlorophyllgehalt im See und seine photosynthetische Valenz als geophysikalisches Problem. Schweiz. Zschr. Hydrol. 11.—(4) 1952. Der Druck in seiner Bedeutung für das Wachstum submerser Wasserpflanzen. Planta, 49.—(5) 1960. Limnologische Untersuchungen am Zusammenfluss des Rio Negro und des Amazonas (Solimoes). Int. Rev. 45.

GOLDMAN, C. R. 1956. Primary production and limiting factors in the lakes of the Alaska Peninsula. Ecol. Monogr. 30.

GOSSLER, O. 1950. Funktionsanalysen am Räderorgan von Rotatorien durch optische Verlangsamung. Oesterr. zool. Zschr. 2.

GÖTZINGER, G. 1912. Die Lunzer Seen I. Physik und Geomorphologie d. Lunzer Seen. Int. Rev. Suppl. zu Bd. 3.

GRIM, J. 1950. Versuche zur Ermittlung der Produktionskoeffizienten einiger Planktophyten in einem flachen See. Biol. Zentralbl. 69.

HARRIS, I. E., and U. K. WOLFE. 1955. A laboratory study of vertical migration. Proc. Roy. Soc. London Ser. B, 144.

HASLER, A. D., and W. EINSELE. 1948. Fertilization for increasing productivity of natural inland lakes. Trans. Thirteenth North Amer. Wildlife Conf.

HAYES, F. R., B. L. REID, and M. L. CAMERON. 1958. Lake water and sediment II. Oxidation-reduction relations at the mud-water interface. Limnol. and Oceanogr. 3.

HAYES, F. R., and collaborators. 1958–59. Lake waters and sediment I–IV. Limnol. and Oceanogr. 3, 4.

HENSEN, V. 1895. Ergebnisse der Planktonexpedition der Humboldtstiftung. Kiel.

HOPPE-SEYLER. 1895. Ueber die Verteilung der absorbierten Gase im Wasser des Bodensees und ihre Beziehungen zu den in ihm lebenden Tieren und Pflanzen. Schr. d. Ver. f. Geschichte des Bodensees u. s. Umgeb. 24.

HUSTEDT, F. 1938. Systematische und ökologische Untersuchungen über die Diatomeenflora von Java, Bali und Sumatra. Arch. Hydrobiol. Suppl. XV, "Tropische Binnengewässer," 7. (In addition numerous fundamental papers on the systematic-morphological and ecological character of diatoms.)

HUTCHINSON, G. E. (1) 1937. A contribution to the limnology of arid regions. Trans. Connecticut Acad. Arts Sci. 33.—(2) 1938. On the relation between the oxygen deficit and the productivity and typology of lakes. Int. Rev. 36.—(3) 1941. Limnological studies in Connecticut. IV. Mechanism of intermediary metabolism in stratified lakes. Ecol. Monogr. 11.—(4) 1944. Critical examination of the supposed relationship between phytoplankton periodicity and chemical changes in lake waters. Ecology, 25.

HUTCHINSON, G. E., and V. T. BOWEN. (1) 1947. A direct demonstration of the phosphorus cycle in a small lake. Proc. Nat. Acad. Sci. 33.—(2) 1950. Limnological studies in Connecticut. IX. A quantitative radiochemical study of the phosphorus cycle in Linsley Pond. Ecology, 31.

HUTCHINSON, G. E., and H. LÖFFLER. 1956. The thermal classification of lakes. Proc. Nat. Acad. Sci. 42.

ILLIES, J. (1) 1952. Die Mölle, faunistisch-ökologische Untersuchungen an einem Forellenbach im Lipper Bergland. Arch. Hydrobiol. 46.—(2) 1956. Seeausfluss-Biozönosen lappländischer Waldbäche. Entomolog. Tidskr. 77.—(3) 1961. Versuch einer allgemeinen biozönotischen Gliederung der Fliesswässer. Int. Rev. (in press).

JAAG, O. 1938. Die Kryptogamenflora des Rheinfalles und des Hochrheins von Stein bis Eglisau. Mitt. Naturf. Ges. Schaffhausen, 14.

JACOBS, W. 1935. Das Schweben der Wasserorganismen. Erg. d. Biol. 11.

JAMES, H. R., and E. A. BIRGE. 1938. A laboratory study of the absorption of light by lake waters. Trans. Wisc. Acad. Sci. 31.

JÄRNEFELT, H. (1) 1925–58. Zur Limnologie einiger Gewässer Finnlands. I–XVIII. Ann. Soc. Zool.-Bot. Fenn.—(2) 1949. Der Einfluss der Stromschnellen auf den Sauerstoff- und Kohlensäuregehalt und den pH des Wassers im Flusse Vuokosi. IVL-Verh. 10.

JOHANNSEN, O. A. 1932. Ceratopogoninae from the Malayan subregion of the Dutch East Indies. Arch. Hydrobiol. Suppl. IX. "Tropische Binnengewässer," 2.

JUDAY, C., and E. A. BIRGE. 1933. The transparency, the color, and the specific conductivity of the lake waters of northeastern Wisconsin. Trans. Wis. Acad. Sci. 29.

KANN, E. (1) 1933. Zur Oekologie des litoralen Algenaufwuchses im Lunzer Untersee. Int. Rev. 28.—(2) 1940. Oekologische Untersuchungen an Litoralalgen ostholsteinischer Seen. Arch. Hydrobiol. 37.—(3) 1959. Die eulitorale Algenzone im Traunsee (Oberösterreich). Arch. Hydrobiol. 55.

KARCHER, F. H. 1939. Untersuchungen über den Stickstoffhaushalt in ostpreussischen Waldseen. Arch. Hydrobiol. 35.

KLEEREKOPER, H. 1957. Une étude limnologique de la chimie des sédiments de fond des lacs de l'Ontario méridional, Canada. s'Gravenhage.

KLEIN, G., and M. STEINER. 1929. Bakteriologisch-chemische Untersuchungen am Lunzer Untersee I. Oesterr. Bot. Zschr. 78.

KOFOID, C. A. 1897. On some important sources of error in the plankton method. Science, N.S. 6.

KOHLRAUSCH and HOLBORN. 1916. Das Leitvermögen der Elektrolyte. 2 Aufl. Leipzig and Berlin.

KOLKWITZ, R. (1) 1911. Ueber das Kammerplankton des Süsswassers und der Meere. Ber. Dtsch. Bot. Ges. 29.—(2) 1935. Pflanzenphysiologie. Versuche und Beobachtungen an höheren und niederen Pflanzen, einschliesslich Bakteriologie und Hydrobiologie mit Planktonkunde, 3. Aufl. Jena. Abschnitt B/VI, Oekologie der Gewässer.

KOLKWITZ, R., and M. MARSSON. 1902. Grundsätze für die biologische Beurteilung des Wassers nach seiner Flora und Fauna. Mitt. Kgl. Prüfungsanstalt f. Wasserversorgung und Abwässerbeseitigung, 1.

KORDE, N. W. 1960. Biostratifikation und Typologie der russischen Sapropele. Akad. U.S.S.R. Moskau. (Russian.)

KRAWANY, H. 1928–37. Trichopterenstudien im Gebiete der Lunzer Seen 1–12. Int. Rev.: 1. Die Verbreitung einiger Bachformen und ihre Abhängigkeit von der Temperatur, 1928.

KRISS, A. E. 1954. Die Rolle der Mikroorganismen in der biologischen Produktivität des Schwarzen Meeres. Erg. d. zeitgen. Biologie, 38. (Russian.)

KRISS, A. E., and E. A. RUKINA. 1953. Purpur-Schwefelbakterien in den Schwefelwasserstofftiefen des Schwarzen Meeres. Ber. Akad. Wiss. U.S.S.R. 93.

KUISEL, H. F. 1936. Wissenschaftliche Erforschung des Zürichsees in den Jahren 1929–1933. Zürich.

KUZNETSOV, N. J. 1910. Ueber einige interessante Seen des Gouvernements Wladimir. Tr. Wladimir, ob-wa ljubit jestestw. 3. (Russian.)

KUZNETSOV, S. J. (1) 1942. Der Kreislauf des Schwefels in Seen. Mikrobiologija, 11. (Russian.)—(2) 1958. A study of the size of bacterial populations and of organic matter formation due to photo- and chemosynthesis in water bodies of different types. IVL-Verh. 13.

KUZNETSOV, S. J. and G. S. KARZINKIN. 1930. Eine Methode der quantitativen Bestimmung der Wasserbakterien. Russk. gidrobiol. Jour. 9. (Russian.)

LAUTERBORN, R. (1) 1900. Der Formenkreis von *Anuraea cochlearis.* Verh. nat. med. Ver., Heidelberg.—(2) 1922. Die Kalksinterbildungen an den unterseeischen Felswänden des Bodensees und ihre Biologie. Mitt. Bad. Landesver. Naturk. N.F. 1.—(3) 1915. Die sapropelische Lebewelt. Verh. naturh.-med. Ver. Heidelberg, N.F. 13.—(4) 1916–18. Die geographische und biologische Gliederung des Rheinstroms. Sitz. Ber. Heidelbg. Akad. Wiss., Math. nat. Kl., Abt. 3, 1, 2, 3.

LIEBMANN, H. (1) 1938. Biologie und Chemismus der Bleilochsperre. Arch. Hydrobiol. 33.—(2) 1950. Zur Biologie der Methanbakterien. "Gesundheitsingenieur," 71.

LINDEMAN, R. L. 1942. The trophic-dynamic aspect of ecology. Ecology, 23.

LINDROTH, A. 1957. Abiogenic gas supersaturations of river water. Arch. Hydrobiol. 53.

LINDSTRØM, T. 1957. Sur les planctons crustacés de la zone littorale. Inst. Freshwater Res. Drottningholm, Rep. 38.

LIVINGSTONE, D. A. 1955. A lightweight piston sampler for lake deposits. Ecology, 36.

LÖFFLER, H. 1958. Die Klimatypen des holomiktischen Sees und ihre Bedeutung für geographische Fragen. Sitzber. Oesterr. Akad. d. Wiss., math. naturw. Kl. I, 167.

LOHAMMAR, G. 1938. Wasserchemie und höhere Vegetation schwedischer Seen. Symbolae Bot. Upsal. III: 1.

LOHMANN, H. 1908. Untersuchungen zur Festellung des vollständigen Gehaltes des Meeres an Plankton. Wissensch. Meeresunters. 10, Kiel.

LOHUIS, D., V. W. MELOCHE and C. JUDAY. 1938. Sodium and potassium content of Wisconsin lake waters and their residues. Trans. Wis. Acad. Sci. 31.

LUNDEGÅRDH, H. 1925. Klima und Boden in ihrer Wirkung auf das Pflanzenleben. Jena.

LUNTZ, A. 1928. Ueber die Sinkgeschwindigkeit einiger Rädertiere. Zool. Jahrb., Abt. Allg. Zool. 44.

MCCARTER, J. A., F. R. HAYES, L. H. JODREY, and M. L. CAMERON. 1952. Movement of materials in the hypolimnion of lakes as studied by addition of radioactive phosphorus. Canad. Jour. Zool. 30.

MCCOMBIE, A. M. 1953. Factors influencing the growth of phytoplankton. Jour. Fish. Res. Bd. Can. 10.

MCCONNELL, W. J., and W. F. SIGLER. 1959. Chlorophyll and productivity in a mountain river. Limnol. and Oceanogr. 4.

MACAN, T. T. (1) 1958. Methods of sampling the bottom fauna in stony streams. IVL-MITT. 8.—(2) 1961. Factors that limit the range of freshwater animals. Biol. Rev. 36.

MACKERETH, F. J. 1953. Phosphorus utilization by *Asterionella formosa.* J. Exper. Bot. 4.

MAHRINGER, W. 1958. Der Oberflächeneffekt bei der Messung der Lichtdurchlässigkeit des Wassers mit dem Lunzer Unterwasser-Photometer. Wetter u. Leben, 10.

MARGALEF, R. 1958. "Trophic" typology versus biotic typology as exemplified in the regional limnology of Northern Spain. IVL-Verh. 13.

MERIAN, J. R. 1828. Ueber die Bewegung tropfbarer Flüssigkeiten in Gefässen. Basel.

MERKER, E. 1931. Die Fluoreszenz und die Lichtdurchlässigkeit der bewohnten Gewässer. Zool. Jahrb. 49.

MEVIUS, W. 1924. Wasserstoffionenkonzentration und Permeabilität bei "kalkfeindlichen" Gewächsen. Zschr. Bot. 16.

MICHAELIS, L. 1933. Oxydations-Reduktionspotentiale mit besonderer Berücksichtigung ihrer physiologischen Bedeutung. Berlin.

MINDER, L. (1) 1922. Ueber biogene Entkalkung im Zürichsee. IVL-Verh. 1.—(2) 1926. Biologisch-chemische Untersuchungen im Zürichsee. Zschr. Hydrol. 3.—(3) 1943. Der Zürichsee im Lichte der Seetypenlehre: Zürich.

MOLISCH, H. (1) 1926. Pflanzenbiologie in Japan auf Grund eigener Beobachtungen. Jena.—(2) 1901. Ueber den Goldglanz von *Chromophyton Rosanoffii*. Sitz. Ber. Akad. Wiss. Wien, 110.

MORTIMER, C. H. (1) 1941–42. The exchange of dissolved substances between mud and water in lakes. Jour. Ecol. 29, 30.—(2) 1950. The use of models in the study of water movement in stratified lakes. IVL-Verh. 11.—(3) 1952. Water movements in lakes during summer. Philos. Trans. Roy. Soc. London, 236.—(4) 1955. Exchange and circulation in lakes. Some effects of the earth rotation on water movements in stratified lakes. IVL-Verh. 12.—(5) 1956. The oxygen content of air-saturated fresh water, and aids in calculating percentage saturation. IVL-Mitt. 6.

MORTIMER, C. H., and W. H. MOORE. 1953. The use of thermistors for the measurement of lake temperatures. IVL-Mitt. 2.

MÜLLER, H. (1) 1933. Limnologische Feldmethoden. Int. Rev. 28.—(2) 1934. Ueber das Auftreten von Nitrit in einigen Seen der oesterr. Alpen. Int. Rev. 30.—(3) 1938. Beiträge zur Frage der biochemischen Schichtung im Lunzer Ober- und Untersee. Int. Rev. 36.—(4) 1937–38. Auswirkungen des Schneedruckes auf die Schwingrasen usw. Int. Rev. 35, 36.

MÜNSTER-STRØM, K. (1) 1932. Nordfjord Lakes. Norske Videnskaps-Akademi i Oslo I, Math.-Naturv.-Kl.—(2) 1936. Land-locked waters. *Ibid.*—(3) 1932. Tyrifjord. *Ibid.*—(4) 1931. Feforvatn. A physiological and biological study of a mountain lake. Arch. Hydrobiol. 22.—(5) 1945. The temperature of maximum density in fresh waters. Geofys. Publik. 16.

NAUMANN, E. (1) 1921. Einige Grundlinien der regionalen Limnologie. Lunds Universitets Årsskrift, N. F., Bd. 2, 17.—(2) 1927. Zur Kritik des Planktonbegriffes. Ark. Bot. 21.—(3) 1930. Die Eisenorganismen. Grundlinien der limnologischen Fragestellung. Int. Rev. 24. (Besides numerous other works on the problem of iron.)

NEEL, J. K. 1948. A limnological investigation of the psammon in Douglas Lake, Michigan, with especial reference to shoal and shoreline dynamics. Trans. Amer. Micros. Soc. 67.

NELSON, P. R., and W. T. EDMONDSON. 1955. Limnological effects of fertilizing Bare Lake, Alaska. Fish. Bull. 102, U.S. Fish and Wildlife Service, vol. 56.

NIPKOW, F. (1) 1920. Vorläufige Mitteilungen über Untersuchungen des

Schlammabsatzes im Zürichsee. Rev. d. Hydrol. 1.—(2) 1950. Ruheformen planktischer Kieselalgen im geschichteten Schlamm des Zürichsees. Schweiz. Zschr. Hydrol. 12.

NÜMANN, W. (1) 1936. Die Leitfähigkeit des Calciumbicarbonates und die Bestimmung der Sulfate und Gesamthärte in natürlichen Gewässern mit Hilfe der elektrischen Leitfähigkeit. "Die Naturwiss." 24.—(2) 1941. Der Stickstoffhaushalt eines mässig eutrophen Sees. Zschr. Fisch. 39.

NYGAARD, G. N. (1) 1955. On the productivity of five Danish waters. IVL-Verh. 12.—(2) 1958. On the productivity of the bottom vegetation in Lake Grane Langsø. IVL-Verh. 13.

ODÉN, S. 1922. Die Huminsäuren. Kolloidchemische Beihefte, 11, 2. Aufl.

ODUM, H. T. 1956. Primary production in flowing waters. Limnol. and Oceanogr. 1.

OHLE, W. (1) 1934. Chemische und physikalische Untersuchungen in norddeutschen Seen. Arch. Hydrobiol. 26.—(2) 1934. Roströhren und verwandte Konkretionen. Geol. Rundschau, 25.—(3) 1934. Ueber organische Stoffe in Binnenseen. IVL-Verh. 6—(4) 1935. Organische Kolloidgele in ihrer Wirkung auf den Stoffhaushalt der Gewässer. Naturwissensch. 23.—(5) 1936. Der schwefelsaure Tonteich bei Reinbeck. Arch. Hydrobiol. 30.—(6) 1938. Die Bedeutung der Austauschvorgänge zwischen Schlamm und Wasser für den Stoffkreislauf der Gewässer. "Vom Wasser," 13.—(7) 1940. Ueber den Kaliumgehalt der Binnengewässer. "Vom Wasser," 14.—(8) 1952. Die hypolimnische Kohlendioxyd-Akkumulation als produktionsbiologischer Indikator. Arch. Hydrobiol. 46.—(9) 1954. Sulfat als "Katalysator" des limnischen Stoffkreislaufes. "Vom Wasser," 21.—(10) 1955. Beiträge zur Produktionsbiologie der Gewässer. Arch. Hydrobiol. Suppl. 22.—(11) 1958. "Bioaktivität." IVL-Verh. 13.

OSTWALD, Wo. 1902. Zur Theorie des Planktons. Biol. Zbl. 22.

PEARSALL, W. H. 1932. Phytoplankton in the English lakes. II. The composition of the phytoplankton in relation to dissolved substances. Jour. Ecol. 20.

PENNAK, R. W. (1) 1940. Ecology of the microscopic metazoa inhabiting the sandy beaches of some Wisconsin lakes. Ecol. Monogr. 10.—(2) 1957. Species composition of limnetic zooplankton communities. Limnol. and Oceanogr. 2.

PERFILIEV, B. W. 1927. Die Mikrobiologie der Süsswasserablagerungen. IVL-Verh. 4.

PETTERSSON, H. 1936. Das Licht im Meer. Bioklim. Beiblätter.

PLESKOT, G. (1) 1949. Der Stand der biologischen Fliesswasserforschung. Verh. d. Deutschen Zool. in Mainz.—(2) 1951. Wassertemperatur und Leben im Bach. "Wetter und Leben," 3.—(3) 1953. Beiträge zur Limnologie der Wienerwaldbäche. "Wetter und Leben," Sonderh. 2.—(4) 1953. Zur Oekologie der Leptophlebiiden. Oesterr. Zoolog. Zschr. 4.

PRINTZ, H. 1939. Ueber die Kohlensäureassimilation der Meeresalgen in verschiedenen Tiefen. Skr. Norske Videnskaps Akad. i. Oslo I, math. naturw. Kl.

PÜTTER, A. 1909. Die Ernährung der Wassertiere und der Stoffhaushalt der Gewässer. Jena.

RAWSON, D. S. (1) 1939. Some physical and chemical factors in the metabolism of lakes. Publ. No. 10 of the American Association for the Advancement of Sciences.—(2) 1951. The total mineral content of lake waters. Ecology, 32.—(3) 1960. A limnological comparison of twelve lakes in Northern Saskatchewan. Limnol. and Oceanogr. 5.—(4) 1957. Species composition of limnetic zooplankton communities. Limnol. and Oceanogr. 2.

RAZUMOV, A. S. (1) 1932. Eine direkte Methode der Zählung von Wasserbakterien. Ihr Vergleich mit der Kochschen Methode. Mikrobiologija, 1. —(2) 1947. Methoden der mikrobiologischen Wasseruntersuchung. Isd. Ministerstwa str. predpr. tjesh. ind. i instituta WODGEO. (Russian.)

REDEKE, H. C. 1933. Ueber den jetzigen Stand unserer Kenntnisse der Flora und Fauna des Brackwassers. IVL-Verh. 6.

REDINGER, K. 1934. Zur Oekologie der Schlenken in den Mooren des Lunzer Untersees. Beih. Bot. Cbl. 52.

REIF, C. B. 1939. The effect of stream conditions on lake plankton. Trans. Amer. Micros. Soc. 58.

RICHTER, E. 1891. Die Temperaturverhältnisse der Alpenseen. Verh. d. IX. deutschen Geographentages Wien.

RICKER, W. E. (1) 1934. A critical discussion of various measures of oxygen saturation in lakes. Ecology, 15.—(2) 1937. Physical and chemical characteristics of Cultus Lake, British Columbia. Jour. Biol. Bd. Can. 3(4).—(3) 1937. Statistical treatment of sampling processes useful in the enumeration of plankton organisms. Arch. Hydrobiol. 31.

RILEY, G. A. 1940. Limnological studies in Connecticut. III. The plankton of Linsley Pond. Ecol. Monogr. 10.

RODHE, W. (1) 1948. Environmental requirements of freshwater plankton algae. Symb. Bot. Upsal. X: 1.—(2) 1949. The ionic composition of lake waters. IVL-Verh. 10.—(3) 1958. Primärproduktion und Seetypen. IVL-Verh. 13.—(4) 1958. The primary production in lakes: some results and restrictions of the C^{14} method. Rapp. et. Proc.-Verb. vol. 144, Cons. Intern. Explor. de la Mer.—(5) 1961. Die Dynamik des limnischen Stoff- und Energiehaushaltes. IVL-Verh. 14.

RODHE, W., R. A. VOLLENWEIDER, and A. NAUWERK. 1956. The primary production and standing crop of phytoplankton. Perspectives in marine biology. Symp. Scripps Inst. Oceanogr., Univ. Calif.

ROSSOLIMO, L. 1932. Ueber die Gasausscheidung im Beloje-See. Arb. Biol. Stat. Kossino.

RUTTNER, F. (1) 1914. Die Verteilung des Planktons in Süsswasserseen. Abderhalden Fortschr. 10.—(2) 1914. Das elektrolytische Leitvermögen des Wassers der Lunzer Seen. Int. Rev. Suppl. 6.—(3) 1921. Das elektrolytische Leitvermögen verdünnter Lösungen unter dem Einfluss submerser Gewächse. Sitz. Ber. Akad. d. Wiss. Wien, math.-naturw. Kl. I/130.—(4) 1926. Bemerkungen über den Sauerstoffgehalt der Gewässer und dessen respiratorischen Wert. Naturwissenschaften.—(5) 1929. Das Plankton des Lunzer Untersees, seine Verteilung im Raum und Zeit. Int. Rev. 23.—(6) 1931. Hydrographische und hydrochemische Beobachtungen auf Java, Sumatra und Bali. Arch. Hydrobiol. Suppl. 8.—(7) 1933. Ueber metalimnische Sauerstoffminima. Naturwissenschaften, 21.—(8) 1937. Limnologische Studien an einigen Seen der Ostalpen. Arch. Hydrobiol. 32.

—(9) 1943. Ueber die Anreicherung des Eisens im Tiefensediment eutroph geschichteter Seen. Zschr. Fisch. 41.—(10) 1947–48. Zur Frage der Karbonatassimilation der Wasserpflanzen, I and II. Oesterr. Bot. Zschr. 94, 95.—(11) 1948. Methoden der quantitativen Planktonforschung. Mikroskopie 3.—(12). 1952. Planktonstudien der Deutschen Limnologischen Sundaexpedition. Arch. Hydrobiol. Suppl. 21.—(13) 1955. Der Lunzer Mittersee, ein Quellsee mit zeitweise meromiktischer Schichtung. Arch. Hydrobiol. Suppl. 22.—(14) 1956. Einige Beobachtungen über das Verhalten des Planktons in Seeabflüssen. Oesterr. Bot. Zschr. 103.—(15) 1960. Das Vorkommen von Kohlendioxyd und Kohlensäure im Süsswasser. Handb. d. Pflanzenphysiologie V.—(16) 1960. Ueber die Kohlenstoffaufnahme bei Algen aus der Rhodophyceen-Gattung *Batrachospermum*. Schweiz. Zschr. Hydrol. 22.

RUTTNER, F., and F. SAUBERER. 1938. Durchsichtigkeit des Wassers und Planktonschichtung. Int. Rev. 37.

RUTTNER-KOLISKO, A. (1) 1938. Die Nahrungsaufnahme bei *Anapus testudo*. Int. Rev. 37.—(2) 1949. Formwechsel und Artproblem von *Anuraea aculeata*. Hydrobiologia, 1.—(3) 1956. Der Lebensraum des Limnopsammals. Verh. D. Zool. Ges.

RYLOV, M. W. 1931. Ueber das Tripton-Problem. IVL-Verh. 5.

RYTHER, J. H. (1) 1956. The measurement of primary production. Limnol. and Oceanogr. 1.—(2) 1956. Interaction between photosynthesis and respiration in the marine flagellate, *Dunaliella euchlora*. Nature, 178.

SAUBERER, F. (1) 1938. Zur Methodik der Durchsichtigkeitsmessung im Wasser und deren Anwendung in der Limnologie. Arch. Hydrobiol. 33.—(2) 1939. Beiträge zur Kenntnis des Lichtklimas einiger Alpenseen. Int. Rev. 39.— 1942. Oberflächenseiches am zugefrorenen Lunzer Untersee. Bioklim. Beibl. 9.—(4) 1950. Die spektrale Strahlungsdurchlässigkeit des Eises. "Wetter und Leben," 2.—(5) 1961. Strahlungsmessungen unter Wasser. I. V. L.-Mitt. (in press).

SAUBERER, F., and O. ECKEL. 1938. Zur Methodik der Strahlungsmessungen unter Wasser. Int. Rev. 37.

SCHMASSMANN, H. J. 1951. Untersuchungen über den Stoffhaushalt fliessender Gewässer. Schweiz. Zschr. Hydrol. 13.

SCHMIDT, H. 1935. Die bionomische Einteilung der fossilen Meeresböden. Fortschr. d. Geologie und Paläontologie XII.

SCHMIDT, W. (1) 1908. Absorption der Sonnenstrahlung im Wasser. Sitzber. Akad. d. Wiss. Wien.—(2) 1928. Ueber den Energiegehalt der Seen. Int. Rev. 6, Suppl. 1915. Also in: Geographiska Annaler, 102.—(3) 1925. Der Massenaustausch in freier Luft und verwandte Erscheinungen. Hamburg.—(4) 1934. Ein Jahr Temperaturmessungen an 17 österr. Alpenseen. Sitz. Ber. Akad. d. Wiss. Wien, math.-naturw. Kl. IIa, 153.

SCHMITZ, W. (1) 1957. Die Bergbach-Zoozönosen und ihre Abgrenzung, dargestellt am Beispiel der oberen Fulda. Arch. Hydrobiol. 53.—(2) Lichtmessungen in Fliessgewässern des deutschen und österreichischen Donaugebietes. Wetter und Leben 12.—(3) 1961. Fliessgewässerforschung, Hydrographie und Botanik. IVL-Verh. 14.

SCHMITZ, W., and E. VOLKERT. 1959. Die Messung von Mitteltemperaturen auf reaktionskinetischer Grundlage mit dem Kreispolarimeter und ihre

Anwendung in Klimatologie und Bioökologie, speziell in Forst- und Gewässerkunde. Zeiss-Mitteilungen I.

SCHOMER, H., and C. JUDAY. 1935. Photosynthesis of algae at different depths in some lakes of northeastern Wisconsin. Trans. Wis. Acad. Sci. 29.

SCHREIBER, E. 1927. Die Reinkultur von marinem Phytoplankton und deren Bedeutung für die Erforschung der Produktivität des Meerwassers. Wiss. Meeresunters. Abt. Helgoland 16, No. 10.

SCHRÖDER, R. (1) 1959. Die Vertikalwanderungen des Crustaceenplanktons der Seen des südlichen Schwarzwaldes. Arch. Hydrobiol. Suppl. 25.—(2) 1961. Untersuchungen über die Planktonverteilung mit Hilfe der Unterwasser-Fernsehanlage und des Echographen. Arch. Hydrobiol. Suppl. 25, Falkau 4.

SCHÜTT, FR. 1893. Das Pflanzenleben der Hochsee. Kiel and Leipzig.

SEBESTYÉN, O. 1960. Horizontale Planktonuntersuchungen im Balaton, I. Orientierende Untersuchungen über die horizontale Verbreitung der Planktonkrebse. Ann. Inst. Biol. Hung. (Tihany), 27.

SHAPIRO, J. 1957. Chemical and biological studies on the yellow organic acids of lake water. Limnol. and Oceanogr. 2.

SIEBECK, O. (1) 1960. Die Bedeutung von Alter und Geschlecht für die Horizontalverteilung planktischer Crustaceen im Lunzer Obersee. Int. Rev. 45.—(2) 1960. Untersuchungen über die Vertikalwanderung planktischer Crustaceen unter Berücksichtigung der Strahlungsverhältnisse. Int. Rev. 45.

SIOLI, H. 1954, 1955, 1957. Beiträge zur regionalen Limnologie des Amazonasgebietes II–IV. Arch. Hydrobiol. 49, 50, 53.

SOROKIN, J. I. (1) 1955. Ueber die bakterielle Chemosynthese in Schlammablagerungen. Mikrobiologija 24. (Russian.)—(2) 1958. Die primäre Produktion organischer Substanz im Rybinskoje-Staubecken. Arb. Inst. f. Biol. der Staubecken, 3. (Russian.)

STANGENBERG, M. 1959. Der biochemische Sauerstoffbedarf des Seewassers. Mem. Ist. Ital. Idrobiol. 11.

STARKEY, R. L. 1945. Precipitation of ferric hydrate by iron bacteria. Science, 120.

STEEMANN NIELSEN, E. (1) 1939. Ueber die Verbreitung der Phytoplanktonten im Meere. Int. Rev. 38.—(2) 1947. Photosynthesis of aquatic plants with special reference to the carbon sources. Dansk. Bot. Ark. 12.—(3) 1955. The interaction of photosynthesis and respiration and its importance for the determination of C^{14}-discrimination in photosynthesis. Physiologia plantarum, 8.—(4) 1959. Untersuchungen über die Primärproduktion des Planktons in einigen Alpenseen Oesterreichs. "Oikos" 10.—(5) 1960. Dark fixation of CO_2 and measurement of organic productivity. With remarks on chemosynthesis. Physiologia plantarum, 13.—(6) 1960. Productivity of the oceans. Ann. Rev. Plant Physiol. 11.—(7) 1952. The use of radioactive carbon (C^{14}) for measuring organic production in the sea. J. Cons. int. Expl. de la Mer 18.

STEINBÖCK, O. 1958. Grundsätzliches zum "kryoeutrophen" See. IVL-Verh. 13.

STEINER, M. (1) 1931. Beiträge zur Kenntnis des Zellulose- und Chitinabbaues durch Mikroorganismen in stehenden Binnengewässern. "Stella matutina," Festschr. II.—(2) 1938. Zur Kenntnis des Phosphatkreislaufes in Seen. Naturwissenschaften, 26.

STEINMANN, P. (1) 1907. Die Tierwelt der Gebirgsbäche. Annal. Biol. Lacustre.—(2) 1915. Praktikum der Süsswasserbiologie, I. Teil: Die Organismen des fliessenden Wassers. Berlin.

ŠTĚPÁNEK, M. 1959. Limnological study of the reservoir Sedlice near Zeliv. IX. Transmission and transparency of water. Sci. Pap. Inst. Chem. Technol., Prague, 3.

STORCH, O. (1) 1924. Morphologie und Physiologie der Fangapparate der Daphniden. Erg. u. Fortschr. d. Zool. 6.—(2) 1925. Der Fangapparat von *Diaptomus*. Ztschr. vgl. Physiol. 3.

TALLING, J. P. 1957. Photosynthetic characteristics of some freshwater plankton diatoms in relation to underwater radiation. The New Phytologist, 56.

TEAL, J. M. 1957. Community metabolism in a temperate cold spring. Ecol. Monogr. 27.

THIENEMANN, A. (1) 1918. Lebensgemeinschaft und Lebensraum. Naturwiss. Wochenschr. 33.—(2) 1913. Die Salzwassertierwelt Westfalens. Verh. d. dtsch. zool. Ges.—(3) 1915. Das Ulmener Maar. Festschr. med. naturf. Ges., Münster.—(4) 1918. Untersuchungen über die Beziehungen zwischen dem Sauerstoffgehalt des Wassers und der Zusammensetzung der Fauna in norddeutschen Seen. Arch. Hydrobiol. 12.—(5) 1928. Die Reliktenkrebse *Mysis relicta*, *Pontoporeia affinis*, *Pallasea quadrispinosa* und die von ihnen bewohnten norddeutschen Seen. Arch. Hydrobiol. 19. —(6) 1911/12. Die Tierwelt der Bäche des Sauerlandes. 40. Jahresber. d. westfäl. Provinzialver. f. Wiss. u. Kunst, Münster.—(7) 1939. Die Chironomidenforschung in ihrer Bedeutung für Limnologie und Biologie. Biol. Jaarb. "Dodonea" 6, Gent.—(8) 1931. Der Produktionsbegriff in der Biologie. Arch. Hydrobiol. 22.

THOMAS, E. A. (1) 1950. Beitrag zur Methodik der Produktionsforschung in Seen. Schweiz. Zschr. Hydrol. 12.—(2) 1955. Sedimentation in oligotrophen und eutrophen Seen als Ausdruck der Produktivität. IVL-Verh. 12.—(3) 1955. Stoffhaushalt und Sedimentation im oligotrophen Aegerisee und im eutrophen Pfäffikersee und Greifensee. Mem. Ist. Ital. Idrobiol. Suppl. 8, Colloque IUBS, 19.—(4) 1956. Sedimentation und Stoffhaushalt im Türlersee. Monatsbull. Schweiz. Ver. v. Gas- u. Abwasserfachmännern, 12.

THORADE, H. 1930. Probleme der Wasserwellen. Leipzig.

TILLMANS, J., and W. SUTTHOFF. 1911. Ein einfaches Verfahren zum Nachweis und zur Bestimmung der Salpetersäure im Wasser. Zschr. analyt. Chemie, 50.

TÖDT, F. 1953. Elektrochemische O_2-Messungen. Berlin.

TONOLLI, V. 1949. Struttura speziale del popolamento mesoplanctico. Eterogenità delle densità dei popolamenti orizontali e sua variazione in funzione della quota. Mem. Ist. Ital. Idrobiol. 5.

ULLYOT, P. 1939. Die täglichen Wanderungen der planktischen Süsswasser-Crustaceen. Int. Rev. 28.

UTERMÖHL, H. (1) 1925. Limnologische Phytoplanktonstudien. Arch. Hydrobiol. Suppl. 5.—(2) 1931. Neue Wege in der quantitativen Erfassung des Planktons. IVL-Verh. 5.

VAN NIEL, C. B. 1936. On the metabolism of the Thiorhodaceae. Arch. Mikrobiol. 7.

VAN NORMAN, R. W., and A. H. BROWN. 1952. The relative rates of photosynthetic assimilation of isotopic forms of carbon dioxide. Plant Physiol. 27.

VERDUIN, J. (1) 1956. Primary production in lakes. Limnol. and Oceanogr. 1.—(2) 1957. Daytime variation in phytoplankton photosynthesis. Limnol. and Oceanogr. 2.

VINOGRADSKY, S. 1888. Beiträge zur Morphologie und Physiologie der Bakterien. 1. Zur Morphologie und Physiologie der Schwefelbakterien. Leipzig.

VOLLENWEIDER, R. A. (1) 1950. Oekologische Untersuchungen von planktischen Algen auf experimenteller Grundlage. Schweiz. Zschr. Hydrol. 12. —(2) 1956. Das Strahlungsklima des Lago Maggiore und seine Bedeutung für die Photosynthese des Phytoplanktons. Mem. Ist. Ital. Idrobiol. 9.—(3) 1960. Beiträge zur Kenntnis optischer Eigenschaften der Gewässer und Primärproduktion. Mem. Ist. Ital. Idrobiol. 12.

WASMUND, E. (1) 1926. Biozönose und Thanatozönose. Biosoziologische Studie über Lebensgemeinschaften und Totengesellschaften. Arch. Hydrobiol. 17.—(2) 1927/28. Die Strömungen im Bodensee. Int. Rev. 18, 19. —(3) 1930. Bitumen, Sapropel und Gyttja. Geol. Föreningens i Stockholm Förhandlingar.

WATSON, E. R. 1904. Movements of the waters of Loch Ness as indicated by temperature observations. Geogr. Jour. 26.

WEDDERBURN, E. M. 1910. Current observations in Loch Garry. Proc. Roy. Soc. Edinburgh, 30.

WELTEN, M. 1944. Pollenanalytische, stratigraphische und geochronologische Untersuchungen aus den Faulenseemoos bei Spiez. Veröff. d. Geobotan. Inst. Rübel in Zürich, 21.

WESENBERG-LUND, C. (1) 1900. Von dem Abhängigkeitsverhältnis zwischen dem Bau der Planktonorganismen und dem spez. Gewichte des Süsswassers. Biol. Cbl. 20.—(2) 1908. Plankton investigations of the Danish lakes. Kopenhagen.—(3) 1917. Furesøestudier (Avec un résumé en français). Det Kgl. Danske Vedensk. Selsk. Skrifter, Nat. og Mathem. Afd., 8 Räkke, III. 1.

WILLER, A. 1920. Ueber den Aufwuchs der Unterwasserpflanzen. Schr. Physikökonom. Ges., Königsberg, 61/62.

WILLER, A., and E. HEINEMANN. 1936. Untersuchungen über Trübungs- und Sedimentationsvorgänge in Gewässern. Schr. Physikökonom. Ges. Königsberg, 69.

WOLTERECK, R. (1) 1908. Die natürliche Nahrung pelagischer Cladoceren und die Rolle des "Zentrifugenplanktons" im Süsswasser. Int. Rev. 1.—(2) 1909. Weitere experimentelle Untersuchungen über die Artveränderung, speziell über das Wesen quantitativer Artunterschiede bei Daphniden. Verh. D. Zool. Ges. 19.—(3) 1913. Ueber Funktion, Herkunft und Entstehungsursachen der sogennanten Schwebefortsätze pelagischer Cladoceren. Zoologica, 67.—(4) 1911. Beitrag zur Analyse der "Vererbung erworbener Eigenschaften": Transmutation und Präinduktion bei *Daphnia.* Verh. D. Zool. Ges.—(5) 1928. Ueber die Spezifität des Lebensraumes, der Nahrung und der Körperformen bei pelagischen Cladoceren und über "ökologische Gestaltsysteme." Biolog. Zbl. 48.—(6) 1928. Bemerkungen über die Begriffe "Reaktionsnorm" und "Klon." Biol. Zbl. 48.

WRIGHT, J. C. 1960. The limnology of Canyon Ferry reservoir: III. Some observations on the density dependence of photosynthesis and its cause. Limnol. and Oceanogr. 5.

WUHRMANN, K. 1945. Beitrag zur Kenntnis der Physiologie von Schmutzwasserorganismen. Verh. Schweiz. Naturf. Ges. Zürich.

YOSHIMURA, S. (1) 1933. Kata-numa, a very strong acid water lake in Volcano Katanuma, Miyagi prefecture, Japan. Arch. Hydrobiol. 24.—(2) 1936. A contribution to the knowledge of deep water temperature of Japanese lakes I. Jap. Journ. Astronomy and Geophysics XIII.

YOUNT, J. L. 1956. Factors that control species numbers in Silver Springs, Florida. Limnol. and Oceanogr. 1.

ZHADIN, W. I. 1950. Das Leben in den Flüssen. In: Das Leben der süssen Gewässer der USSR, Bd. 3. Moskow und Leningrad. (Russian.)

ZÜLLIG, H. 1956. Sedimente als Ausdruck des Zustandes eines Gewässers. Schweiz. Zschr. Hydrol. 18.

INDEX